Structure	Class of Compound	Specific Example	Name	Use
C. Containing nitrogen				
$-NH_2$	primary amine	$CH_3CH_2NH_2$	ethylamine	intermediate for dyes, medicinals
$-NHR$	secondary amine	$(CH_3CH_2)_2NH$	diethylamine	pharmaceuticals
$-NR_2$	tertiary amine	$(CH_3)_3N$	trimethylamine	insect attractant
$-C\equiv N$	nitrile (cyanide)	$CH_2=CHCN$	acrylonitrile	orlon manufacture
D. Containing oxygen and nitrogen				
$-\overset{+}{N}\overset{O}{\underset{O^-}{\diagdown}}$	nitro compounds	CH_3NO_2	nitromethane	rocket fuel
$-\overset{O}{\overset{\|}{C}}-NH_2$	primary amide	NH_2CNH_2	urea	fertilizer
E. Containing halogen				
$-X$	alkyl or aryl halide	CH_3Cl	methyl chloride	refrigerant, local anesthetic
$-\overset{O}{\overset{\|}{C}}-X$	acid (acyl) halide	CH_3CCl	acetyl chloride	acetylating agent
F. Containing sulfur				
$-SH$	thiol	CH_3CH_2SH	ethanethiol	odorant to detect gas leaks
$-S-$	thioether	$(CH_2=CHCH_2)_2S$	allyl sulfide	odor of garlic
$-\overset{O}{\underset{O}{\overset{\|}{\underset{\|}{S}}}}-OH$	sulfonic acid	$CH_3-\!\!\langle\!\bigcirc\!\rangle\!\!-SO_3H$	*para*-toluenesulfonic acid	manufacture of oral antidiabetic drug

Organic Chemistry
A Short Course
Sixth Edition

Organic Chemistry
A Short Course
Sixth Edition

Harold Hart Michigan State University

Houghton Mifflin Company   Boston
Dallas   Geneva, Illinois   Hopewell, New Jersey   Palo Alto

A word about the cover

The cover shows a photomicrograph taken with polarized light (Section 5.2) of ordinarily colorless crystals of 4-*t*-butylbenzoic acid. The compound is manufactured commercially by oxidizing 4-*t*-butyltoluene (Section 10.8b) and is used as a modifier in the formulation of alkyd resins, a type of polyester (Section 11.4) used for protective surface coatings.

Cover photograph by Manfred Kage—Peter Arnold, Inc.

Photo credits for "word abouts"

Page 57, Bureau of Mines; page 87, U.S. Forest Service; page 92, Bechtel Petroleum, Inc.; page 114, American Cancer Society; page 164, courtesy of Alpha Therapeutic Corp.; page 165, USDA photo from Acme; page 196, engraving from *Trials of a Public Benefactor* by Nathan P. Rice, 1859; page 198, USDA photo; page 200, "Scotch" 5-minute Epoxy Adhesive by 3M; page 225, Corning Glass Works; page 257, Tennessee Valley Authority; page 277, Fundamental Photographs, New York; page 306, courtesy of Du Pont Magazine; page 399, Pfizer, Inc.

Printed in the U.S.A.

Library of Congress Catalog Card Number: 82-84391

ISBN: 0-395-32611-7

Contents

twelve
Amines and Related Nitrogen Compounds   290

fifteen

Amino Acids, Peptides, and Proteins

sixteen
Nucleotides and Nucleic Acids

Preface

This text provides a brief introduction to modern organic chemistry. It is designed for use in a one-semester or one-quarter course taken by students who, for the most part, will not major in chemistry. The primary interest of these students may be in agriculture, biology, medical technology, nursing, home economics, health sciences, human and veterinary medicine, dentistry, or pharmacy—areas of science that require varying degrees of background in organic chemistry. To assure these students that what they study in this course is worthwhile, relevant to their goals, and interesting, a special effort has been made to illustrate the practical applications of organic chemistry both to everyday life and to biological processes.

WHAT REMAINS THE SAME IN THE SIXTH EDITION

Thoroughly, yet carefully, revised, the sixth edition builds on the strengths of the five previous editions that are responsible for the text's wide acceptance and successful use over thirty years. A very deliberate effort has been made to retain the authoritativeness that has come to characterize and define this presentation of fundamental principles for the short organic chemistry course. The text continues to offer a selection of topics that are appropriate for the nonmajor, soundly organized, logically developed, and accurately explained.

WHAT HAS CHANGED IN THE SIXTH EDITION

Even the soundest presentation of principles must be continually adapted to reflect a modern treatment of facts and an up-to-date approach to the teaching of the science. With the intention of providing coverage of the necessary topics in a way that makes the material easier for instructors to teach today's students, we have instituted several important changes:

1. Selective refinement of the presentation and sequencing of principles coverage Although the general plan of the text has been retained, instructors familiar with previous editions will note some newly initiated changes in the treatment of specific topics. Chapter 1 now provides an integrated treatment of bonding and isomerism (combining material from Chapters 1 and 2 in the fifth edition). Organic halogen compounds now follow the chapter on stereoisomerism, so that the consequences of stereochemistry can be immediately applied to reactions. The sections on difunctional acids have been excised from the chapter on carboxylic acids and combined with the material on polyesters and fats. In this way, the effect that one functional group exerts on the chemistry of another group in the same molecule becomes the theme of this chapter. Finally, the chapter on spectroscopy (an appendix in the fifth edition) has been placed as Chapter 13, immediately following the last of the functional group chapters. Although this location makes good sense because the student is at this point fully familiar with all the main functional groups and can appreciate how spectra can be used to obtain structural information, all or parts of this chapter can be introduced earlier if desired. Students will be familiar with all the structures in the in-text problems if infrared spectroscopy is introduced after Chapter 8, visible-ultraviolet and mass spectroscopy after Chapter 9, and NMR spectroscopy after Chapter 10. At the other extreme, the spectroscopy chapter can be omitted altogether, since it is not essential for the remaining biochemically oriented chapters that follow it.

2. Development of a modernized problem-solving program New to this revision is the addition of examples with carefully worked-out solutions, which appear at appropriate places within each chapter to help students develop problem-solving skills. Also incorporated throughout each of the chapters are substantial numbers of short, unsolved problems that will provide immediate reinforcement of learning. The abundant end-of-chapter problems round out a problem-solving program that will really offer students ample opportunity to learn by doing.

3. Expansion of the coverage of relevant and applied topics While relevant applications are stressed throughout the text, thirty short sections under the general rubric, *A Word About*, emphasize these applications to other branches of science and to human life. These sections appear within the text rather than as isolated essays. However, since these sections are numbered and printed in color, they stand out from the text clearly so that instructors can easily choose to require these sections or not, as desired.

4. Enhancement of the readability of the narrative Large portions of the text have been rewritten to simplify and humanize the writing style and improve the text's teachability. The subject matter is subdivided into smaller packages, with attention being given to making appropriate transitions that will lead the reader logically from one section to another and improve the natural flow of ideas. Because of the small number of new concepts or facts introduced in each section, this arrangement will assist students in assimilating the material. It will also make it easier for the instructor to omit sections if necessary, in addressing the time demands of the course, the particular student audience, and the instructor's preferences.

5. The presentation of an even more complete learning package

Study Guide and Solutions Book contains answers to all text problems, a guide on how to reason out the answers, a summary of each chapter, a summary of the new reactions in each chapter, a list of learning objectives for each chapter, a summary of important reaction mechanisms, and sample test questions.

Laboratory Manual contains 27 experiments, time-tested with thousands of students. The latest edition has been carefully combed to avoid hazardous chemicals on the OSHA list and to minimize contact with solvents, etc. The student and instructor are clearly warned whenever caution or special care is required. The manual has tear-out, perforated report sheets convenient for student and instructor. It is also a convenient size for the non-major lab. Most experiments can be completed in the relatively short two- or three-hour lab for non-majors. The manual contains appendices giving atomic weights, other properties of common reagents, instructions for the teacher on how to make or obtain special reagents, and a list of chemicals and equipment required for each experiment to simplify ordering and stocking the labs. Experiments are a good mix of techniques, preparations, tests, and applications.

Instructor's Manual to Text and Laboratory Manual is available for the first time with the sixth edition. It includes recommended sections of the text to omit if necessary (and which problems to also omit, *if* those sections are skipped). It also includes expected responses to the questions on the lab reports in the laboratory manual. (These can be used by the instructor or teaching assistants for grading reports.)

A STATEMENT OF THE AUTHOR'S TEACHING APPROACH

A word about my general philosophy regarding the short organic chemistry course may be pertinent. This book is not a "watered-down" text. Superficiality has been avoided whenever possible. Students are entitled to clear and accurate explanations for chemical properties and chemical reactions. Mechanisms can be overdone for this group of students, but to discard mechanisms altogether only enhances the archaic view that organic chemistry is a rote memory subject. To be sure, some memorizing is necessary and I see nothing wrong with that. But mechanisms properly understood can minimize memory work and can provide the student with a framework for facing new, previously unencountered chemistry. For these reasons I have included mechanistic discussions of major reaction types at appropriate places in the text.

ACKNOWLEDG-MENTS

For their frankness and diligence in reviewing the entire manuscript, I would like to thank the following professors:

R. A. Abramovitch (Clemson University)

Raylene A. Coad (El Camino College, California)

J. D. Gettler (New York University)

D. Larson (Oklahoma State University)

R. K. Murray, Jr. (University of Delaware)

D. J. Rislove (Winona State University, Minnesota)

P. M. Warner (Iowa State University)

I incorporated at least 80% of their many recommendations and know that the book is very much improved as a consequence.

My special thanks to Professor D. J. Hart (Ohio State University) for having the courage to act as critic for his father with detachment and professionalism, and for assisting with the proofreading. Thanks also to my secretary, Denise McCune, for her skill and effervescent cheerfulness during work on "the book." Not least, I thank my wife, Gerry, for withstanding once again the seige of a revision.

A Special Tribute All earlier editions of this text carried the name of my former co-author, the late Professor Robert D. Schuetz. Although this edition is largely rewritten, it is a logical successor to previous editions and still carries remnants of his contributions. I believe that Bob, a sincere teacher, would approve of the direction I have taken.

I will be happy to hear from users and nonusers, faculty and students, with suggestions for further improvements.

Harold Hart
Professor of Chemistry
Michigan State University
East Lansing, MI 48824

introduction
To the Student

WHAT IS ORGANIC CHEMISTRY ABOUT?

The term *organic* suggests that this kind of chemistry has something to do with organisms or living things. Originally organic chemistry did deal only with substances obtained from living matter. In the late seventeen and early eighteen hundreds, many chemists spent much of their time extracting, purifying, and analyzing substances from animals and plants. They were motivated in part by curiosity about living matter and in part by the desire to obtain useful ingredients for medicines, dyes, and other purposes.

It gradually became clear that most compounds that occur in plants and animals differ in several respects from those that occur in nonliving matter, such as minerals. In particular, most compounds from living matter are made up of the same few elements: carbon, hydrogen, oxygen, nitrogen, and sometimes sulfur, phosphorus, and a few others. Carbon is virtually always present. This fact led to our present definition: *Organic chemistry is the chemistry of carbon compounds.* This definition broadens the scope of the subject to include not only compounds from nature but synthetic compounds as well—compounds prepared for the first time in the laboratory.

SYNTHETIC ORGANIC COMPOUNDS

For years scientists thought compounds that occurred in living matter were different from other substances in that they contained some sort of intangible **vital force.** This idea discouraged chemists from trying to make organic compounds in the laboratory. But in 1828 the German chemist Friedrich Wöhler, then 28 years old, accidentally prepared urea, a well-known constituent of urine, by heating the inorganic (or mineral) substance ammonium cyanate. He was quite excited about this result, and in a letter to his former teacher, the Swedish chemist J. J. Berzelius, he wrote, "I can make urea without the necessity of a kidney, or even of an animal, whether man or dog." This experiment and others like it gradually discredited the vital-force theory and opened the way for modern synthetic organic chemistry.

Synthesis usually consists of piecing together small, relatively simple molecules into larger, more complex ones. To make a molecule that contains many atoms from molecules that contain fewer atoms, one must know how to link atoms to each other—that is, one must know how to make and to break chemical bonds. Although Wöhler's synthesis of urea was accidental, synthesis is much more effective if it can be carried out in a controlled and rational way, so that, when all the atoms are assembled, they will be connected to one another in the correct manner to give the desired product.

Chemical bonds are made and broken via chemical reactions. During this course you will learn about quite a few different reactions that can be used to form new chemical bonds. Thus you will learn how to link molecules together in specific ways that are useful for synthesis.

WHY SYNTHESIS?

At present, the number of organic compounds that have been synthesized in research laboratories and by the chemical industry is far greater than the number isolated from nature—from plants or animals. Why is it important to know how to synthesize molecules? There are several reasons. For one, it might be important to be able to synthesize a natural product in the laboratory in order to make the substance more widely available at lower cost than would be required if the compound had to be extracted from its natural source. Examples of compounds first isolated from nature but now produced synthetically for commercial use are the vitamins, the amino acids, the dye indigo, the moth repellent camphor, the antibiotic penicillin, and many others. Although the term *synthetic* is sometimes frowned on as implying something artificial or unnatural, these synthetic natural products are in fact identical to the same compounds extracted from nature.

Another reason for synthesis is to create new substances that may have more useful properties than the natural products. Synthetic fibers such as nylon and Orlon, for example, have properties that make them superior for some uses to natural fibers such as silk, cotton, and hemp. Many compounds used in medicine are synthetic (including aspirin, ether, Novocain, and barbiturates). The list of synthetic products that we almost take for granted in our industrial society is long indeed—plastics, detergents, insecticides, and oral contraceptives are just a few. All of these are compounds of carbon, organic compounds.

Finally, organic chemists sometimes synthesize new compounds to test chemical theories, and sometimes just for the fun of it. Certain geometric structures, for example, are aesthetically pleasing, and it can be a challenge to make a molecule in which the carbon atoms are arranged in this way. An example is the hydrocarbon cubane, C_8H_8. First synthesized in 1964, it has eight carbons at the corners of a cube, each carbon with one hydrogen and three other carbons connected to it.

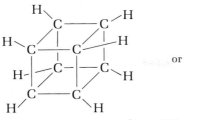

cubane, C_8H_8
mp 130–131°C
P. E. Eaton (U. of Chicago), 1964

Related three-dimensional figures whose corresponding hydrocarbons (compounds with only carbon and hydrogen) have been synthesized in recent years include the following:

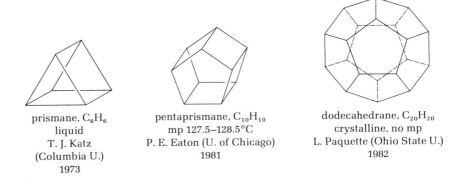

prismane, C_6H_6
liquid
T. J. Katz
(Columbia U.)
1973

pentaprismane, $C_{10}H_{10}$
mp 127.5–128.5°C
P. E. Eaton (U. of Chicago)
1981

dodecahedrane, $C_{20}H_{20}$
crystalline, no mp
L. Paquette (Ohio State U.)
1982

ORGANIC CHEMISTRY IN EVERYDAY LIFE

Organic chemistry touches our daily lives as much as or more than any other branch of science. Almost all the reactions in living matter involve organic substances, and it is impossible to understand life, at least from the physical point of view, without knowing organic chemistry. The major constituents of living matter—proteins, carbohydrates, lipids (fats), nucleic acids (DNA, RNA), cell membranes, enzymes, hormones—are organic, and toward the latter part of this book their chemical structures will be described. Their structures are quite complex. To understand them we will have to discuss simpler molecules first.

Other organic substances with which we have contact almost daily include the gasoline, oil, and tires for our cars, the clothing we wear, the wood and paper of our furniture and books, the medicines we take when ill, plastic containers, camera film, perfume, carpeting, and other fabrics. You name it, and the chances are good that it is organic. Daily in the papers or on TV we encounter mention of polyethylene, epoxy, Styrofoam, nicotine, polyunsaturated fat, cholesterol, and octane numbers. All of these terms refer to organic substances, and we will describe their structures and many more in this book.

In short, organic chemistry is more than just a branch of science for the professional chemist or for the physician, dentist, veterinarian, pharmacist, nurse, or agriculturist. It is part of our technological culture.

Let us now begin our study of organic chemistry with a brief review of chemical bonding, placing special emphasis on carbon.

one

Bonding and Isomerism

In this first chapter we will discuss some rather abstract concepts concerning the way bonds are formed between atoms. Although abstract, the ideas are important because they help to explain the structures of molecules, and they also explain why particular molecules react as they do with other molecules. It may be that you have already studied some of these ideas in a beginning chemistry course. Browse through each section of the chapter to see whether it is familiar, and try to work the problems. If you have no trouble with them, you can safely skip that section. But if you have any difficulty with the problems within or at the end of the chapter, be sure to study the entire chapter carefully, because we will use the ideas developed here frequently throughout the rest of the book.

1.1
HOW ELECTRONS ARE ARRANGED IN ATOMS

Atoms contain a small, dense **nucleus** surrounded by **electrons.** The nucleus is positively charged and contains most of the mass of the atom. The nucleus consists of **protons,** which are positive, and **neutrons,** which are neutral. The only exception is hydrogen, whose nucleus consists of only a single proton. The positive charge of the nucleus is exactly balanced by the negative charge of the electrons that surround it. The **atomic number** of an element is equal to the number of protons in the nucleus (or the number of electrons that surround the nucleus—the two numbers are the same). The **atomic weight** is approximately equal to the sum of the number of protons and the number of neutrons in the nucleus, because the electrons are exceedingly light by comparison.

We will be concerned mainly with the electrons, because their number and arrangement are the key to how a particular atom reacts with other atoms to form molecules. We will also deal with electron arrangements in only the lighter elements, because these elements are most important in organic chemistry.

Electrons are concentrated in certain regions of space around the nucleus, called **orbitals.** Each orbital can contain a maximum of two electrons. The orbitals, which differ in shape, are designated by the letters s, p, and d. Orbitals are grouped in **shells** designated by the numbers 1, 2, 3, and so on. Each shell can contain different types and numbers of orbitals corresponding to the shell number. For example, shell 1 contains only one type of orbital, designated the 1s orbital. Shell 2 may contain two types of orbitals, 2s and 2p, and the third shell may contain three types, 3s, 3p, and 3d. Finally, within any particular shell, the numbers of s, p, and d orbitals are 1, 3, and 5, respectively.

These rules permit us to determine how many electrons each shell will contain when it is filled, as shown in Table 1.1. Table 1.2 shows how the electrons for the first eighteen elements are arranged in the various shells and orbitals.

TABLE 1.1	Shell number	Number of orbitals of each type			Total number of electrons when shell is filled
The orbital arrangements in electron shells		s	p	d	
	1	1	0	0	2
	2	1	3	0	8
	3	1	3	5	18

TABLE 1.2	Atomic number	Element	Number of electrons in the orbitals				
Electronic arrangement of the first eighteen elements			1s	2s	2p	3s	3p
	1	H	1				
	2	He	2				
	3	Li	2	1			
	4	Be	2	2			
	5	B	2	2	1		
	6	C	2	2	2		
	7	N	2	2	3		
	8	O	2	2	4		
	9	F	2	2	5		
	10	Ne	2	2	6		
	11	Na	2	2	6	1	
	12	Mg	2	2	6	2	
	13	Al	2	2	6	2	1
	14	Si	2	2	6	2	2
	15	P	2	2	6	2	3
	16	S	2	2	6	2	4
	17	Cl	2	2	6	2	5
	18	Ar	2	2	6	2	6

TABLE 1.3	Group	1	2	3	4	5	6	7	8
Valence electrons of the first eighteen elements		H·							He:
		Li·	Be·	·B·	·C·	·N:	·O:	:F:	:Ne:
		Na·	Mg·	·Al·	·Si·	·P:	·S:	:Cl:	:Ar:

Note that the first shell is filled for helium (He) and all elements beyond, and the second shell is filled for neon (Ne) and all elements beyond. Filled shells play almost no role in chemical bonding. It is the outer shells, or **valence shells,** that are mainly involved in chemical bonding, and we will focus our attention on these shells.

Table 1.3 shows the **valence electrons,** the electrons in the outermost shell, for the first eighteen elements. The symbol stands for the **kernel** of the element (the nucleus plus the filled electron shells), and the dots represent the valence electrons. The elements are arranged in groups according to the periodic table, and (except for helium) these group numbers correspond to the number of valence electrons.

Armed with this information about atomic structure, we are now ready to tackle the problem of how elements combine to form chemical bonds.

1.2 IONIC AND COVALENT BONDING

An early and still useful theory of chemical bonding was proposed in 1916 by Gilbert Newton Lewis, then a professor at the University of California in Berkeley. Lewis noticed that the **inert gas** helium had only two electrons surrounding its nucleus and that the inert gas with the next higher atomic weight, neon, had ten such electrons (2 + 8; see Table 1.2). He concluded that atoms of these gases, because they were then thought not to combine with other atoms, must have very stable electron arrangements. He further suggested that other atoms might react in such a way as to achieve these stable arrangements. This stability could be achieved in one of two ways —by complete transfer of electrons from one atom to another or by sharing electrons between atoms.

1.2a Ionic Compounds *Ionic bonds are formed by the transfer of one or more valence electrons from one atom to another.* The atom that gives up the electron(s) becomes positively charged, a **cation.** The atom that receives the electron(s) becomes negatively charged, an **anion.** The reaction between sodium and chlorine atoms to form sodium chloride (ordinary table salt) is a typical electron-transfer reaction.

$$\text{Na·} \quad + \quad \text{·Cl:} \quad \rightarrow \quad \text{Na}^+ \quad + \quad \text{:Cl:}^- \tag{1.1}$$

| sodium | chlorine | sodium | chloride |
| atom | atom | cation | anion |

FIGURE 1.1   Sodium chloride, Na$^+$Cl$^-$, is an ionic crystal. The colored spheres represent
sodium ions, Na$^+$, and the gray spheres are chloride ions, Cl$^-$. Each ion is
surrounded by six oppositely charged ions, except for those ions that are at the
surface of the crystal.

The sodium atom has only one valence electron (in its third shell). By
giving up that electron, it achieves the electron arrangement of neon and
becomes positively charged, a sodium cation. The chlorine atom has seven
valence electrons. By accepting an additional electron, it achieves the electron arrangement of argon and becomes negatively charged, a chloride
anion. *Atoms,* such as sodium, *that tend to give up electrons are said to
be* **electropositive.** *Atoms,* such as chlorine, *that tend to accept electrons
are said to be* **electronegative.**

The product of eq. 1.1 is sodium chloride, an ionic compound made up
of equal numbers of sodium and chloride ions. In general, ionic compounds are formed when strongly electropositive atoms and strongly electronegative atoms combine. The ions in a crystal of an ionic substance
are held together by the attractive force between the opposite charges on
the ions, as shown in Figure 1.1 for a sodium chloride crystal.

In a sense, the ionic bond is not really a bond at all. Being oppositely
charged, the ions are attracted to one another like the opposite poles of
a magnet. In the crystal the ions are packed in a definite arrangement, but
we cannot say that any particular ion is bonded or connected to another
particular ion. And, of course, when the substance is dissolved, the ions
are able to move about in solution relatively freely.

Problem 1.1   **Using Table 1.3 as a guide, tell what charges the ion would carry when
each of the following elements reacts to form an ionic compound: Al,
F, Li, Mg, S, H.**

Problem 1.2   **Using Table 1.3 as a guide, tell which is the more electropositive element,
sodium or aluminum? lithium or beryllium?**

Problem 1.3   **Using Table 1.3 as a guide, tell which is the more electronegative element,
oxygen or fluorine? oxygen or nitrogen?**

Problem 1.4 **Judging from its position in Table 1.3, do you expect a carbon atom to be electropositive or electronegative?**

1.2b The Covalent Bond Elements that are neither strongly electronegative nor strongly electropositive tend to form bonds by sharing electron pairs instead of by complete transfer of electrons. A **covalent bond** *involves the mutual sharing of one or more electron pairs between atoms.* When the two atoms are identical or have equal electronegativities, the electron pairs are shared equally. The hydrogen molecule is an example.

$$H\cdot \;+\; H\cdot \;\rightarrow\; H\colon H \;+\; heat \tag{1.2}$$

hydrogen   hydrogen
atoms   molecule

Each hydrogen atom can be considered as having filled its first electron shell by the sharing process. That is, each atom is considered to "own" all the electrons it shares with another atom, as shown by the loops in these structures:

$$\boxed{H\colon}H \qquad\qquad H\boxed{\colon H}$$

When two hydrogen atoms combine to form a molecule, a rather large amount of heat is liberated. Conversely, this same amount of heat (energy) would have to be supplied to a hydrogen molecule if we were to try to cleave it into atoms. To cleave 1 mol of hydrogen molecules (one gram molecular weight, in this case 2 g) into atoms requires 104 kcal (or 435 kJ*) of heat.

The H—H bond is a very strong bond. The main reason for this is that the shared electron pair is attracted to *both* positively charged hydrogen nuclei, whereas, in a hydrogen atom, the valence electron is associated with only one nucleus. But other forces in the hydrogen molecule tend to counterbalance the attraction between the electron pair and the nuclei. They are the repulsion between the two like-charged nuclei and the repulsion between the two like-charged electrons. A balance is struck between these attractive and repulsive forces. The hydrogens neither fly apart nor fuse together. Instead, they remain connected, or bonded, and vibrate at some equilibrium distance that we call the **bond length.** For a hydrogen molecule the bond length (that is, the average distance between the two hydrogen nuclei) is 0.74 Å.**

1.3 CARBON AND THE COVALENT BOND

Now let us look at carbon and its bonding. We represent atomic carbon by the symbol $\cdot\dot{C}\cdot$, where the letter C stands for the kernel (the nucleus plus first-shell electrons) and the dots represent the valence electrons.

With four valence electrons, the valence shell of carbon is half filled (or half empty). Carbon atoms have neither a strong tendency to lose all

---

*Although most organic chemists use kilocalories as a unit of heat, the currently used international unit is the kilojoule. 1 kcal = 4.184 kJ
**1 Å, or angstrom unit, is 10^{-8} cm, so the H—H bond length is 0.74×10^{-8} cm.

their electrons (and become C^{4+}) nor a strong tendency to gain four electrons (and become C^{4-}). Being in the middle of the periodic table, carbon is neither strongly electropositive nor strongly electronegative. Instead, it usually forms covalent bonds with other atoms by sharing electrons. For example, carbon combines with four hydrogen atoms (each of which supplies one valence electron) by sharing electron pairs. The substance formed is known as **methane.** Carbon can do the same thing with four chlorine atoms, forming **carbon tetrachloride.**

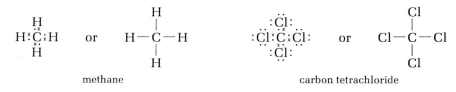

methane   carbon tetrachloride

By sharing electron pairs, the atoms complete their valence shells. In both examples, carbon has eight valence electrons surrounding it. In methane, the hydrogen completes its first valence shell with two electrons, and in carbon tetrachloride, the chlorine atoms fill their third valence shell with eight electrons. In this way all valence shells are filled and the compounds are quite stable.

The shared electron pair is called a **covalent bond,** because it bonds or links the atoms together (by mutual attraction for their nuclei). The single bond is usually represented by a dash or a single line, as shown in the second formulas for methane and carbon tetrachloride.

Example 1.1   **Draw the formula for chloromethane (also called methyl chloride), CH_3Cl.**

Solution
$$H:\overset{\textstyle H}{\underset{\textstyle H}{C}}:\overset{..}{\underset{..}{Cl}}: \qquad or \qquad H-\overset{\textstyle H}{\underset{\textstyle H}{C}}-Cl$$

Problem 1.5   **Draw the formulas for dichloromethane (also called methylene chloride), CH_2Cl_2, and for trichloromethane (chloroform), $CHCl_3$.**

**1.4
CARBON–CARBON
SINGLE BONDS**

The unique property of carbon atoms, the property that makes it possible for chemists to construct literally millions of organic compounds, is their almost unlimited ability to share electrons not only with different elements but also with other carbon atoms. For example, two carbon atoms may be bonded to one another, and each of these carbon atoms may be linked to other atoms. In ethane and hexachloroethane, each carbon is connected to three hydrogen atoms or to three chlorine atoms.

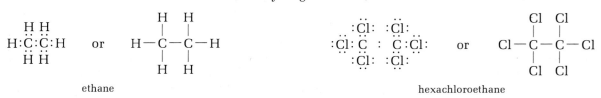

ethane   hexachloroethane

Although they have two carbon atoms instead of one, these compounds have chemical properties similar to those of methane and carbon tetrachloride, respectively.

The carbon–carbon bond in ethane, like the hydrogen–hydrogen bond in a hydrogen molecule, is a purely covalent bond, with the electrons shared equally between the two identical carbon atoms. As with the hydrogen molecule, heat is required to break the carbon–carbon bond into two CH_3 fragments (called **methyl radicals**). *A* **radical** *is a fragment with an odd number of unshared electrons.*

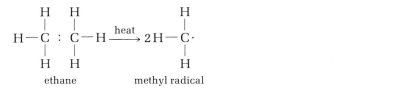

$$\text{(1.3)}$$

ethane   methyl radical

However, less heat is required to break the carbon–carbon bond in ethane than is required to break the hydrogen–hydrogen bond in a hydrogen molecule. The actual amount is 88 kcal (or 368 kJ) per mole of ethane. The carbon–carbon bond in ethane is longer (1.54 Å) than the hydrogen–hydrogen bond (0.74 Å) and also somewhat weaker. Breaking carbon–carbon bonds by heat, as represented in eq. 1.3, is the first step in the "cracking" of petroleum, an important commercial process in the manufacture of gasoline (see A Word About Petroleum, page 91).

Example 1.2 **What do you expect the bond length of a C—H bond (as in methane or ethane) to be?**

Solution **Probably somewhere between the H—H bond length (0.74 Å) and the C—C bond length in ethane (1.54 Å). The actual value is about 1.09 Å.**

Problem 1.6 **Which bond will be longer, H—H or C—Cl? C—C or C—Cl? The Cl—Cl bond length is 1.98 Å.**

Problem 1.7 **Using the structure of ethane as a guide, draw the structure for propane, C_3H_8.**

There is almost no limit to the number of carbon atoms that can be linked together, and some molecules contain as many as 100 or more carbon–carbon bonds in a row. This ability of an element to form chains as a result of bonding with itself is called **catenation.** Carbon atoms may be linked not only in continuous chains, but also in branched chains and rings of endless variety, as we shall see.

1.5
POLAR COVALENT
BONDS
As we have seen, covalent bonds can be formed not only between identical atoms (H—H, C—C), but also between different atoms (C—H, C—Cl), provided that the atoms do not differ too greatly in electronegativity. In the latter cases, the electron pair will not be shared equally between the

two atoms. Such a bond is sometimes called a **polar covalent bond,** because the atoms that are linked carry partial negative and partial positive charges.

The hydrogen chloride molecule provides an example of a polar covalent bond. Chlorine atoms are more electronegative than hydrogen atoms, but the electronegativity difference between them is such that they still form a covalent bond rather than an ionic bond. However, the electron pair is attracted more toward the chlorine, which therefore is slightly negative with respect to the hydrogen. This bond polarization may be indicated by an arrow whose head is negative and whose tail is marked with a plus sign. Alternatively, a partial charge written as $\delta+$ and $\delta-$ (to be read "delta plus" or "delta minus") may be shown:

$$\overset{\longmapsto}{\text{H}:\ddot{\text{C}}\text{l}:} \qquad \text{or} \qquad \overset{\delta+}{\text{H}}:\overset{\cdot\cdot\;\delta-}{\ddot{\text{C}}\text{l}:} \qquad \text{or} \qquad \overset{\delta+}{\text{H}}-\overset{\cdot\cdot\;\delta-}{\ddot{\text{C}}\text{l}:}$$

The bonding electron pair, which is shared unequally, is displaced toward the chlorine.

You can usually rely on the periodic table (Table 1.3) to determine which end of a polar covalent bond is more negative and which end is more positive. As we proceed from left to right across the table within a given period, the elements become *more* electronegative, owing to their increasing atomic number, or charge on the nucleus. As we proceed from the top to the bottom of the table within a given group (down a column), the elements become *less* electronegative, because the valence electrons are shielded from the nucleus by an increasing number of inner-shell electrons. From these generalizations we can safely predict that the atom on the right in each of the following bonds will be negative with respect to the atom on the left:

$$\overset{\longmapsto}{\text{C}-\text{N}} \qquad \overset{\longmapsto}{\text{C}-\text{Cl}} \qquad \overset{\longmapsto}{\text{H}-\text{O}} \qquad \overset{\longmapsto}{\text{Br}-\text{Cl}}$$
$$\overset{\longmapsto}{\text{C}-\text{O}} \qquad \overset{\longmapsto}{\text{C}-\text{Br}} \qquad \overset{\longmapsto}{\text{H}-\text{S}} \qquad \overset{\longmapsto}{\text{Si}-\text{C}}$$

Problem 1.8   **Which element is more electropositive, sodium or lithium? Which element is more electronegative, oxygen or sulfur?**

Problem 1.9   **Predict the polarity of the N—Cl and S—O bonds.**

The carbon–hydrogen bond, which is so common in organic compounds, requires special mention. Carbon and hydrogen have nearly identical electronegativities, so the C—H bond is almost purely covalent. The C—H bond length in molecules such as methane or ethane is 1.09 Å, intermediate between H—H and C—C bond lengths.

Problem 1.10   **Indicate any bond polarization in the structure of carbon tetrachloride.**

Problem 1.11   **Draw the formula for methyl alcohol, CH_3OH, and (where appropriate) indicate bond polarity with an arrow, $\longmapsto$.**

**1.6
MULTIPLE
COVALENT BONDS**

To complete their valence shells, atoms may sometimes share more than one electron pair. Carbon dioxide, CO_2, is an example. The carbon atom has four valence electrons, and each oxygen has six valence electrons. A formula that allows each atom to complete its valence shell with eight electrons is

$$\overset{+}{\underset{+}{\text{:}}}\text{O}\text{::}\text{C}\text{::}\overset{+}{\underset{+}{\text{O}}}\text{:} \quad \text{or} \quad \overset{\times\times}{\text{O}}=\text{C}=\overset{\times\times}{\underset{\times\times}{\text{O}}} \quad \text{or} \quad \text{O}=\text{C}=\text{O}$$

$$(1) \qquad\qquad\qquad (2) \qquad\qquad\qquad (3)$$

In (1), the dots show the electrons from carbon; the x's are electrons from the oxygens; (2) shows the bonds and oxygen's unshared electrons, and (3) shows only the covalent bonds. That each atom is surrounded by eight valence electrons is shown by the "loops" in the following sketches:

Two electron pairs are shared between carbon and oxygen. Consequently, the bond is called a **double bond.** Each oxygen atom also has two pairs of **nonbonding electrons,** or **unshared electron pairs.**

Hydrogen cyanide, HCN, is an example of a simple compound with a triple bond, a bond in which three electron pairs are shared.

$$\text{H:C:::N:} \quad \text{or} \quad \text{H}-\text{C}\equiv\text{N:} \quad \text{or} \quad \text{H}-\text{C}\equiv\text{N}$$

hydrogen cyanide

Problem 1.12 **Show with "loops" how each atom in hydrogen cyanide completes its valence shell.**

Problem 1.13 **Show what, if anything, is wrong with each of the following electronic arrangements for carbon dioxide.**

a. :O:::C:::O: **b.** :Ö:C:Ö: **c.** :Ö:C:::O:

Problem 1.14 **Formaldehyde has the formula H_2CO. Draw a formula that shows how the valence electrons are arranged.**

Carbon atoms can be connected to one another by double bonds or triple bonds, as well as single bonds. Thus there are three **hydrocarbons** (compounds with just carbon and hydrogen atoms) that have two carbon atoms per molecule: ethane, ethylene, and acetylene.

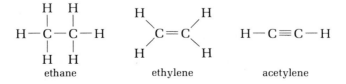

ethane   ethylene   acetylene

They differ in having carbon–carbon single, double, and triple bonds. As we shall see later, these compounds differ quite a bit from one another in chemical reactivity because of the different types of bonds between the carbon atoms.

Example 1.3   Draw the formula for C_3H_6 with one carbon–carbon double bond.

Solution   First draw the three carbons with one double bond.

$$C{=}C{-}C$$

Then add the hydrogens in such a way that each carbon has eight electrons around it (or, in such a way that each carbon has four bonds).

$$\begin{array}{c c c} H & H & H \\ | & | & | \\ H{-}C{=}C{-}C{-}H \\ & & | \\ & & H \end{array}$$

Problem 1.15   Draw at least three different structures that have the formula C_4H_8 and have one carbon–carbon double bond.

1.7
VALENCE

The word *valence* comes from the Latin word *valentia*, meaning power or capacity, and it deals with the combining power of an element. *The valence of an element is simply the number of bonds that the element can form.* This number is usually equal to the number of electrons needed to fill the valence shell. Table 1.4 gives the common valences of several elements.

TABLE 1.4	**Element**	H·	·C·	·N:	·O:	:F:	or	:Cl:
Valences of common elements	**Valence**	1	4	3	2	1		

Note the distinction between valence electrons and valence. Oxygen, for example, has six valence electrons but a valence of 2. Only for hydrogen and carbon, in Table 1.4, are the two numbers the same. In all cases, the *sum* of the two numbers is equal to the number of electrons in the filled shell.

The valences in Table 1.4 apply whether the bonds are single, double, or triple. For example, carbon has four bonds in all of the formulas we have written so far: methane, ethane, ethylene, acetylene, carbon dioxide, carbon tetrachloride, and so on. These valences are worth remembering, because they will help you to write correct formulas.

Example 1.4   Using dashes for bonds, draw a formula for C_3H_4 that has the proper valences of 1 for hydrogen and 4 for carbon.

Solution   There are three possibilities.

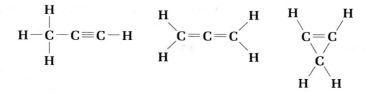

Compounds corresponding to each of these three different arrangements of the atoms are known.

Problem 1.16 **Using dashes for bonds, and using the valences given in Table 1.4, write structures for:**
a. CH_5N b. CH_4O

Problem 1.17 **Does C_2H_5 represent a stable molecule?**

In Example 1.4, we saw that three carbon atoms and four hydrogen atoms could be connected to one another in three different ways, each of which satisfies the valences of both kinds of atoms. Let us take a closer look at this phenomenon.

1.8 ISOMERISM
The **molecular formula** of a substance tells us only the number and kinds of atoms present, but the **structural formula** tells us how those atoms are arranged. For example, H_2O is the molecular formula for water. It tells us that each water molecule contains two hydrogen atoms and one oxygen atom. But the structural formula $H—O—H$ tells us more than that. It tells us that the hydrogens are connected to the oxygen (and not to each other).

As we just saw in Sec. 1.7, it is sometimes possible to arrange the same atoms in more than one way and still satisfy their valences. Molecules of this type are called **isomers,** a term that comes from the Greek (*isos,* equal, and *meros,* part). **Structural isomers** *are compounds that have the same molecular formula but different structural formulas.* Let us look at a particular pair of isomers.

Two very different chemical substances are known that have the molecular formula C_2H_6O. One of these substances is a colorless liquid that boils at 78.5°C, whereas the other is a colorless gas at ordinary temperatures (bp −23.6°C). The only possible explanation is that the atoms are arranged differently in the molecules of each substance, and that these arrangements are somehow responsible for the fact that one substance is a liquid and the other a gas.

For the molecular formula C_2H_6O, two (and only two) structural formulas are possible that satisfy the valence requirements of 4 for carbon, 2 for oxygen, and 1 for hydrogen. They are

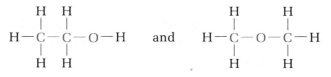

In one formula the two carbons are connected by a single covalent bond; in the other formula each carbon is connected to the oxygen.

We are left with the problem of figuring out which arrangement corresponds to the liquid and which to the gas. There are many ways to solve this problem; in our example, a simple chemical test gives the answer. The liquid C_2H_6O (called ethyl alcohol, or ethanol) reacts with sodium metal to produce hydrogen gas and a new compound, C_2H_5ONa. On the other hand, the gaseous C_2H_6O (called dimethyl ether) does not react with sodium metal at all. The most reasonable interpretation of this experimental result is that ethyl alcohol is represented by the structural formula in which one hydrogen is different from the other five. Apparently the hydrogen on oxygen can be replaced by sodium. In dimethyl ether all six hydrogens are alike, connected to carbon, and nonreplaceable by sodium. Thus we can associate names and properties with each structural formula:

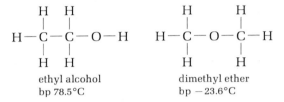

ethyl alcohol　　　　　　　dimethyl ether
bp 78.5°C　　　　　　　　　bp −23.6°C

We leave until later (Chapters 7 and 8) an explanation of why these arrangements of atoms produce substances that are so different and boil at temperatures over 100° apart.

Ethyl alcohol and dimethyl ether are structural isomers. They have the same molecular formula but different structural formulas. Isomers of this type differ in physical and chemical properties as a consequence of their different molecular structures.

Problem 1.18　　**Draw structural formulas for all possible isomers of C_3H_8O (there are three).**

**1.9
WRITING
STRUCTURAL
FORMULAS**

You will be writing structural formulas throughout this course. Perhaps a few hints about how to do so will be helpful. Let's look at another case of isomerism. Suppose we want to write out all possible structural formulas that correspond to the molecular formula C_5H_{12}. Begin by writing all five carbons in a **continuous chain**.

C—C—C—C—C

a continuous chain

This chain uses up one valence for each of the "end" carbons and two valences for the carbons in the "middle" of the chain. Each end carbon therefore has three valences left for bonds to hydrogens. The middle carbons have only two bonds each for hydrogens. The structural formula is therefore

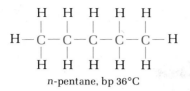

n-pentane, bp 36°C

(The n stands for *normal*.) To find structural formulas for the other isomers, we must consider **branched chains**. For example, we could reduce
the longest chain to only four carbons but connect the fifth carbon to one
of the middle carbons, as in

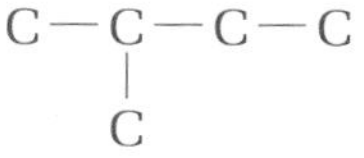

C—C—C—C
      |
      C

a branched chain

If we add the remaining bonds for each carbon to have a valence of 4, we
see that three of the carbons have three hydrogens attached, but the other
carbons have only one or two hydrogens.

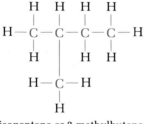

isopentane or 2-methylbutane,
      bp 28°C

Suppose we keep the chain of four carbons and try to connect the fifth
carbon somewhere else. Consider the following chains:

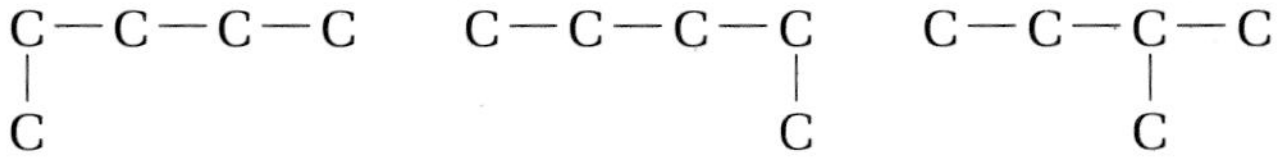

C—C—C—C   C—C—C—C   C—C—C—C
| | |
C   C   C

Do we have anything new here? NO! The first two structures have consecutive carbon chains, exactly as in the formula for n-pentane, and the
third structure is identical to the branched chain we have already drawn
for isopentane—a four-carbon chain with a one-carbon branch attached to
the second carbon in the chain (counting now from the right instead of
from the left).

But there *is* a third isomer of C_5H_{12}. We can find it by reducing the
longest chain to only three carbons and connecting two one-carbon branches
to the middle carbon.

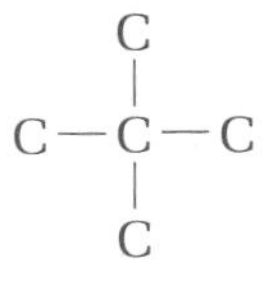

            C
            |
      C—C—C
            |
            C

If we fill in the hydrogens, we see that the middle carbon has no hydrogens attached to it.

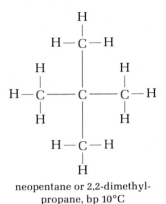

neopentane or 2,2-dimethyl-
propane, bp 10°C

So we can draw three and only three different structural formulas that correspond to the molecular formula C_5H_{12}, and in fact we find that only three different chemical substances with this formula exist. They are commonly called *n*-pentane, isopentane, and neopentane.

Problem 1.19   **To which isomer of C_5H_{12} does each of the following structural formulas correspond?**

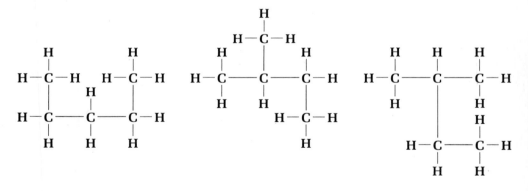

1.10
ABBREVIATED
STRUCTURAL
FORMULAS

Structural formulas such as the ones we have written so far are somewhat cumbersome. They take up a lot of space on the page, and they are tiresome to write out. Consequently, we often take some short cuts that nevertheless still convey the meaning of structural formulas. For example, we may abbreviate the structural formula of ethyl alcohol from

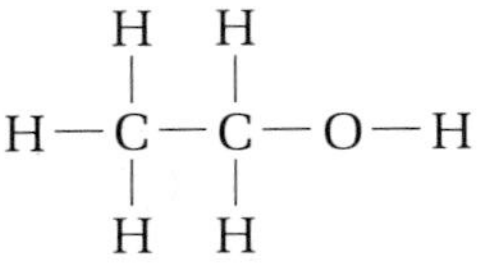

$$H-\underset{\underset{\displaystyle H}{|}}{\overset{\overset{\displaystyle H}{|}}{C}}-\underset{\underset{\displaystyle H}{|}}{\overset{\overset{\displaystyle H}{|}}{C}}-O-H \qquad \text{to} \qquad CH_3-CH_2-OH \qquad \text{or} \qquad CH_3CH_2OH$$

Each of these formulas conveys that two carbon atoms are joined to one another, that one carbon has three hydrogens attached to it, and that the other carbon has two hydrogens and an OH (hydroxyl) group attached to it. Each formula clearly differentiates ethyl alcohol from dimethyl ether, which can be represented by any of the following structures:

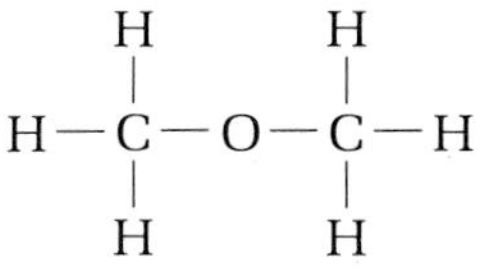

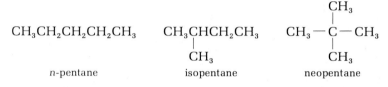

The structural formulas for the three pentanes can be abbreviated in a similar fashion.

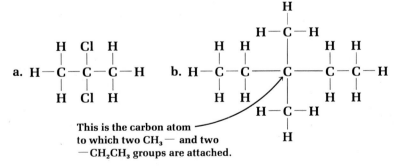

| n-pentane | isopentane | neopentane |

Sometimes these formulas are abbreviated even further. For example, they can be printed on a single line in the following ways:

$CH_3(CH_2)_3CH_3$ $(CH_3)_2CHCH_2CH_3$ $(CH_3)_4C$

    n-pentane   isopentane   neopentane

Example 1.5   **Write a structural formula that shows all the bonds for:**
a. $CH_3CCl_2CH_3$ b. $(CH_3)_2C(CH_2CH_3)_2$

Solution

a.
```
     H  Cl  H
     |  |   |
 H — C — C — C — H
     |  |   |
     H  Cl  H
```

This is the carbon atom
to which two CH_3— and two
—CH_2CH_3 groups are attached.

b.
```
              H
              |
          H — C — H
              |
   H  H       |       H  H
   |  |       |       |  |
H—C — C ————— C ————— C — C — H
   |  |       |       |  |
   H  H       |       H  H
          H — C — H
              |
              H
```

Problem 1.20   **Write a structural formula that shows all the bonds for**
a. $(CH_3)_2CHCH_2OH$ b. $CCl_2 = CCl_2$

Perhaps the ultimate abbreviation of structures is the use of lines to represent the carbon framework:

| n-pentane | isopentane | neopentane |

Each line *segment* is understood to have a carbon atom at each end. The hydrogens are omitted, but we can quickly find the number of hydrogens on each carbon by subtracting from four the number of line segments that emanate from any point. Multiple bonds are represented by multiple line segments. For example, the hydrocarbon with a chain of five carbon atoms and a double bond between the second and third carbon atoms (that is, $CH_3CH = CHCH_2CH_3$) would be represented as follows:

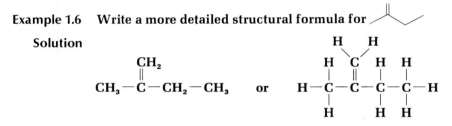

Three line segments emanate from this point; therefore this carbon has one $(4-3=1)$ hydrogen attached to it.

Two line segments emanate from this point; therefore this carbon has two $(4-2=2)$ hydrogens attached to it.

Example 1.6   **Write a more detailed structural formula for**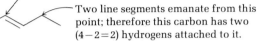

Solution

$$
\begin{array}{c}
CH_2 \\
\| \\
CH_3 - C - CH_2 - CH_3
\end{array}
\quad \text{or} \quad
\begin{array}{c}
\;\;\; H \quad H \\
\;\;\; \backslash \;\; / \\
H \quad C \quad H \quad H \\
| \quad | \quad | \quad | \\
H - C - C - C - C - H \\
| \quad \| \quad | \quad | \\
H \quad \quad H \quad H
\end{array}
$$

Problem 1.21   **Write a line-segment formula for $(CH_3)_2CHCH(CH_3)_2$.**

So far we have considered molecules whose atoms are neutral. But in some compounds one or more atoms may carry a charge, either plus or minus. Because these charges usually influence the chemical reactions of such molecules, it is important to know how to tell where that charge is located.

1.11   Consider the hydronium ion, H_3O^+, the product of reaction of a water
FORMAL CHARGE   molecule with a proton.

$$
H - \overset{..}{\underset{..}{O}} - H + H^+ \rightarrow
\left[\begin{array}{c} H \\ | \\ H - \underset{..}{O} - H \end{array} \right]^+
\tag{1.4}
$$

hydronium ion

The structure has eight electrons around the oxygen and two electrons around each hydrogen, so that all valence shells are complete. (Note that there are eight valence electrons altogether—oxygen contributes six and each hydrogen contributes one for a total of nine, but the ion has a single positive charge, so one electron must have been given away, leaving eight. These eight electrons form three O—H single bonds, leaving one unshared electron pair on the oxygen.)

Although the entire hydronium ion carries a positive charge, we can ask, "Which atom, in a formal sense, bears the charge?" To determine **formal charge,** we consider that each atom "owns" *all* of its unshared electrons plus only *half* of its shared electrons. We then subtract this total from the number of valence electrons in the neutral atom to get the formal charge. This definition can be expressed in equation form as follows:

$$\text{Formal charge} = \frac{\text{number of valence electrons}}{\text{in the neutral atom}} - \left(\substack{\text{unshared} \\ \text{electrons}} + \substack{\text{half the shared} \\ \text{electrons}}\right)$$

or, in a simplified form,   **(1.5)**

$$\text{Formal charge} = \frac{\text{number of valence electrons}}{\text{in the neutral atom}} - (\text{dots} + \text{bonds})$$

Let us apply this definition to the hydronium ion. For each hydrogen the formal charge $= 1 - (0 + 1) = 0$; for the oxygen, the formal charge $= 6 - (2 + 3) = 1$. The oxygen atom formally carries the $+1$ charge of the hydronium ion.

Example 1.7   **Where is the formal charge in the hydroxide ion, OH^-?**

Solution   **The electron-dot formula is $\left[:\overset{..}{\underset{..}{O}}:H\right]^-$ (oxygen contributes six electrons, hydrogen one, and there is one more for the negative charge, for a total of eight electrons). The formal charge on oxygen is $6 - (6 + 1) = -1$, so the oxygen carries the negative charge.**

Problem 1.22   **Calculate the formal charge on the nitrogen atom in ammonia, NH_3; ammonium ion, NH_4^+; and amide ion, NH_2^-.**

Now let us look at a slightly more complex situation involving electron-dot formulas and formal charge.

1.12 RESONANCE   Sometimes electrons are not as associated with one particular bond as the Lewis electron-dot formulas we have been writing would suggest. As an example, let us consider the carbonate ion, CO_3^{2-}.

The total number of valence electrons in the carbonate ion is 24 (4 from the carbon, $3 \times 6 = 18$ from the three oxygens, *plus* 2 more electrons that give the ion its negative charge—these 2 electrons presumably come from some metal, such as 2 sodium atoms). An electron-dot formula that completes the octet around the carbon and each oxygen is

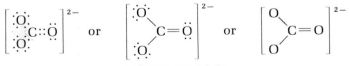

carbonate ion, CO_3^{2-}.

The structure contains 2 carbon–oxygen single bonds and 1 carbon–oxygen double bond. Application of the definition for formal charge shows that the carbon is formally neutral, each singly bound oxygen has a formal charge of −1, and the doubly bound oxygen is formally neutral.

Problem 1.23   **Verify the last sentence.**

When we wrote the electron-dot formula for carbonate ion, our choice of which oxygen atom was doubly bound to the carbon was purely arbitrary. There are in fact *three exactly equivalent* structures that we might write:

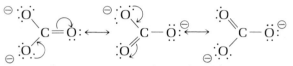

equivalent resonance contributors for carbonate ion

In each structure there is one C=O bond and there are two C—O bonds. The curved arrows show how one contributing structure is related to another.

Physical measurements tell us that none of the foregoing structures accurately describes the "real" carbonate ion. For example, we find experimentally that all three carbon–oxygen bond lengths are identical: 1.31 Å. This distance is intermediate between the normal C=O (1.20 Å) and C—O (1.41 Å) bond lengths. To get around this dilemma we usually say that the "real" carbonate ion has a structure that is a **resonance hybrid** of the three contributing resonance structures. It is as though we could take an average of the three structures. In the "real" carbonate ion, the two formal negative charges are spread out over the three oxygen atoms, so that each oxygen atom carries about two-thirds of a negative charge.

Resonance *arises whenever we can write two or more structures for a molecule with different arrangements of the electrons but identical arrangements of the atoms.* Resonance is very different from isomerism, wherein the atoms themselves are arranged differently. When resonance is possible, the substance (such as carbonate ion) is said to have a structure that is a resonance hybrid of the various **contributing structures.** We use a double-headed arrow ↔ between contributing structures to distinguish resonance from an equilibrium, for which we use ⇌.

Problem 1.24   **Draw the three equivalent contributing electron-dot structures for the nitrate ion, NO_3^{1-}. What is the formal charge on the nitrogen atom and on each oxygen atom in the individual structures? What is the charge on the oxygens and on the nitrogen in the resonance hybrid structure? Show with curved arrows how the structures can be interconverted.**

Each carbon–oxygen bond in the carbonate ion is neither single nor double, but something in between—perhaps a one-and-one-third bond (any particular carbon–oxygen bond is single in two contributing structures and double in one). There is no entirely satisfactory way of representing the resonance hybrid with one electron-dot formula, but sometimes it is done by writing a solid line for the full bond and a dotted line for the partial (in this case, one-third) bond:

carbonate ion
resonance hybrid

Although electron-dot formulas are often useful, they have some limitations. The Lewis theory of bonding itself has some limitations, especially in explaining the three-dimensional geometries of molecules. For this purpose in particular, another theory of bonding, involving orbitals, is more useful.

1.13
THE ORBITAL VIEW
OF BONDING;
THE SIGMA BOND

The atomic orbitals we named in Sec. 1.1 have definite shapes. The s orbitals are spherical. The electrons that fill s orbitals confine their movement to a spherical region of space around the nucleus. The p orbitals are dumbbell-shaped, with their axes along the three coordinates x, y, and z. Figure 1.2 shows the shapes of these orbitals.

In the orbital view of bonding, atoms approach each other in such a way that their atomic orbitals can *overlap* to form a bond. For example,

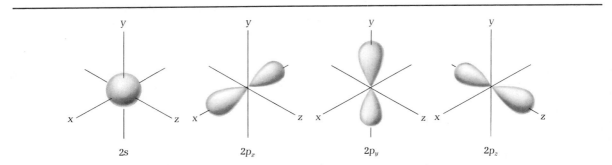

FIGURE 1.2   The shapes of s and p orbitals used by the valence electrons of carbon. The nucleus is at the origin of the three coordinate axes.

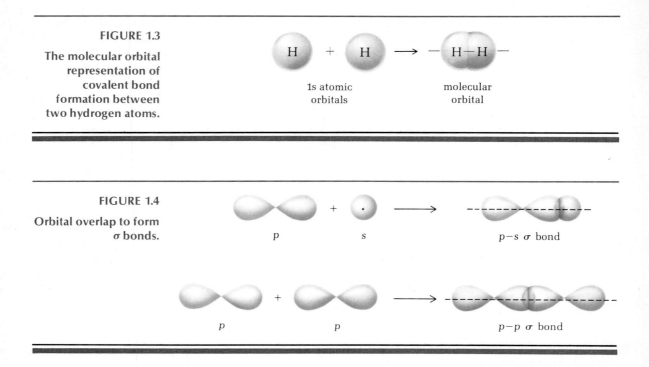

FIGURE 1.3

The molecular orbital representation of covalent bond formation between two hydrogen atoms.

1s atomic orbitals

molecular orbital

FIGURE 1.4

Orbital overlap to form σ bonds.

p s $p-s$ σ bond

p p $p-p$ σ bond

if two hydrogen atoms combine, their two spherical 1s orbitals combine to form a new orbital that encompasses both of the atoms. This orbital contains both valence electrons (one from each hydrogen). Like atomic orbitals, these orbitals can contain no more than two electrons. In the hydrogen molecule these electrons mainly occupy the space between the two nuclei, as shown in Figure 1.3.

The orbital in the hydrogen molecule is cylindrically symmetric along the H—H internuclear axis. Such orbitals are called **sigma (σ) orbitals,** and the bond is referred to as a **sigma bond.** Sigma bonds could also be formed by the overlap of s and p orbitals or of two p orbitals, as shown in Figure 1.4.

Let us see how these ideas apply to the bonding in carbon compounds.

1.14
CARBON sp^3
HYBRID ORBITALS

In an isolated carbon atom the six electrons are arranged as shown in Figure 1.5. The 1s shell is filled, and the four valence electrons are in the 2s orbital and two different 2p orbitals. There are a few things to notice about Figure 1.5. The energy scale at the left represents the energy of electrons in the various orbitals. The further the electron from the nucleus, the greater its potential energy. The 2s orbital has a slightly lower energy than the three 2p orbitals, which have equal energies (they differ from one another only in direction, as already illustrated in Figure 1.2). The two highest-energy electrons are placed in different $2p$ orbitals rather than in the same orbital, because this keeps them further apart and reduces the repulsion between the like-charged particles.

We might get a false idea about the bonding of carbon from Figure 1.5. For example, we might think that carbon should form only two bonds (to complete the partially filled $2p$ orbitals), or perhaps three bonds (if some atom donated two electrons to the empty $2p$ orbital). But we know from experience that this picture is wrong. Carbon usually forms *four* single bonds, and often these bonds are all equivalent, as in CH_4 or CCl_4. How can this discrepancy between theory and fact be resolved?

One way is illustrated in Figure 1.6. First, one of the $2s$ electrons is "promoted" to the vacant $2p$ orbital. This electron arrangement is shown in the middle of the figure. Although it costs some energy to raise an electron from the $2s$ level to the $2p$ level, some of this energy is recovered by reducing electron repulsion (there are no longer two electrons in any one orbital).

The picture in the middle of Figure 1.6 is still not quite satisfactory. There are four half-filled orbitals, so we can now expect to form four bonds (with, say, four hydrogen atoms). But the bonds would not have identical energies, and we know that all four C — H bonds in methane are

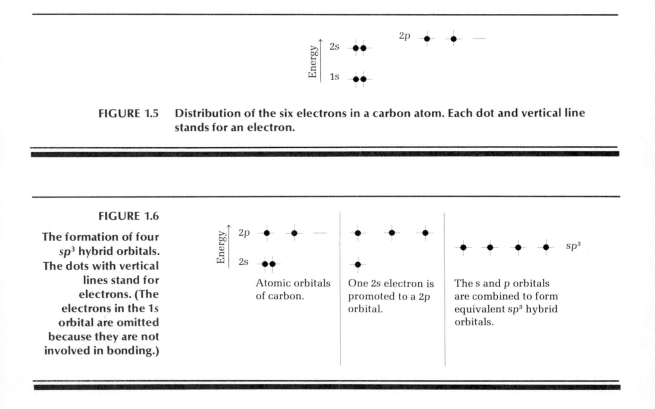

FIGURE 1.5 Distribution of the six electrons in a carbon atom. Each dot and vertical line stands for an electron.

FIGURE 1.6

The formation of four *sp*³ hybrid orbitals. The dots with vertical lines stand for electrons. (The electrons in the 1*s* orbital are omitted because they are not involved in bonding.)

Atomic orbitals of carbon.

One 2s electron is promoted to a 2p orbital.

The s and p orbitals are combined to form equivalent sp³ hybrid orbitals.

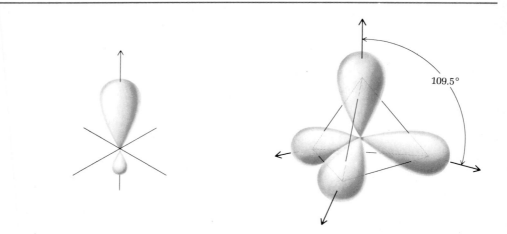

FIGURE 1.7   An *sp*³ orbital extends mainly in one direction from the nucleus and forms
bonds with other atoms in that direction. The four *sp*³ orbitals of any particular
carbon atom are directed toward the corners of a regular tetrahedron, as shown
in the right-hand part of the figure (in this part of the drawing, the small "back"
lobes of the orbital have been omitted for simplification).

FIGURE 1.8

**Sigma (σ) bonds
formed from *sp*³
hybrid orbitals.**

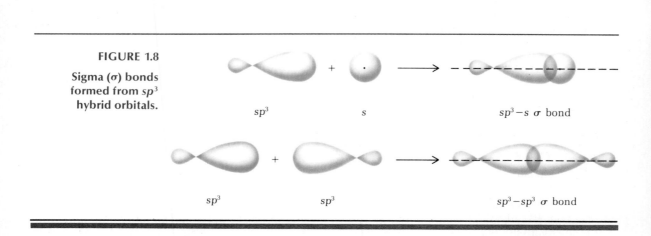

identical. The way out of the dilemma is to allow the s and p orbitals
to "mix" or "combine" and form four identical **sp³ hybrid orbitals.** These
hybrid orbitals are called *sp*³ because they are made by combining one
s orbital and three p orbitals. As shown in Figure 1.6, their energy is a little less than that of the 2*p* orbitals but somewhat greater than that of the
2*s* orbital. The shape of *sp*³ orbitals resembles the shape of p orbitals,
except that the dumbbell is lop-sided, and the electrons are more likely to
be found in the space that extends out the greatest distance in one direction from the nucleus, as shown in Figure 1.7. The four *sp*³ hybrid orbitals

of a single carbon atom are directed toward the corners of a regular tetrahedron, as shown in the figure. This particular geometry puts each orbital as far from the other three orbitals as it can be and thus minimizes repulsion when the orbitals are filled with electron pairs. The angle between bonds formed from four sp^3 orbitals is approximately 109.5°, the angle made by lines drawn from the center to the corners of a regular tetrahedron.

Hybrid orbitals can form sigma bonds by overlap with other orbitals of the same type or of a different type. Figure 1.8 shows some examples.

**1.15
TETRAHEDRAL
CARBON; THE
BONDING IN
METHANE**

We can now describe the way in which a carbon atom combines with four hydrogen atoms to form methane. This process is pictured in Figure 1.9. The carbon atom is joined to each hydrogen atom by a σ bond, which is formed by the overlap of a carbon sp^3 orbital with a hydrogen 1s orbital. The four σ bonds are directed from the carbon nucleus to the corners of a regular tetrahedron. In this way, the electron pair in any one bond experiences minimum repulsion from the electrons in the other bonds. Each H—C—H **bond angle** is the same, 109.5°. To summarize, in methane there are four sp^3–s C—H sigma bonds, each directed from the carbon atom to one of the four corners of a regular tetrahedron.

Problem 1.25   **Considering the repulsion between electrons in different bonds, give a reason why a planar geometry for methane would be less stable than the tetrahedral geometry.**

Because the tetrahedral geometry of carbon plays such an important role in organic chemistry, it is a good idea to become familiar with the

FIGURE 1.9

A molecule of methane, CH_4, formed by the overlap of the four sp^3 carbon orbitals with the 1s orbitals of four hydrogen atoms. The resulting molecule has the geometry of a regular tetrahedron, and contains four σ bonds of the sp^3–s type.

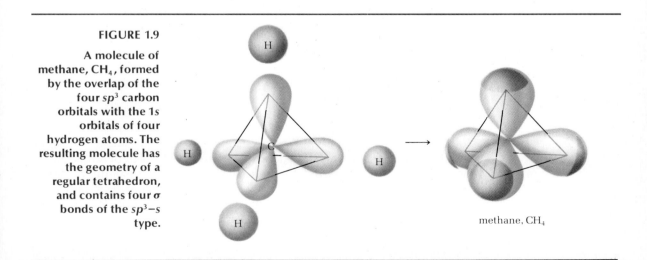

methane, CH_4

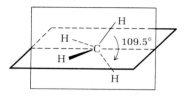

FIGURE 1.10   The carbon and two of the hydrogens in methane form a plane that is the perpendicular bisector of the plane formed by the carbon and the other two hydrogens.

features of a regular tetrahedron. One feature is that the center and any two corners of a tetrahedron form a plane that is the perpendicular bisector of a similar plane formed by the center and the other two corners. In methane, for example, any two hydrogens and the carbon form a plane that perpendicularly bisects the plane formed by the carbon and the other two hydrogens. These planes are illustrated in Figure 1.10.

The geometry of carbon with four single bonds, as in methane, is often represented as

$$
\begin{array}{c}
\text{H} \\
| \\
\text{H} -\!\!\!-\!\!\!- \text{C} \quad 109.5° \\
\text{H} \quad\quad \text{H}
\end{array}
$$

where the solid bonds lie in the plane of the page, the dashed bond goes behind the plane of the paper, and the wedge bond extends out of the plane of the paper toward you.

Now that we have described single covalent bonds and their geometry in carbon compounds,* we are ready to tackle, in the next chapter, the structure and chemistry of saturated hydrocarbons. But before we do that, it will be a good idea to present a brief overview of organic chemistry so that you can see how the subject will be organized for study.

Because carbon atoms can be linked to one another or to other atoms in so many different ways, the number of possible organic compounds is almost limitless. At present, over 2 million organic compounds have been characterized, and the number grows daily. How, then, can we hope to study this vast subject systematically? Fortunately, organic compounds can be classified according to their structures into a relatively small number of groups. Structures can be classified both according to the carbon framework (sometimes called the carbon "skeleton") and according to the groups that are attached to that framework.

*We will deal with the orbital description and geometry of double and triple bonds in Chapter 3.

**1.16
CLASSIFICATION
ACCORDING TO
MOLECULAR
FRAMEWORK**

There are three main classes of molecular or carbon frameworks for organic structures.

1.16a Acyclic Compounds By *acyclic* we mean *not* cyclic. Acyclic organic molecules have chains of carbon atoms but no rings. As we have seen, the chains may be unbranched or branched.

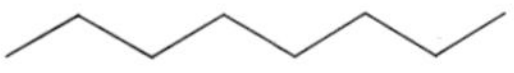

unbranched chain of
eight carbon atoms

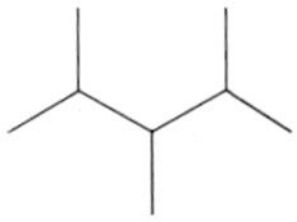

branched chain of
eight carbon atoms

Pentane is an example of an acyclic compound with an unbranched carbon chain, whereas isopentane and neopentane are also acyclic but have branched carbon frameworks (Sec. 1.9). Ethyl alcohol, dimethyl ether, ethane, ethylene, acetylene, methane, and carbon tetrachloride, all of whose structures have been illustrated in this chapter, are acyclic organic compounds. Figure 1.11 shows the structures of a few acyclic compounds that occur in nature.

1.16b Carbocyclic Compounds Carbocyclic compounds contain rings of carbon atoms. The smallest possible carbocyclic ring has three carbon atoms, but rings may come in many sizes and shapes. Rings may also have chains of carbon atoms attached to them and may contain multiple bonds. Many compounds with more than one carbocyclic ring are known. Figure 1.12 shows the structures of a few carbocyclic compounds that occur in nature. Five- and six-membered rings are most common in nature, but smaller and larger rings are also found.

1.16c Heterocyclic Compounds Heterocyclic compounds make up the third and perhaps largest class of molecular frameworks for organic compounds. In heterocyclic compounds at least one atom in the ring must be

FIGURE 1.11

Examples of natural acyclic compounds, their sources (in parentheses), and selected characteristics.

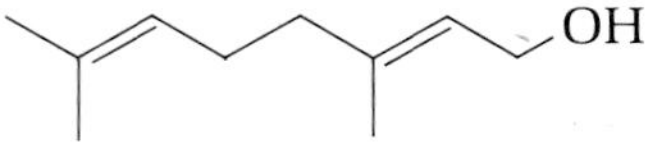

geraniol
(oil of roses)
bp 229–230°C

A branched chain
compound used in
perfumes

$CH_3(CH_2)_5CH_3$

heptane
(petroleum)
bp 98.4°C

A hydrocarbon
present in petroleum,
used as a standard in
testing the knock of
gasoline engines

$CH_3\overset{\displaystyle O}{\overset{\|}{C}}(CH_2)_4CH_3$

2-heptanone
(oil of cloves)
bp 151.5°C

A colorless liquid
with a fruity odor,
in part responsible
for the "peppery"
odor of blue cheese

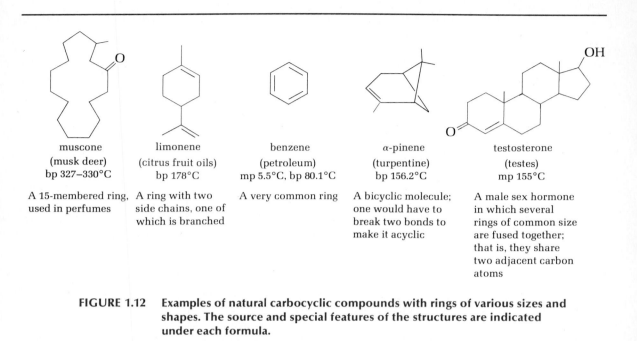

muscone (musk deer) bp 327–330°C	limonene (citrus fruit oils) bp 178°C	benzene (petroleum) mp 5.5°C, bp 80.1°C	α-pinene (turpentine) bp 156.2°C	testosterone (testes) mp 155°C
A 15-membered ring, used in perfumes	A ring with two side chains, one of which is branched	A very common ring	A bicyclic molecule; one would have to break two bonds to make it acyclic	A male sex hormone in which several rings of common size are fused together; that is, they share two adjacent carbon atoms

FIGURE 1.12   **Examples of natural carbocyclic compounds with rings of various sizes and shapes. The source and special features of the structures are indicated under each formula.**

a heteroatom, an atom that is *not* carbon. The most common heteroatoms are oxygen, nitrogen, and sulfur, but many heterocycles with other elements are also known. More than one heteroatom may be present and, if so, they may be alike or different. Heterocyclic rings come in many sizes, may contain multiple bonds, may have carbon chains or rings attached to them, and in short may exhibit a great variety of structures. Figure 1.13 illustrates the structures of a few natural products that contain heterocyclic rings. In these abbreviated structural formulas the symbols for the heteroatoms are shown, but the carbons are indicated in the usual way.

The formulas in Figures 1.11–1.13 show not only the molecular frameworks, but also various groups of atoms that may be attached to the framework. Fortunately these groups can also be classified in a way that helps simplify the study of organic chemistry.

**1.17
CLASSIFICATION
ACCORDING TO
FUNCTIONAL
GROUP**

Certain groups of atoms have chemical properties that depend only moderately on the particular molecular framework to which they are attached. These groups of atoms, whose chemical behavior is similar regardless of what molecule they are a part of, are called **functional groups.** The **hydroxyl group, —OH,** is an example of a functional group, and compounds with only this group are called **alcohols.** In most organic reactions, some chemical change occurs at the functional group, but the rest of the molecule keeps its original structure. For example, in the reaction of ethyl alcohol with sodium (Sec. 1.8), the hydrogen of the hydroxyl is replaced by sodium.

$$2 \text{ CH}_3\text{CH}_2\text{OH} + 2 \text{ Na} \rightarrow 2 \text{ CH}_3\text{CH}_2\text{O}^-\text{Na}^+ + \text{H}_2 \tag{1.6}$$

ethyl alcohol   sodium   sodium ethoxide   hydrogen

The other atoms (2 carbons, 5 hydrogens, and the oxygen) have the same arrangement in the product (sodium ethoxide) that they had in the starting material (ethyl alcohol). This maintenance of most of the structural

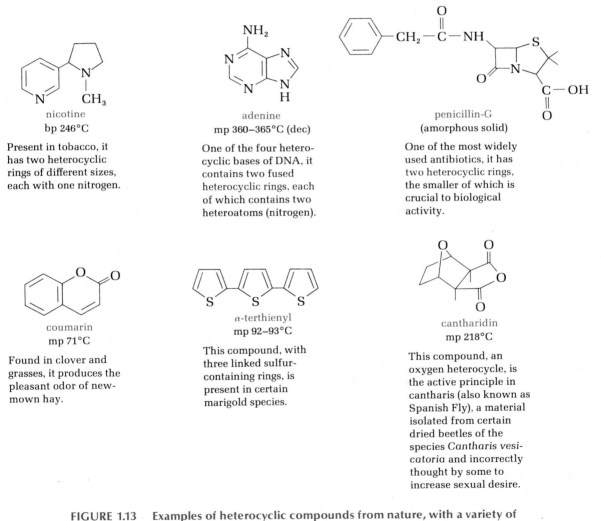

nicotine
bp 246°C

Present in tobacco, it has two heterocyclic rings of different sizes, each with one nitrogen.

adenine
mp 360–365°C (dec)

One of the four heterocyclic bases of DNA, it contains two fused heterocyclic rings, each of which contains two heteroatoms (nitrogen).

penicillin-G
(amorphous solid)

One of the most widely used antibiotics, it has two heterocyclic rings, the smaller of which is crucial to biological activity.

coumarin
mp 71°C

Found in clover and grasses, it produces the pleasant odor of new-mown hay.

α-terthienyl
mp 92–93°C

This compound, with three linked sulfur-containing rings, is present in certain marigold species.

cantharidin
mp 218°C

This compound, an oxygen heterocycle, is the active principle in cantharis (also known as Spanish Fly), a material isolated from certain dried beetles of the species *Cantharis vesicatoria* and incorrectly thought by some to increase sexual desire.

FIGURE 1.13   Examples of heterocyclic compounds from nature, with a variety of heteroatoms and ring sizes.

	Structure	Class of compound	Specific example	Name and use of specific example
A. Functional groups that are a part of the molecular framework	$\diagdown C = C \diagup$ (with slashes)	alkene	$CH_2 = CH_2$	ethylene, used to make polyethylene
	$-C \equiv C-$	alkyne	$HC \equiv CH$	acetylene, used in welding
B. Functional groups containing oxygen				
1. With *one* carbon–oxygen bond	$-\overset{\mid}{\underset{\mid}{C}}-OH$	alcohol	CH_3CH_2OH	ethyl alcohol, in beer, wines, and liquors
	$-\overset{\mid}{\underset{\mid}{C}}-O-\overset{\mid}{\underset{\mid}{C}}-$	ether	$CH_3CH_2OCH_2CH_3$	diethyl ether, the common anesthetic
2. With *two* carbon–oxygen bonds*	$-\overset{\overset{O}{\|}}{C}-H$	aldehyde	$CH_2 = O$	formaldehyde, used to preserve biological specimens
	$\diagdown C \diagdown \overset{\overset{O}{\|}}{C} \diagup C \diagup$	ketone	$CH_3\overset{\overset{O}{\|}}{C}CH_3$	acetone, a solvent for varnish and rubber cement
3. With *three* carbon–oxygen bonds	$-C\overset{\diagup O}{\diagdown OH}$	carboxylic acid	$CH_3\overset{\overset{O}{\|}}{C}-OH$	acetic acid, a component of vinegar
	$-C\overset{\diagup O}{\diagdown O-\overset{\mid}{\underset{\mid}{C}}-}$	ester	$CH_3\overset{\overset{O}{\|}}{C}-OCH_2CH_3$	ethyl acetate, a solvent for nail polish and model airplane dope
C. Functional groups containing nitrogen[†]	$-\overset{\mid}{\underset{\mid}{C}}-NH_2$	primary amine	$CH_3CH_2NH_2$	ethylamine, smells like ammonia
	$-C \equiv N$	cyanide or nitrile	$CH_2 = CH - C \equiv N$	acrylonitrile, raw material for making Orlon
D. Functional group with oxygen and nitrogen	$-\overset{\overset{O}{\|}}{C}-NH_2$	primary amide	$H_2N-\overset{\overset{O}{\|}}{C}-NH_2$	urea, a fertilizer and odorless component of urine

*The $\diagdown C = O$ group, present in several functional groups, is called a **carbonyl group.** The $\overset{\overset{O}{\|}}{C}-OH$ group of acids is called a **carboxyl group** (a contraction of *carb*onyl and hydr*oxyl*).

[†]The $-NH_2$ group is called an **amino group.**

	Structure	Class of compound	Specific example	Name and use of specific example
E. Functional group containing sulfur[‡]	$-\overset{\textstyle\vert}{\underset{\textstyle\vert}{C}}-SH$	thiol (also called mercaptan)	CH_3SH	methanethiol, has the odor of rotten cabbage
	$-\overset{\textstyle\vert}{\underset{\textstyle\vert}{C}}-S-\overset{\textstyle\vert}{\underset{\textstyle\vert}{C}}-$	thioether (also called sulfide)	$(CH_2=CHCH_2)_2S$	allyl sulfide, has the odor of garlic

[‡]Thiols and sulfides are the sulfur analogs of alcohols and ethers.

TABLE 1.5   The main functional groups

formula throughout a chemical reaction simplifies our study of organic chemistry. It allows us to focus our attention on the chemistry of the various functional groups. We can study classes of compounds (such as alcohols) instead of having to learn the chemistry of each individual compound.

Example 1.8 **Predict the product of the reaction of isopropyl alcohol, $CH_3\overset{\textstyle\vert}{\underset{\textstyle OH}{C}}HCH_3$,**

with sodium.

Solution $2\ CH_3\overset{\textstyle\vert}{\underset{\textstyle OH}{C}}HCH_3 + 2\ Na \rightarrow 2\ CH_3\overset{\textstyle\vert}{\underset{\textstyle O^-Na^+}{C}}HCH_3 + H_2$

isopropyl alcohol   sodium isopropoxide

In the product, only the H of the OH group is replaced by Na, just as in eq. 1.6. The oxygen in the product remains attached to the middle carbon of the three-carbon chain.

Problem 1.26 **Write the structure of the products of reaction of sodium metal with:**
a. $CH_3CH_2CH_2OH$ b. $CH_3CH(OH)CH_2CH_3$

Some of the main functional groups that we will study are listed in Table 1.5, together with typical compounds of each type.

Although we will describe these classes of compounds in greater detail in separate chapters, it would be a good idea for you to become familiar with their names and structures now. If a particular functional group is mentioned before its chemistry is discussed in detail, and you forget what it is, you can refer to Table 1.5 or to the inside front cover of the book.

Problem 1.27 **What functional groups can you find in the following natural products? (their formulas are given in Figures 1.11 and 1.12).**
a. geraniol   b. muscone   c. limonene   d. testosterone

ADDITIONAL **1.28.** Show the number of valence electrons in each of the following atoms. Let
PROBLEMS the element's symbol represent its kernel, and use dots for the valence electrons.
a. carbon   **b.** fluorine   **c.** silicon
d. boron   **e.** sulfur   **f.** phosphorus

1.29. When a solution of salt (sodium chloride) in water is treated with a silver nitrate solution, a white precipitate forms immediately. When carbon tetrachloride is shaken with aqueous silver nitrate, no such precipitate is produced. Explain these facts in terms of the types of bonds in the two chlorides.

1.30. Use the relative positions of the elements in the periodic table (Table 1.3) to classify the following substances as ionic or covalent.
a. NaF   **b.** F_2 **c.** $MgCl_2$ **d.** P_2S_5
e. S_2Cl_2 **f.** LiCl   **g.** ClF   **h.** $SiCl_4$

1.31. For each of the following elements, tell (a) how many valence electrons it has and (b) what its common valence is.
a. oxygen   **b.** hydrogen   **c.** chlorine
d. nitrogen   **e.** sulfur   **f.** carbon

1.32. Write structural formulas for each of the following compounds, using a line to represent each single bond and dots for any unshared electron pairs.
a. CH_3Cl **b.** C_3H_8 **c.** C_2H_5F
d. CH_3NH_2 **e.** CH_3CH_2OH **f.** CH_2O

1.33. Draw structural formulas for each of the following covalent molecules. Which bonds are polar? Indicate the polarity by properly placing the symbols $\delta+$ and $\delta-$.
a. Cl_2 **b.** CH_3F **c.** CO_2 **d.** HBr
e. SF_6 **f.** CH_4 **g.** SO_2 **h.** CH_3OCH_3

1.34. Considering bond polarity, which hydrogen in acetic acid, $CH_3\overset{\displaystyle O}{\overset{\displaystyle \|}{C}}-OH$, do you expect to be most acidic? Write an equation for the reaction between acetic acid and metallic sodium.

1.35. Draw structural formulas for all possible isomers with the following molecular formulas.
a. C_3H_8 **b.** C_3H_7Cl **c.** $C_2H_4Cl_2$ **d.** $C_3H_6Br_2$
e. C_4H_9F **f.** $C_2H_2Cl_2$ **g.** C_3H_6 **h.** $C_4H_{10}O$

1.36. Draw structural formulas for the five isomers of C_6H_{14}. Try to be systematic in your thinking.

1.37. For each of the following abbreviated structural formulas, write a structural formula that shows all the bonds.
a. $CH_3(CH_2)_4CH_3$ **b.** $(CH_3)_3CCH_2CH_3$ **c.** $(CH_3)_2CHOH$
d. $(CH_3CH_2)_2S$ **e.** CH_2ClCH_2OH **f.** $(CH_3)_2NCH_2CH_3$

1.38. Write structural formulas that show the correct number of hydrogens on each carbon and correspond to the following abbreviated structures.

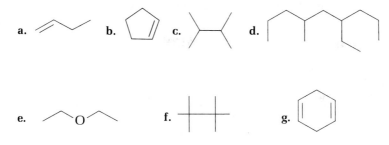

1.39. An abbreviated formula of geraniol is shown in Figure 1.11.
a. How many carbons does geraniol have?
b. What is its molecular formula?
c. Write a more detailed structural formula for it.

1.40. What is the *molecular formula* for each of the following compounds? Consult Figures 1.12 and 1.13 for their abbreviated structural formulas.
a. muscone   **b.** benzene   **c.** testosterone
d. nicotine   **e.** limonene   **f.** adenine

1.41. Write electron-dot formulas for each of the following species. Show where the formal charges, if any, are located.
a. nitrous acid, HONO   **b.** nitric acid, $HONO_2$
c. formaldehyde, H_2CO **d.** ammonium ion, $NH_4{}^+$
e. cyanide ion, CN^- **f.** carbon monoxide
g. sulfate ion, $SO_4{}^{2-}$ **h.** boron trifluoride, BF_3
i. hydrogen peroxide, H_2O_2 **j.** bicarbonate ion, $HCO_3{}^-$

1.42. Draw electron-dot formulas for the two contributors to the resonance hybrid nitrite ion, $NO_2{}^-$. (Both oxygens are connected to the nitrogen.) What is the charge on each oxygen in each contributor and in the hybrid structure? Show by curved arrows how the electron pairs can relocate to interconvert the two structures.

1.43. Consider each of the following highly reactive carbon species. What is the formal charge on carbon in each species?

$$
\begin{array}{cccc}
\text{H} & \text{H} & \text{H} & \text{H} \\
| & | & | & | \\
\text{H}-\text{C} & \text{H}-\text{C}\cdot & \text{H}-\text{C}: & \text{H}-\text{C}\cdot \\
| & | & | & \cdot \\
\text{H} & \text{H} & \text{H} &
\end{array}
$$

1.44. The following substances contain both ionic and covalent bonds. Draw electron-dot formulas for each.
a. CH_3ONa **b.** NH_4Cl

1.45. Fill in any unshared electron pairs that are missing from the following formulas.

$$
\begin{array}{c}
\text{O} \\
||
\end{array}
$$

a. CH_3CH_2OH **b.** CH_3C-OH
c. $(CH_3)_2NH$ **d.** $CH_3OCH_2CH_2OH$

1.46. Write the contributors to the resonance hybrid structures of:
a. azide ion, a linear ion with three connected nitrogens, N_3^-
b. acetate ion, $CH_3CO_2^-$

1.47. Make a drawing (similar to the right-hand part of Figure 1.6) of the electron distribution in nitrogen atoms that would be expected if the s and p orbitals were hybridized to sp^3. Based on this model, predict the geometry of the ammonia molecule, NH_3.

1.48. The ammonium ion, NH_4^+, has a tetrahedral geometry exactly analogous to that of methane. Explain this structure in terms of atomic and molecular orbitals.

1.49. Silicon is just below carbon in the periodic table (Table 1.3). Predict the geometry of silane, SiH_4.

1.50. Use solid, dashed, and wedged lines to show the geometry of CCl_4 and CH_3OH.

1.51. Write a structural formula that corresponds to the molecular formula $C_5H_{10}O$ and is:
a. acyclic   **b.** carbocyclic   **c.** heterocyclic

1.52. Divide the following compounds into groups that might be expected to exhibit similar chemical reactions.
a. CH_3OH **b.** CH_3OCH_3 **c.** $CH_2(OH)CH(OH)CH_2(OH)$
d. C_5H_{12} **e.** C_4H_9OH **f.** C_8H_{18}
g. C_3H_7OH **h.** $(CH_3CH_2)_2O$ **i.** $CH_3OCH_2CH_2OCH_3$

1.53. Using eq. 1.6 as a guide, write an equation for the reaction of cyclohexanol with sodium metal.

H OH

cyclohexanol

1.54. Using Table 1.5 as a guide, write a structural formula for each of the following.
a. an alcohol, $C_4H_{10}O$ **b.** an ether, C_3H_8O
c. an aldehyde, C_3H_6O **d.** a ketone, C_4H_8O
e. a carboxylic acid, $C_3H_6O_2$ **f.** an ester, $C_5H_{10}O_2$
g. an ester that is an isomer   **h.** an amine with the molecular
    of your answer to part f   formula C_3H_9N

two

Alkanes and Cycloalkanes; Conformational and Geometric Isomerism

2.1
INTRODUCTION

The main components of petroleum and natural gas, resources that supply a large fraction of our energy, are **hydrocarbons.** As the name implies, hydrocarbons contain only carbon and hydrogen. There are three main classes of hydrocarbons: **saturated, unsaturated,** and **aromatic.** These classifications are based on the types of carbon–carbon bonds present. Saturated hydrocarbons contain only carbon–carbon *single* bonds. Unsaturated hydrocarbons contain carbon–carbon *multiple* bonds, either double bonds or triple bonds or both. Aromatic hydrocarbons are a special class of cyclic unsaturated compounds related in structure to benzene. In this chapter we discuss saturated hydrocarbons. The other two classes of hydrocarbons will be discussed in the next two chapters.

Saturated hydrocarbons may be acyclic or cyclic and are referred to as **alkanes** or **cycloalkanes,** respectively. An older term for alkanes, still sometimes used, is *paraffins.* The term is derived from paraffin wax, which is a mixture of long-chain saturated hydrocarbons.

Let us look first at the structure and properties of alkanes.

2.2
THE STRUCTURE OF ALKANES

The simplest alkane is methane. Its tetrahedral three-dimensional structure was described in Sec. 1.15 (see Figure 1.9). Additional members of the alkane series are constructed by lengthening the carbon chain and adding an appropriate number of hydrogens to complete the carbon valences (Table 2.1 and Figure 2.1*). Alkanes have an acyclic molecular framework (Sec. 1.16a).

*Molecular models can help you visualize organic structures in three dimensions. They will be extremely useful to you throughout this course, especially when we consider various types of isomerism. Relatively inexpensive sets (about $10) are usually available at stores that sell textbooks, and your instructor can suggest which kind to buy. If you cannot locate or afford a set, you can create models that are adequate for some purposes from toothpicks (for bonds) and marshmallows, gum drops, or jelly beans (for atoms).

FIGURE 2.1

Three-dimensional
models of ethane,
propane, and butane.
The stick-and-ball
models at the left
show the way in
which the atoms are
connected and depict
the correct bond
angles. The space-
filling models at the
right are constructed
to scale and give a
better idea of the
molecular shape.

ethane

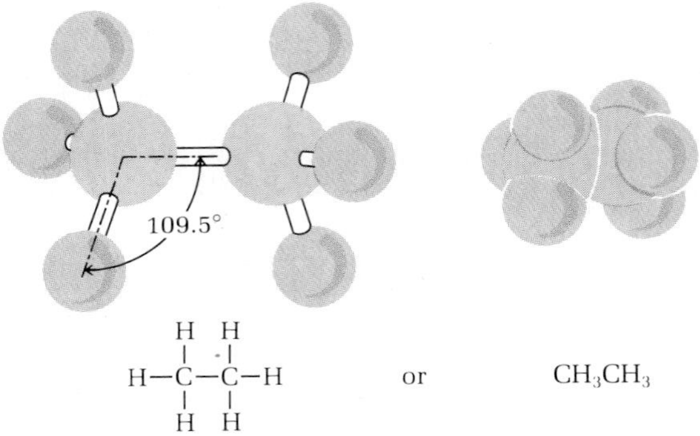

109.5°

$$H-\overset{\displaystyle H}{\underset{\displaystyle H}{C}}-\overset{\displaystyle H}{\underset{\displaystyle H}{C}}-H \qquad \text{or} \qquad CH_3CH_3$$

propane

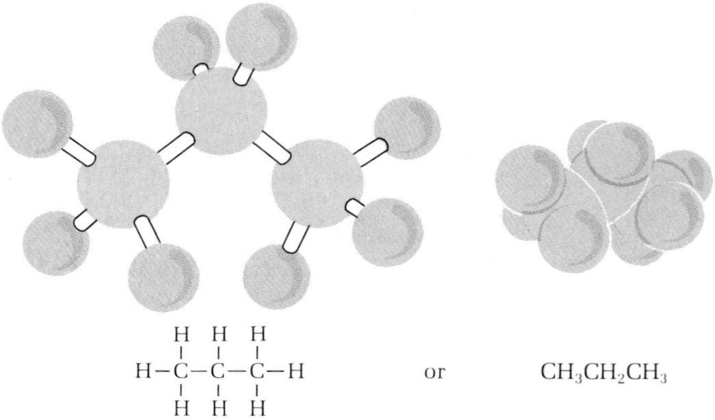

$$H-\overset{\displaystyle H}{\underset{\displaystyle H}{C}}-\overset{\displaystyle H}{\underset{\displaystyle H}{C}}-\overset{\displaystyle H}{\underset{\displaystyle H}{C}}-H \qquad \text{or} \qquad CH_3CH_2CH_3$$

butane

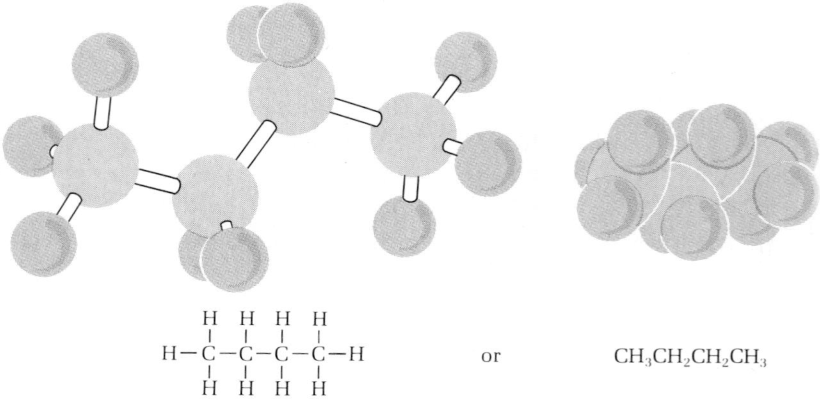

$$H-\overset{\displaystyle H}{\underset{\displaystyle H}{C}}-\overset{\displaystyle H}{\underset{\displaystyle H}{C}}-\overset{\displaystyle H}{\underset{\displaystyle H}{C}}-\overset{\displaystyle H}{\underset{\displaystyle H}{C}}-H \qquad \text{or} \qquad CH_3CH_2CH_2CH_3$$

Name	Number of carbons	Molecular formula	Structural formula	Number of possible isomers
methane	1	CH_4	CH_4	1
ethane	2	C_2H_6	CH_3CH_3	1
propane	3	C_3H_8	$CH_3CH_2CH_3$	1
butane	4	C_4H_{10}	$CH_3CH_2CH_2CH_3$	2
pentane	5	C_5H_{12}	$CH_3(CH_2)_3CH_3$	3
hexane	6	C_6H_{14}	$CH_3(CH_2)_4CH_3$	5
heptane	7	C_7H_{16}	$CH_3(CH_2)_5CH_3$	9
octane	8	C_8H_{18}	$CH_3(CH_2)_6CH_3$	18
nonane	9	C_9H_{20}	$CH_3(CH_2)_7CH_3$	35
decane	10	$C_{10}H_{22}$	$CH_3(CH_2)_8CH_3$	75

TABLE 2.1   **Names and formulas of the first ten unbranched alkanes**

All alkanes fit the general molecular formula C_nH_{2n+2}, where n is the number of carbon atoms. Alkanes with carbon chains that are unbranched (Table 2.1) are referred to as **normal alkanes.** Each member of this series differs from the next higher and the next lower member by a $-CH_2-$ group (called a **methylene group**). A series of compounds related in this way is sometimes referred to as a **homologous series.** Members of such a series have similar chemical properties and physical properties (such as boiling point and density), which change gradually as we add carbon atoms to the chain.

Example 2.1   What is the molecular formula of an alkane with 6 carbon atoms?

Solution   If $n = 6$, then $2n + 2 = 2(6) + 2 = 14$. The formula is C_6H_{14}.

Problem 2.1   What is the molecular formula of an alkane with 20 carbon atoms?

Problem 2.2   Which of the following are alkanes?
a. C_8H_{16} b. C_7H_{16} c. C_7H_{18} d. $C_{27}H_{56}$

2.3
NOMENCLATURE
OF ORGANIC
COMPOUNDS

In the early days of organic chemistry each new compound was given a name that was usually based on the source or use of the compound. Most of the structures shown in Figures 1.11–1.13 were named this way. Examples include limonene (from lemons), α-pinene (from pine trees), coumarin (from the tonka bean, known to South American natives as cumaru), and penicillin (from the mold that produces it, *Penicillium notatum*). Even today this method of naming may be used to give a short, simple name to a molecule with a complex structure.

Other names may be given for more trivial reasons. Barbituric acid, from which the sedatives known as barbiturates are derived, is thought to have been named by its discoverer (Adolph von Baeyer) after a friend, Barbara. Cubane and the prismanes (page 3) were named after their shapes.

It became clear to chemists many years ago that one could not rely only on common or trivial names and that a systematic method for naming compounds was needed. Ideally, the rules of the system should lead to a unique name for each compound. Knowing the rules and seeing a structure, one should be able to write out the systematic name. Seeing the name, one should be able to write out the correct structure.

Eventually, a system of nomenclature was devised that is recognized and used by organic chemists throughout the world. The system is recommended by the International Union of Pure and Applied Chemistry and is known as the IUPAC (pronounced "eye-you-pack") System.

If chemists used only IUPAC names, we would have to learn only the one system. Unfortunately, some common names have been in use for such a long time that they seem likely to be used indefinitely. For example, almost no one uses the systematic name, ethanoic acid, for acetic acid. Other common names, though they are relatively new (such as cubane), are too catchy to be replaced by their systematic name (which in the case of cubane is pentacyclo[4.2.0.0.^{2,5}0.^{3,8}0.^{4,7}]octane). In this book we will emphasize IUPAC names when they are used, but we will also use common names when they are used more often by practicing chemists.

2.4
THE IUPAC RULES
FOR ALKANES

1. The general name for acyclic saturated hydrocarbons is **alkanes.** The -*ane* ending is used for all saturated hydrocarbons.

2. Alkanes without branches are named (after the first four, which have common names) according to the *number of carbon atoms.* Greek roots (*pent-*, *hex-*, and so on) are used to designate the length of the chain (see Table 2.1).

3. For alkanes with branched chains, *the root name is that of the longest continuous chain of carbon atoms.* For example, in the structure

$$CH_3 \quad CH_3$$
$$CH_3-CH-CH-CH_2-CH_3$$

the longest continous chain (colored C's) has five carbon atoms. The compound is therefore named as a substituted *pentane,* even though there are seven carbon atoms altogether.

4. Groups attached to the main chain are called **substituents.** Saturated substituents that contain only carbon and hydrogen are called **alkyl groups.** They are named from the alkane with the same number of carbon atoms by changing the -*ane* ending to -*yl.*

In the example, each branch has only one carbon. Derived from the hydrocarbon methane by removing one of the hydrogens, it is called the **methyl group.**

$$H-\overset{\overset{\displaystyle H}{|}}{\underset{\underset{\displaystyle H}{|}}{C}}-H \qquad H-\overset{\overset{\displaystyle H}{|}}{\underset{\underset{\displaystyle H}{|}}{C}}- \qquad \text{or} \qquad CH_3- \qquad \text{or} \qquad Me-$$

methane   methyl group

5. Groups are located by a name and a number. *The main chain is numbered in such a way that the first substituent encountered along the chain receives the lowest possible number.* When two identical groups are attached to the main chain, prefixes such as *di-, tri-,* and *tetra-* are used. Every substituent must be named and numbered, even if two identical substituents are attached to the same carbon of the main chain. The compound

$$\underset{1}{CH_3}-\underset{2}{\overset{\overset{\displaystyle CH_3}{|}}{CH}}-\underset{3}{\overset{\overset{\displaystyle CH_3}{|}}{CH}}-\underset{4}{CH_2}-\underset{5}{CH_3}$$

is correctly named 2,3-dimethylpentane (the name tells us that the compound has two methyl substituents, one attached to carbon-2 and one attached to carbon-3 of a five carbon chain).

6. *Punctuation is important in writing IUPAC names.* The names are written as one word. Numbers are separated from each other by commas and from letters by hyphens. If two or more different types of substituents are present, they are listed alphabetically (to aid indexing), except that prefixes such as *di-* and *tri-* are not counted for alphabetizing. The last substituent named is prefixed to the parent alkane to form one word.

Example 2.2 **Give the IUPAC name for**

$$CH_3-\overset{\overset{\displaystyle CH_3}{|}}{\underset{\underset{\displaystyle CH_3}{|}}{C}}-CH_2CH_2CH_3$$

Solution Number the longest chain from left to right (so that the substituents receive a low number—rule 5). The correct name is **2,2-dimethylpentane.**

Problem 2.3 **Give the IUPAC name for the following compounds.**

a. $CH_3\overset{\overset{\displaystyle }{}}{\underset{\underset{\displaystyle CH_3}{|}}{CH}}CH_2CH_3$ b. $CH_3CH_2\overset{\overset{\displaystyle }{}}{\underset{\underset{\displaystyle CH_3}{|}}{CH}}CH_3$ c. $CH_3-\overset{\overset{\displaystyle CH_3}{|}}{\underset{\underset{\displaystyle CH_3}{|}}{C}}-CH_3$

2.5
ALKYL AND
HALOGEN
SUBSTITUENTS

As illustrated for the methyl groups, under IUPAC rule 4, the names of alkyl substituents are derived by changing the -*ane* ending of alkanes to -*yl*. Thus the two-carbon alkyl group is called an ethyl group, from ethane.

$$CH_3CH_3 \qquad CH_3CH_2 \text{—} \qquad \text{or} \qquad C_2H_5 \text{—} \qquad \text{or} \qquad Et \text{—}$$

ethane   ethyl group

However, when we come to propane there are two possible alkyl groups, depending on which type of hydrogen is removed. If a terminal hydrogen is removed, the group is called a normal or *n*-propyl (or 1-propyl) group.

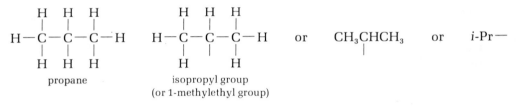

propane   *n*-propyl or 1-propyl group

But if a hydrogen is removed from the central carbon atom, we get a different propyl group called the isopropyl group.

$$\underset{\text{propane}}{H-\overset{\displaystyle H}{\underset{\displaystyle H}{C}}-\overset{\displaystyle H}{\underset{\displaystyle H}{C}}-\overset{\displaystyle H}{\underset{\displaystyle H}{C}}-H} \qquad \underset{\substack{\text{isopropyl group} \\ \text{(or 1-methylethyl group)}}}{H-\overset{\displaystyle H}{\underset{\displaystyle H}{C}}-\overset{\displaystyle H}{\underset{\displaystyle}{C}}-\overset{\displaystyle H}{\underset{\displaystyle H}{C}}-H} \qquad \text{or} \qquad CH_3CHCH_3 \qquad \text{or} \qquad i\text{-Pr}\text{—}$$

There are four different butyl groups:

$$CH_3CH_2CH_2CH_2\text{—} \qquad CH_3CHCH_2CH_3 \qquad \overset{CH_3}{\underset{CH_3}{\diagdown}}CH\text{—}CH_2\text{—} \qquad CH_3\text{—}\overset{CH_3}{\underset{CH_3}{\overset{|}{\underset{|}{C}}}}\text{—}$$

| *n*-butyl | *sec*-butyl | isobutyl | *tert*-butyl |
| (or butyl) | (or 1-methylpropyl) | (or 2-methylpropyl) | (or 1,1-dimethylethyl) |

These names for the alkyl groups with 1–4 carbon atoms are very commonly used, so you must memorize them.

The symbol R— is used as a general symbol for an alkyl group. The formula R—H therefore represents any alkane, and the formula R—Cl stands for any alkyl chloride (methyl chloride, ethyl chloride, and so on).

Halogen substituents are named by changing the -*ine* ending of the element to -*o*.

$$F\text{—} \qquad Cl\text{—} \qquad Br\text{—} \qquad I\text{—}$$

fluoro-   *chloro-*   *bromo-*   *iodo-*

Example 2.3   **Give the common and IUPAC names for $CH_3CH_2CH_2Br$.**

Solution   The common name is *n*-propyl bromide (the common name of the alkyl group is followed by the name of the halide). The IUPAC name is 1-bromopropane, the halogen being named as a substituent on the three-carbon chain.

Problem 2.4   Write the formula for each of the following: a 1-pentyl group; a 2-pentyl group.

Problem 2.5   Write the formula for each of the following compounds.
   a. *n*-propyl bromide
   b. isopropyl chloride
   c. 2-chloropropane
   d. *tert*-butyl iodide
   e. isobutyl alcohol
   f. any alkyl fluoride

**2.6
USE OF THE
IUPAC RULES**

The examples given in Table 2.2 illustrate how the IUPAC rules are applied to name a particular structure.

   When writing a structural formula from a name, first write the longest carbon chain, number it, add the substituents to the correct carbon atoms, and then fill in the formula with the correct number of hydrogens at each

---

TABLE 2.2

Examples of the use of the IUPAC rules

$$\overset{5}{C}H_3\overset{4}{C}H_2\overset{3}{C}H_2\overset{2}{C}H\overset{1}{C}H_3$$
$$\underset{\displaystyle CH_3}{|}$$

2-methylpentane
(*not* 4-methylpentane)

The *-ane* tells us that all the carbon–carbon bonds are single; *pent-* indicates five carbons in the longest chain, and we number from right to left to give the methyl substituent the lowest possible number.

$$\overset{3}{C}H_3\overset{4}{C}HCH_2\overset{5}{C}H_2\overset{6}{C}H_3$$
$$\underset{\displaystyle \overset{2}{C}H_2\overset{1}{C}H_3}{|}$$

3-methylhexane
(not 2-ethylpentane)

A six-carbon saturated chain with a methyl group on the third carbon. We would usually write the structure as $CH_3CH_2CHCH_2CH_2CH_3$.
$$\underset{\displaystyle CH_3}{|}$$

$$CH_3$$
$$|$$
$$\overset{1}{C}H_3-\overset{2}{C}-\overset{3}{C}H_2\overset{4}{C}H_3$$
$$|$$
$$CH_3$$

2,2-dimethylbutane
(*not* 2,2-methylbutane,
not 2-dimethylbutane)

There must be a number for each substituent, and the prefix *di-* says that there are two methyl substituents.

$$CH_3$$
$$|$$
$$\overset{1}{C}H_2\overset{2}{C}H_2\overset{3}{C}H-\overset{4}{C}H_3$$
$$|$$
$$Cl$$

1-chloro-3-methylbutane
(not 4-chloro-2-methylbutane)

Number the butane chain to give the first substituent the lowest possible number.

carbon. For example, to write the formula for 2,2,4-trimethylpentane, we go through these steps:

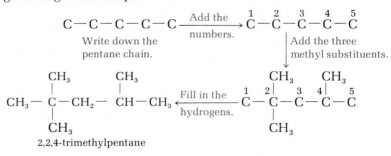

2,2,4-trimethylpentane

Problem 2.6   Name the following compounds by the IUPAC system.
a. CH_3CHFCH_3 b. $(CH_3)_3CCH_2CHClCH_3$

Problem 2.7   Write the structure for 3,3-dimethylpentane.

Problem 2.8   Explain why 1,3-dichlorobutane is a correct IUPAC name, but 1,3-dimethylbutane is *not* a correct IUPAC name.

2.7
PHYSICAL
PROPERTIES
OF ALKANES

Alkanes are insoluble in water, and those that are liquid are less dense than water and float on it. (Hence the saying "oil and water do not mix.") The reason is that water molecules are polar and attract one another, whereas alkanes are nonpolar. To intersperse alkane and water molecules, we would have to break up the attractive force between the water molecules, and this would be difficult to do.

The mutual insolubility of alkanes and water is used to advantage by many plants. Alkanes often form part of the protective coating on leaves and fruits. If you have ever polished an apple, you know that the skin or cuticle contains waxes. Among them are the C_{27} and C_{29} n-alkanes. The leaf wax of cabbage and broccoli is mainly $n\text{-}C_{29}H_{60}$, whereas the main alkane of tobacco leaves is $n\text{-}C_{31}H_{64}$. Similar hydrocarbons are found in beeswax. The major function of these waxes is to prevent water loss.

Alkanes have lower boiling points for a given molecular weight than most other organic compounds. This is because the attractive forces between nonpolar molecules are weak, and the process of separating molecules from one another (which is what we do when we convert a liquid to a gas) requires relatively little energy. Figure 2.2 shows the boiling points of some alkanes. The greater the molecular surface area, the greater the attractive forces between molecules. Therefore boiling points rise as the chains increase in length and fall as the chains become branched and more spherical in shape.

2.8
CONFORMATIONS
OF ALKANES

The shapes of molecules affect their properties, and in recent years chemists have begun to pay careful attention to the fine details of molecular geometry. In a simple molecule like ethane, for example, an infinite num-

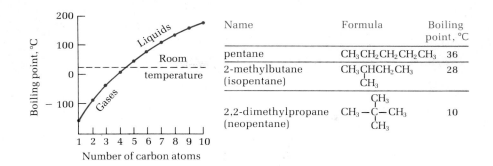

Name	Formula	Boiling point, °C
pentane	$CH_3CH_2CH_2CH_2CH_3$	36
2-methylbutane (isopentane)	$CH_3\underset{\underset{CH_3}{\mid}}{C}HCH_2CH_3$	28
2,2-dimethylpropane (neopentane)	$CH_3-\underset{\underset{CH_3}{\mid}}{\overset{\overset{CH_3}{\mid}}{C}}-CH_3$	10

FIGURE 2.2   As shown by the smooth curve, the boiling points of the normal alkanes rise smoothly as the length of the carbon chain increases. Note from the table, however, that chain branching causes a decrease in boiling point (each compound in the table has the same number of carbons and hydrogens, C_5H_{12}).

ber of structures is possible as a consequence of rotating one carbon atom (and its attached hydrogens) with respect to the other carbon atom. These arrangements are called **conformations.** Two particular possibilities for ethane are shown in Figure 2.3.

The Newman projections (devised by Professor Melvin S. Newman of Ohio State University) are particularly useful for representing conformations. In Newman formulas, one views the carbon–carbon bond end-on. Bonds to the "front" carbon go to the center of the circle, whereas bonds to the "rear" carbon go only to the edge of the circle. As shown in Figure 2.3, conformations can also be illustrated by sawhorse projections and by space-filling models.

In the **staggered conformation** of ethane, each C—H bond on one carbon bisects an H—C—H angle on the other carbon, whereas in the **eclipsed conformation**, C—H bonds on the front and back carbons are aligned. By rotating one carbon with respect to the other by 60°, we can interconvert staggered and eclipsed conformations. Between these two extremes there is an infinite number of intermediate conformations of ethane.

The staggered and eclipsed conformations of ethane can be regarded as **rotational isomers** (or **rotamers**), because they can be converted into one another by rotation about the carbon–carbon bond. Such rotation about a single bond occurs quite easily, however, because overlap of the sp^3 orbitals on each carbon is not affected by rotation about the σ bond (see Figure 1.4). Indeed, there is enough heat available at room temperature for the staggered and eclipsed forms of ethane to interconvert rapidly. For this reason, the two forms cannot be separated at room temperature. We know, however, that both forms are not equally stable. The staggered

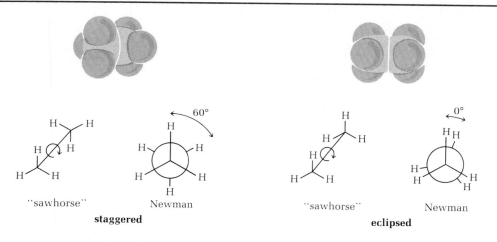

FIGURE 2.3   Two of the possible conformations of ethane, staggered and eclipsed. Interconversion is easily possible via a 60° rotation about the C—C bond, as shown by the curved arrows. The upper formulas show space-filling models. The structure at the lower left in each case is a "sawhorse" drawing. The structure at the lower right in each case is a Newman projection formula, an end-on view down the C—C axis.

conformation is much preferred, and at room temperature more than 99% of the ethane molecules have the staggered arrangement. This is probably because the bonding electrons on adjacent carbons are furthest apart in this arrangement.

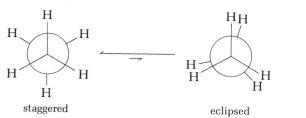

staggered   eclipsed   (2.1)

Example 2.4   Draw the Newman projection formulas for the staggered and eclipsed conformations of propane.

Solution

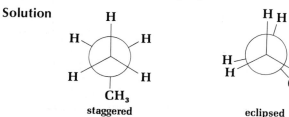

staggered   eclipsed

Problem 2.9   **Draw Newman projection formulas for two different staggered conformations of butane (looking down the carbon-2–carbon-3 bond), and predict which of the two conformations will be more stable.**

   The most important thing to remember about conformational isomers is that they are just different forms of a single molecule that can be interconverted by rotational motions about single (σ) bonds. More often than not, sufficient thermal energy for this rotation is available at room temperature. Consequently, at room temperature it is usually not possible to separate conformers from one another.
   Now let us take a look at cycloalkanes and the effect that the cyclic structure imposes on their possible conformations.

**2.9
CYCLOALKANE
NOMENCLATURE
AND
CONFORMATION**

Cycloalkanes are carbocyclic hydrocarbons that are named by placing the prefix *cyclo-* before the alkane name that corresponds to the number of carbon atoms in the ring. The structures and names of the first six unsubstituted cycloalkanes are

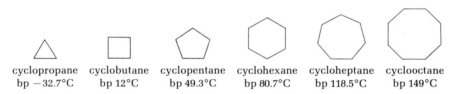

cyclopropane   cyclobutane   cyclopentane   cyclohexane   cycloheptane   cyclooctane
bp −32.7°C   bp 12°C   bp 49.3°C   bp 80.7°C   bp 118.5°C   bp 149°C

When the rings carry alkyl or halogen substituents, they are named in the usual way. If only one substituent is present, no number is needed to locate it. If there are several substituents, numbers are required. One substituent is always assigned the number 1, and the ring carbons are then numbered consecutively in a way that gives the other substituents the lowest possible numbers. The following examples illustrate the system:

methylcyclopentane   1,1-dimethylcyclopentane   1,2-dimethylcyclopentane
(*not* 1-methylcyclopentane)   (*not*-1,5-dimethylcyclopentane)

Problem 2.10   **Draw the structural formulas for:**
                   a. 1,3-dimethylcyclopentane   b. 1,2,3-trichlorocyclopropane

Problem 2.11   **Give IUPAC names for:**

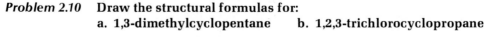
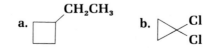

Let us look now at the conformations of cycloalkanes. Cyclopropane, with only three carbon atoms, is necessarily planar (this is because three points determine a plane). The C—C—C bond angle is only 60°, much less than the usual tetrahedral angle of 109.5°. The hydrogens lie above and below the carbon plane, and hydrogens on adjacent carbons are eclipsed.

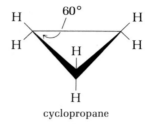

cyclopropane

Because of the strain that results primarily from the small angles, cyclopropane often reacts to give ring-opened products.

All cycloalkanes with more than three carbon atoms are nonplanar and have "puckered" conformations. With cyclobutane and cyclopentane, puckering actually makes the C—C—C angles a little smaller than they would be if the molecules were planar, but less eclipsing of adjacent hydrogens compensates for this.

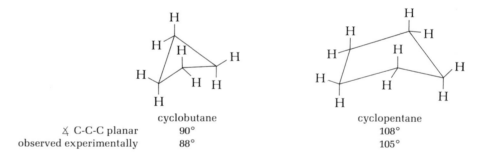

	cyclobutane	cyclopentane
∡ C-C-C planar	90°	108°
observed experimentally	88°	105°

Six-membered rings are rather special and have been studied in great detail because they are so common in nature. If cyclohexane were a planar hexagon, the internal C—C—C angles would be 120°—quite a bit larger than the tetrahedral angle (109.5°). Strain prevents cyclohexane from being planar (or flat). The most favored conformation of cyclohexane is the **chair conformation,** an arrangement in which all the C—C—C angles are the normal 109.5° and all the hydrogens on adjacent carbon atoms are perfectly staggered. Figure 2.4 shows models of the cyclohexane chair conformation.

In the chair conformation, the hydrogens fall into two sets, called **axial** and **equatorial.** Three axial hydrogens lie above and three lie below the average or mean plane of the carbon atoms, whereas the six equatorial hydrogens lie approximately in that plane (see Figure 2.5). Through a motion in which alternate ring carbons (say 1, 3, and 5) move in one direction (down) and the other three ring carbons move in the opposite direction (up), one chair conformation can be converted into another chair

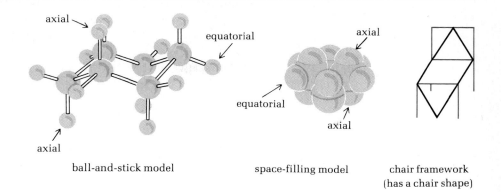

ball-and-stick model   space-filling model   chair framework
                                                           (has a chair shape)

FIGURE 2.4   The chair conformation of cyclohexane, shown in ball-and-stick and
space-filling models. The axial hydrogens lie above or below the mean
plane of the carbons, and the equatorial hydrogens lie approximately
within the mean plane of the carbons. The origin of the "chair"
terminology is illustrated at the right.

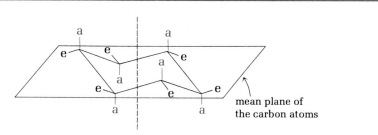

FIGURE 2.5   Axial and equatorial bonds in the chair conformation. The axial bonds (a),
shown in color, are parallel to the vertical axis through the middle of the
ring. Equatorial bonds (e), in black, lie roughly in the ring plane.

conformation in which all axial hydrogens in one conformation become
equatorial in the other, and vice versa.

(2.2)

Axial bonds (in color) in the left structure become
equatorial bonds (in color) in the right structure
when the ring "flips."

At room temperature this flipping process is rapid, but at low temperatures (say −90°C) the flipping process slows down so that the two different types of hydrogens can actually be detected by certain spectroscopic methods.

There is another important feature of cyclohexane conformations to consider. If you look carefully at the space-filling model of cyclohexane (Figure 2.4), you will notice that the three axial hydrogens on the same face of the ring nearly touch each other. If one of these axial hydrogens were replaced by a larger group (such as methyl), the axial crowding would be even worse. Therefore, the preferred conformation is the one in which the methyl substituent is equatorial.

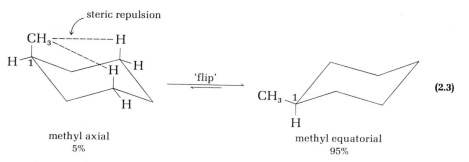

methyl axial
5%

methyl equatorial
95%

(2.3)

Problem 2.12   **tert-Butylcyclohexane exists almost 100% in a single conformation, with the *tert*-butyl group equatorial. Explain why the conformational preference is greater than for methylcyclohexane.**

Problem 2.13   **Another puckered conformation for cyclohexane, one in which all C—C—C angles are the normal 109.5°, is the "boat" conformation. Explain why this conformation is very much less stable than the chair conformation.**

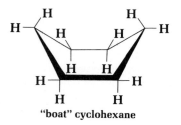

"boat" cyclohexane

Before we proceed to reactions of alkanes and cycloalkanes, we need to consider a kind of isomerism that may arise when substituents are connected to two or more different ring carbon atoms in a cycloalkane.

**2.10
CIS-TRANS
ISOMERISM IN
CYCLOALKANES**

Stereoisomerism deals with molecules that have the same order of attachment of the atoms but different arrangements of the atoms in space. ***Cis-trans isomerism*** (also called geometric isomerism) is one kind of stereoisomerism, and it is most easily seen with a specific case. Consider, for example, the possible structures of 1,2-dimethylcyclopentane. For simplicity, let us neglect the slight puckering of the ring and draw it as if it were planar. The methyl groups may be on the same side of the ring plane or they may be on opposite sides:

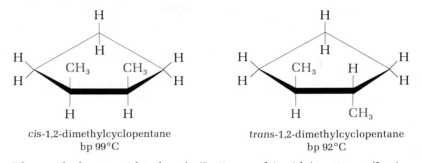

cis-1,2-dimethylcyclopentane   trans-1,2-dimethylcyclopentane
bp 99°C   bp 92°C

The methyls are said to be **cis** (Latin, on this side) or **trans** (Latin, across), respectively.

Cis-trans isomers differ from one another only in the way the atoms or groups are positioned in space. Yet this difference is sufficient to give them different physical and chemical properties (note, for example, the boiling points under the 1,2-dimethylcyclopentane structures). *Cis-trans isomers are separate and unique compounds.* Unlike conformational isomers, they cannot be interconverted by rotation around the carbon–carbon bonds. In the example described, the cyclic structure prevents this rotation. Cis-trans isomers can usually be separated from each other and kept separate.

Problem 2.14   **Draw the structures for the *cis*- and *trans*-isomers of:**
a. 1,2-dichlorocyclopropane
b. 1-bromo-3-chlorocyclobutane

2.11
***CIS-TRANS* AND**
CONFORMATIONAL
ISOMERS OF
CYCLOHEXANES

When we combine *cis-trans* isomerism with conformational isomerism, as is necessary for cyclohexanes, the situation becomes a little more complicated. Let us examine, for example, the *cis*- and *trans*- isomers of 1,2-dimethylcyclohexane. The *cis*-isomer is represented by the following equilibrium:

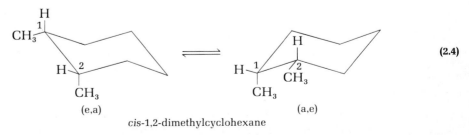

(2.4)

cis-1,2-dimethylcyclohexane

In each chair conformation, the methyl substituents are on the lower or bottom face of the mean ring plane. Another way to look at the structure is to notice that the methyl groups are attached to carbon-1 and carbon-2 by the lower of the 2 bonds at each carbon. Consequently the methyl groups are *cis* to each other in both structures. Now, in the conformation at the left, the methyl group at C-1 is equatorial and the methyl at C-2 is axial (hence, the **e,a**-conformation). When the ring "flips," the methyl at C-1 becomes axial and the methyl at C-2 becomes equatorial (the **a,e** conformation). Both conformations have one equatorial and one axial methyl group, and for this reason have equal stabilities. The equilibrium shown in eq. 2.4 is therefore 50:50.

The situation for *trans*-1,2-dimethylcyclohexane is different.

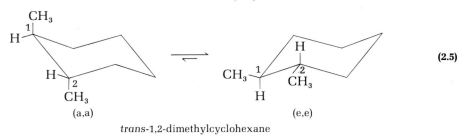

(2.5)

trans-1,2-dimethylcyclohexane

In both conformations the methyl groups are *trans* (the methyl at C-1 is attached to the upper of the two bonds, whereas the methyl at C-2 is attached to the lower of the two bonds). However, in the conformation at the left both methyls are axial, whereas at the right both methyls are equatorial. The conformation at the right (e,e) is therefore the preferred structure for *trans*-1,2-dimethylcyclohexane, and the equilibrium shown in eq. 2.5 is strongly shifted to the right.

Example 2.5   The following structures all represent 1,4-dimethylcyclohexane. For each structure, tell whether the methyl groups are *cis* or *trans,* and also tell whether they are axial or equatorial.

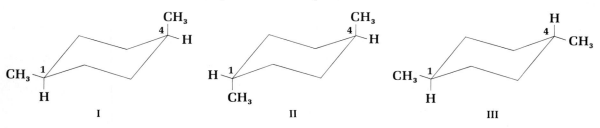

Solution   I *cis* (e,a): both methyls are attached to the upper bond at C-1 and C-4. II *trans* (a,a) and III *trans* (e,e): methyls in II and III are attached to the upper bond of one carbon and to the lower bond of the other carbon.

Problem 2.15   Which form of 1,4-dimethylcyclohexane is expected to be most stable: I, II, or III?

Problem 2.16   Draw the structures obtained when the cyclohexane ring "flips" from one chair conformation to the other (a) for I and (b) for II. Use equilibrium arrows such as those in eq. 2.4 or eq. 2.5 to describe the position of the interconversion equilibrium.

**2.12
REACTIONS OF
ALKANES**

The bonds in alkanes are single, covalent, and nonpolar. Hence alkanes are relatively inert (the old term *paraffins* comes from the Latin *parum affinitas*, little affinity). Alkanes do not react with most common acids, bases, or oxidizing or reducing agents. Because of this inertness, alkanes can be used as solvents for extraction or for carrying out chemical reactions with other substances. However, alkanes do react with some reagents, such as oxygen and the halogens. We will discuss those reactions here.

**2.13
OXIDATION AND
COMBUSTION.
ALKANES AS FUELS**

The most important use of alkanes is as fuels. Alkanes burn in an excess of oxygen to form carbon dioxide and water. Most important, the reaction evolves large quantities of heat (that is, the reactions are **exothermic**).

$$CH_4 + 2\,O_2 \rightarrow CO_2 + 2\,H_2O + 212.8\,\text{kcal/mol} \tag{2.6}$$

$$C_4H_{10} + \tfrac{13}{2}\,O_2 \rightarrow 4\,CO_2 + 5\,H_2O + 688.0\,\text{kcal/mol} \tag{2.7}$$

These combustion reactions are the basis for the use of hydrocarbons for heat (natural gas and heating oil) and for power (gasoline). An initiation step is required—usually ignition by a spark or flame. Once initiated, the reaction proceeds spontaneously and exothermically.

If insufficient oxygen for complete reaction is present, partial combustion may occur. In this case, carbon in the hydrocarbon may be oxidized only to carbon monoxide or even only to carbon.

$$2\,CH_4 + 3\,O_2 \rightarrow 2\,CO + 4\,H_2O \tag{2.8}$$

$$CH_4 + O_2 \rightarrow C + 2\,H_2O \tag{2.9}$$

The effects of incomplete combustion are well known to every motorist. They include the gradual build-up of carbon deposits on the head and pistons of the engine and toxic carbon monoxide in exhaust fumes. Although usually undesirable, incomplete combustion of natural gas is sometimes purposely carried out to manufacture carbon blacks, particularly lampblack, a pigment for ink.

**2.14
HALOGENATION
OF ALKANES**

When a mixture of an alkane and chlorine gas is stored at low temperatures in the dark, no reaction occurs. In sunlight or at high temperatures, however, an exothermic reaction occurs. One or more hydrogen atoms of the alkane are replaced by chlorine atoms. The reaction can be represented by the general equation

$$R-H + Cl-Cl \xrightarrow[\text{heat}]{\text{light or}} R-Cl + H-Cl \tag{2.10}$$

or, specifically for methane,

$$CH_4 + Cl-Cl \xrightarrow[\text{or heat}]{\text{sunlight}} CH_3Cl + HCl \qquad (2.11)$$

methane   chloromethane
                                        (methyl chloride)
                                        bp −24.2°C

The reaction is called **chlorination.** It is a **substitution reaction;** a chlorine is substituted for a hydrogen.

An analogous reaction, **bromination,** occurs when the halogen is bromine.

$$R-H + Br-Br \xrightarrow[\text{heat}]{\text{light or}} R-Br + HBr \qquad (2.12)$$

If excess halogen is present, the reaction can continue further to give polyhalogenated products. Thus methane and excess chlorine can give methylene chloride, chloroform, or carbon tetrachloride.*

$$CH_3Cl \xrightarrow{Cl_2} CH_2Cl_2 \xrightarrow{Cl_2} CHCl_3 \xrightarrow{Cl_2} CCl_4 \qquad (2.13)$$

dichloromethane   trichloromethane   tetrachloromethane
(methylene chloride)   (chloroform)   (carbon tetrachloride)
bp 40°C   bp 61.7°C   bp 76.5°C

By controlling the reaction conditions and the ratio of chlorine to methane, we can favor formation of one or another of the possible products.

Problem 2.17   **Write the names and structures of all possible bromination products of methane.**

With longer-chain alkanes, mixtures of products can be obtained even at the first step. For example, with propane,

$$CH_3CH_2CH_3 + Cl_2 \xrightarrow[\text{or heat}]{\text{light}} CH_3CH_2CH_2Cl + CH_3\underset{\underset{Cl}{|}}{C}HCH_3 + HCl \qquad (2.14)$$

propane   1-chloropropane   2-chloropropane
                               (n-propyl chloride) (isopropyl chloride)

When larger alkanes (such as octane) are halogenated, the mixture of products becomes more complex, so that individual isomers become difficult to separate and obtain pure, and halogenation tends to become less useful as a way of synthesizing alkyl halides.[†] With unsubstituted cycloalkanes, however, where all the hydrogens are equivalent, a single pure organic product can be obtained.

---

*Note that, as in eq. 2.13, we sometimes write the formula of one of the reactants (in this case Cl_2) over the arrow for convenience. We also sometimes omit obvious inorganic products (in this case HCl).

[†]Note that, as in eq. 2.14, we often do not write a balanced equation, especially when more than one product is formed from a single organic reactant. Instead, we show on the right side of the equation the structures of *all* the important organic products.

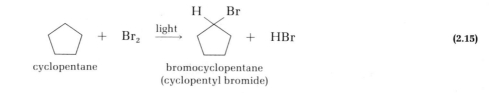

$$\text{cyclopentane} \quad + \quad Br_2 \quad \xrightarrow{\text{light}} \quad \text{bromocyclopentane} \quad + \quad HBr \qquad (2.15)$$

cyclopentane

bromocyclopentane
(cyclopentyl bromide)

Problem 2.18 Write out the structures of all the monobromination products of pentane. Note the complexity of the product mixture, compared to the corresponding reaction with cyclopentane (eq. 2.15).

Problem 2.19 How many organic products can be obtained from the monochlorination of octane? cyclooctane?

**2.15
THE MECHANISM
OF
HALOGENATION**

One may well ask how halogenation occurs. Why is light or heat necessary? Eqs. 2.10 and 2.11 express the *overall* reaction for halogenation. They describe the structures of the reactants and the products, and they indicate necessary reaction conditions or catalysts over the arrow. But they do *not* tell us exactly how the products are formed from the reactants. A **reaction mechanism** is a step-by-step description of the bond-breaking and bond-making processes that occur when reagents react to form products.

In the case of halogenation, various experiments show that the reaction occurs in several steps, not in one step. Halogenation occurs via a **free-radical chain** of reactions.

The **chain-initiating step** is the breaking of the halogen molecule into two halogen atoms.

initiation $:\!\overset{..}{\underset{..}{Cl}}\!-\!\overset{..}{\underset{..}{Cl}}\!: \xrightarrow[\text{heat}]{\text{light or}} 2 \ :\!\overset{..}{\underset{..}{Cl}}\!\cdot$ (2.16)

The halogen–halogen bond is weaker than either the $C\!-\!H$ bond or the $C\!-\!C$ bond, so the halogen, not the alkane, absorbs energy (from light or heat) and gets the reaction going.

The **chain-propagating steps** are

propagation
$\begin{cases} R\!-\!H + :\!\overset{..}{\underset{..}{Cl}}\!\cdot \rightarrow R\!\cdot + HCl \qquad\qquad\qquad (2.17) \\ \qquad\qquad \text{alkyl} \\ \qquad\qquad \text{radical} \\[1em] R\!\cdot + Cl\!-\!Cl \rightarrow R\!-\!Cl + :\!\overset{..}{\underset{..}{Cl}}\!\cdot \qquad\quad (2.18) \\ \qquad\qquad \text{alkyl} \\ \qquad\qquad \text{chloride} \end{cases}$

Chlorine atoms are very reactive. They may either recombine to form chlorine molecules (the reverse of eq. 2.16) or, if they collide with an alkane molecule, abstract a hydrogen atom to form hydrogen chloride and an alkyl radical, $R\cdot$. Recall (eq 1.3, Sec. 1.4) that a radical is a fragment with an odd number of unshared electrons. Note from the space-filling models in Figure 2.1 that alkanes seem to have an exposed surface of hydrogens covering the carbon skeleton. So it is most likely that, if a halogen atom were to collide with an alkane molecule, it would hit the hydrogen end of a C—H bond.

Like a chlorine atom, the alkyl radical formed in the first step of the chain (eq. 2.17) is very reactive. If it were to collide with a chlorine molecule, it could form an alkyl chloride molecule and a chlorine atom (eq. 2.18). The chlorine atom formed in this step can then react as in eq. 2.17 to repeat the sequence. Note that, when you add eq. 2.17 and eq. 2.18, you get the overall equation for chlorination (eq. 2.10). In each chain-propagating step a radical (or atom) is formed and can continue the chain. Almost all of the reactants are consumed, and almost all of the products are formed in these steps.

Were it not for some **chain-terminating steps,** all of the reactants could, in principle, be consumed by initiating a single chain. However, because many chlorine molecules react to form chlorine atoms in the chain-initiating step, many chains are started simultaneously. Quite a few radicals are present as the reaction proceeds. If any two radicals combine, the chain will be terminated. Three possible chain-terminating steps are

$$\textit{termination} \begin{cases} 2 : \overset{..}{\underset{..}{C}} l \cdot \rightarrow Cl-Cl & \text{(2.19)} \\ 2 R \cdot \rightarrow R-R & \text{(2.20)} \\ R \cdot + : \overset{..}{\underset{..}{C}} l \cdot \rightarrow R-Cl & \text{(2.21)} \end{cases}$$

No new radicals are formed in these reactions, so the chain is broken or, as we say, terminated.

Problem 2.20　**Write equations for all the steps (initiation, propagation, termination) in the free-radical chlorination of methane to methyl chloride.**

Problem 2.21　**Account for the experimental observation that small amounts of ethane and chloroethane are produced during the chlorination of methane. (Hint: Consider the possible chain-terminating steps).**

a word about ————————————————————————

1. Methane, Marsh Gas, and Miller's Experiment

Methane is commonly found in nature wherever bacterial decomposition of organic matter can occur in the absence of oxygen (anaerobic conditions), as in marshes, swamps, or the muddy sediment of lakes. Hence its common name, marsh gas. In China methane has been collected from the mud at the bottom of swamps for use in domestic cooking and lighting.

Methane is similarly formed in the digestive tracts of certain animals—ruminants, such as cows.

The scale of methane production by bacteria is considerable. The earth's atmosphere contains an average of 1 part per million of methane. Because our planet is small and because methane is light compared to most other air constituents (O_2, N_2), one would expect most of the methane to escape, and it has been calculated that the equilibrium concentration should be very much less than is observed (perhaps only 1 part in 10^{35} parts of air). The reason for the relatively high observed concentra-

and oxygen as water. Indeed, some of the larger planets (such as Saturn and Jupiter), which have very strong gravitational fields and low surface temperatures that help retain light molecules, still have atmospheres that are rich in methane and ammonia. A now-famous experiment carried out in 1955 by Stanley L. Miller (working in the laboratory of H. C. Urey at Columbia University) supports the idea that life could have arisen in a reducing atmosphere. Miller found that when mixtures of methane, ammonia, water, and hydrogen are subjected to electric discharges to simulate lightning, some organic compounds are formed (amino acids for example) that are important in biology and necessary for life. Similar results have since been obtained using heat or ultraviolet light in place of the electric discharges (it seems likely that the earth's early atmosphere was subjected to much more ultraviolet light than it is now). When oxygen was added to these simulated primeval atmospheres, no amino acids were produced—strong evidence that the earth's original atmosphere did not contain free oxygen. Miller's experiment provided the model for much work in the branch of science now called chemical evolution, the study of chemical events that took place on earth and led to the appearance of the first living cell.

At present, the two most important natural sources of alkanes on earth are petroleum and natural gas. Petroleum is a complex liquid mixture of organic compounds, many of which are alkanes or cyclo-alkanes. All the normal alkanes through $C_{33}H_{68}$ have been isolated from petroleum, and many branched-chain alkanes have also been found there. Natural gas, often found associated with petroleum deposits, consists mainly of methane (about 80%) together with lesser amounts of ethane (5–10%) and higher alkanes. In the United States alone there are over 250,000 miles of natural gas pipelines, distributing this source of energy to all parts of the country. Natural gas is also distributed worldwide via huge tankers. The gas is liquefied ($-160°C$) to conserve space, because 1 m³ of the liquefied gas is equivalent to about 600 m³ of gas at atmospheric pressure. Some tankers are capable of carrying over 100,000 m³ of the liquefied gas. The depletion of these two major energy sources, petroleum and natural gas, and the problem of finding substitutes for them are major concerns of our time.

tion is that methane is constantly being produced by the bacterial decay of plant matter.

In cities the amount of methane in the atmosphere reaches much higher levels, up to several parts per million. The peak concentrations are achieved in the early morning and late afternoon, a direct correlation with automobile traffic. Fortunately methane, which constitutes about 50% of the total hydrocarbons that are urban atmospheric pollutants, seems to have no direct harmful effect on human health.

Methane sometimes accumulates in coal mines, where it is a hazard. Mixed with 5–14% air, methane is explosive. And miners can be asphyxiated by it (due to lack of sufficient oxygen). Fortunately, as they are overcome, they fall to the ground and often revive, because methane is lighter than air and tends to rise to the top of the mine tunnels. Dangerous concentrations of methane are readily detected by a variety of safety devices, including canaries, which succumb to lower concentrations of methane than are harmful to humans and alert miners to the presence of a hazard.

Methane was probably one of the main components of the earth's atmosphere in its early years. Also, hydrogen is the most common element in the solar system (it constitutes about 87% of the sun's mass). It is therefore reasonable that, when the planets were formed, other elements should have been present in reduced (rather than oxidized) forms: carbon as methane, nitrogen as ammonia,

2.22. Write structural formulas for the following compounds.
a. 3-methylpentane
b. 2,3-dimethylbutane
c. 3,3-dimethyl-4-ethylhexane
d. 2-chloro-3-methylpentane
e. 2,2,3-trimethylbutane
f. 2-bromopropane
g. 1,1-dichlorocyclopropane
h. 1,1,3,3-tetrachloropropane
i. 3-bromo-1,1-dimethylcyclopentane
j. 1,4-dichlorocyclohexane

2.23. Write expanded formulas for the following compounds and name them by using the IUPAC system.
a. $CH_3(CH_2)_3CH_3$
b. $CH_3CH(CH_3)CH_2CH_3$
c. $CH_3CH_2C(CH_3)_2CH_2CH_3$
d. $CH_3(CH_2)_2C(CH_3)_3$
e. $CH_3CH_2CHBrCH_3$
f. $CH_3CCl_2CBr_3$
g. $(CH_3CH_2)_4C$
h. CH_2ClCH_2Br
i. $CH_2BrCH(CH_3)CH(CH_3)_2$
j. $(CH_2)_5$
k. MeI
l. i-PrBr

2.24. Give a common and an IUPAC name for the following compounds.
a. CH_3I
b. CH_3CH_2Cl
c. CH_2Cl_2
d. $CHBr_3$
e. $CH_3CH_2CH_2Cl$
f. $(CH_3)_2CHBr$
g. $CHCl_3$
h. CH_2—CH—Cl
                                                                  | |
i. $CH_3CHICH_2CH_3$
j. $(CH_3)_3CCl$
                                                                   CH_2—CH_2

2.25. Write a structure for each of the compounds listed. Explain why the given name is objectionable and give a correct name in each case.
a. 1-methylpentane
b. 2-ethylbutane
c. 2,3-dichloropropane
d. 1,4-dimethylcyclobutane
e. 1,1,3-trimethylpropane
f. 3-bromo-2-methylpropane

2.26. Write the structural formulas for all the isomers (numbers indicated in parentheses) for each of the following compounds, and name each isomer by the IUPAC system.
a. C_4H_{10} (2)
b. C_4H_9Br (4)
c. C_6H_{14} (5)
d. $C_3H_6Br_2$ (4)
e. $C_2H_2BrCl_3$ (3)
f. C_3H_6BrCl (5)

2.27. The general formula for an alkane is C_nH_{2n+2}. What would be the corresponding formula for a cycloalkane?

2.28. Write structural formulas and names for all possible cycloalkanes with each of the following molecular formulas. Be sure to include *cis-trans* isomers, when appropriate. Name each compound by the IUPAC system.
a. C_5H_{10}
b. C_6H_{12} (there are 16).

2.29. Without referring to tables, arrange the following five hydrocarbons in order of increasing boiling point.
a. 2-methylhexane
b. *n*-heptane
c. 3,3-dimethylpentane
d. *n*-hexane
e. 2-methylpentane

2.30. In Problem 2.9 you drew two staggered conformations of butane (looking end-on down the carbon-2–carbon-3 bond). There are also two eclipsed conformations around this bond. Draw Newman projection formulas for them. Arrange all four conformations in order of decreasing stability.

2.31. Draw all possible staggered and eclipsed conformations of 1-bromo-2-chloroethane, using Newman projections. Underneath each, draw the corresponding "sawhorse" projection. Rank the structures in order of decreasing stability.

2.32. Draw formulas for the preferred conformations of:
a. ethylcyclohexane   **b.** *trans*-1,2-dibromocyclohexane
c. *cis*-1-methyl-3-isopropylcyclohexane   **d.** 1,1-dibromocyclohexane

2.33. Name each of the following *cis-trans* pairs.

a.

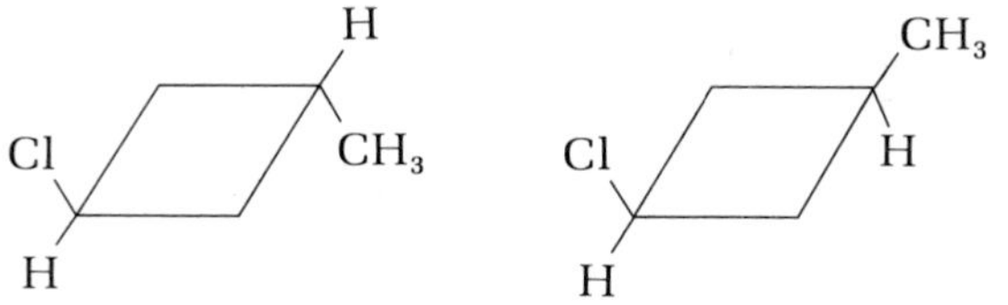

b.

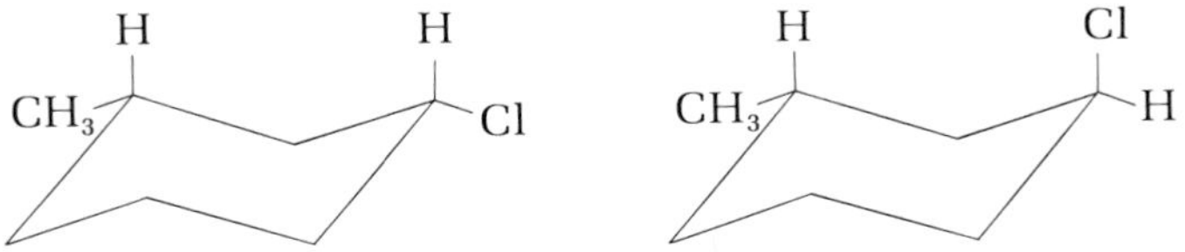

c.

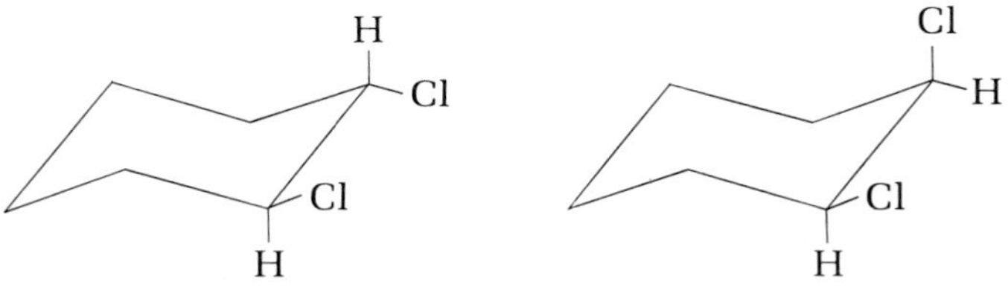

d.

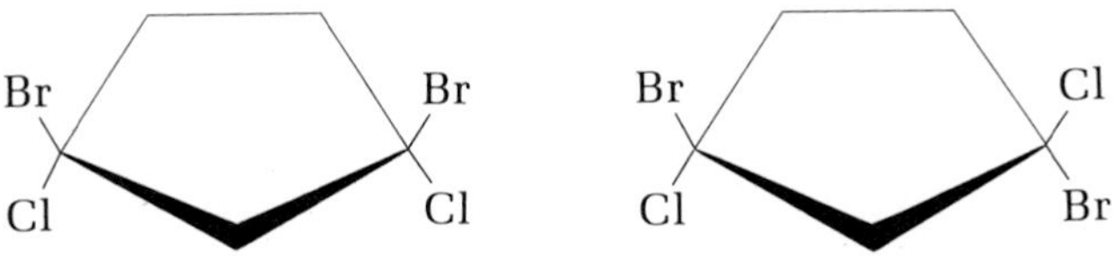

2.34. Explain with the aid of conformational formulas why *cis*-1,3-dimethylcyclo-hexane is more stable than *trans*-1,3-dimethylcyclohexane, whereas the reverse order of stability is observed for the 1,2 and 1,4 isomers.

2.35. Which will be more stable, *cis*- or *trans*-1,4-di-tert-butylcyclohexane? Explain your answer by drawing conformational structures for each compound.

2.36. Draw structural formulas for all possible products of the dichlorination of cyclopentane. Include *cis-trans* isomers.

2.37. Predict which isomer of 1,2,3,4,5,6-hexachlorocyclohexane will be the most stable. (A mixture of isomers of this compound is sold as an insecticide under the trade name of Lindane.)

2.38. Using structural formulas, write equations for the following reactions and name each organic product.
a. the complete combustion of pentane
b. the complete combustion of cyclopentane
c. the monobromination of propane
d. the monochlorination of cyclopentane

2.39. In the dichlorination of propane, four isomeric products with the formula $C_3H_6Cl_2$ were isolated and designated A, B, C, and D. Each was separated and further chlorinated to give one or more trichloropropanes, $C_3H_5Cl_3$. A and B gave three trichloro compounds, C gave one, and D gave two. Deduce the structures of C and D. One of the products from A was identical with the product from C. Deduce structures for A and B.

2.40. Write out all the steps in the free-radical chain mechanism for the mono-chlorination of ethane:

$$CH_3CH_3 + Cl_2 \rightarrow CH_3CH_2Cl + HCl$$

What trace by-products would you expect to be formed as a consequence of the chain-terminating steps?

three _____

Alkenes and Alkynes

3.1
DEFINITION AND CLASSIFICATION

Hydrocarbons that contain a carbon–carbon double bond are called **alkenes;** those with a triple bond are **alkynes.*** Their general formulas are

$$C_nH_{2n} \qquad C_nH_{2n-2}$$
alkenes \qquad alkynes

Both of these classes of hydrocarbons are referred to as **unsaturated,** because they contain fewer hydrogens per carbon than alkanes (C_nH_{2n+2}). Alkanes can be obtained from alkenes or alkynes by adding one or two moles of hydrogen:

$$RCH{=}CHR \xrightarrow[\text{catalyst}]{H_2} RCH_2CH_2R \xleftarrow[\text{catalyst}]{2\,H_2} RC{\equiv}CR \qquad (3.1)$$
alkene \qquad\qquad alkane \qquad\qquad alkyne

Compounds with more than one double or triple bond are known. If two double bonds are present the compounds are called **alkadienes** or, more commonly, **dienes.** There are also trienes, tetraenes, and eventually polyenes (compounds with *many* double bonds, from the Greek *poly*, many). Compounds with more than one triple bond, or with double and triple bonds, are also known.

Example 3.1
What are all the structural possibilities for a compound C_3H_4?

Solution
The formula C_3H_4 corresponds to the general formula C_nH_{2n-2}. The compound could have one triple bond, two double bonds, or one ring and one double bond. For the structures, see the solution to the example in Sec. 1.7, page 15.

Problem 3.1
What are all the structural possibilities for C_4H_6? (There are nine, four acyclic and five cyclic; all nine are known compounds.)

When more than one multiple bond is present in a molecule, it is useful to classify the structure further, depending on the relative positions of the multiple bonds. Double bonds are said to be **cumulated** when they are right next to one another. When multiple bonds alternate with single

---

*An old but still used synonym for alkenes is olefins, which means oil-forming. This name was originally given to ethylene because ethylene formed an oily liquid when treated with chlorine. Alkynes are sometimes called *acetylenes* after the first member of the series.

bonds, they are said to be **conjugated.** When more than one single bond comes between multiple bonds, the latter are said to be **isolated** or **nonconjugated.** Of these three arrangements, conjugated systems are by far the most common in nature and have the most interesting chemistry.

$$C=C=C \qquad\qquad C=C-C=C \qquad C=C-C-C=C$$

$$C=C=C=C \qquad C=C-C\equiv C \qquad C\equiv C-C-C\equiv C$$

| cumulated | conjugated | isolated |

Problem 3.2   **Which of the following compounds have conjugated multiple bonds?**

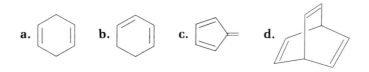

a.   b.   c.   d.

3.2
NOMENCLATURE   The IUPAC rules are similar to those for alkanes (Sec. 2.4), but a few rules must be added to name and locate the multiple bonds. These additional rules are as follows:

1. Carbon–carbon double bonds are designated by the ending **-ene;** when more than one double bond is present, the ending is **-diene, -triene,** and so on. Triple bonds are designated by the ending **-yne (-diyne** for two triple bonds, and so on). Both endings (**-ene** and **-yne**) are used when both types of bonds are present.
2. The numbered chain must include the multiple bond and is numbered so that the carbon atoms in the multiple bond have the lowest possible numbers.
3. The position of the multiple bond(s) is indicated by the number(s) of the **lower-numbered carbon atom** of each multiple bond. These numbers are placed in front of the name of the compound.

Let us see how these rules are applied. The first members of the series are

$$CH_3CH_3 \qquad CH_2=CH_2 \qquad HC\equiv CH$$

   ethane   ethene   ethyne

$$CH_3CH_2CH_3 \qquad CH_2=CHCH_3 \qquad HC\equiv CCH_3$$

   propane   propene   propyne

Note that the root of the name (*-eth-* or *-prop-*) tells us the number of carbons and that the ending (*-ane, -ene,* or *-yne*) tells us whether the bonds are single, double, or triple. No number is necessary in these cases, because in each instance only one structure is possible.

With four carbons, a number is necessary to locate the double or triple bond.

$$\overset{1}{C}H_2=\overset{2}{C}H\overset{3}{C}H_2\overset{4}{C}H_3 \qquad \overset{1}{C}H_3\overset{2}{C}H=\overset{3}{C}H\overset{4}{C}H_3 \qquad H\overset{1}{C}\equiv\overset{2}{C}\overset{3}{C}H_2\overset{4}{C}H_3 \qquad \overset{1}{C}H_3\overset{2}{C}\equiv\overset{3}{C}\overset{4}{C}H_3$$

   1-butene   2-butene   1-butyne   2-butyne

The number used is the lower of the two numbered carbons involved in the double or triple bond.

Branches are named in the usual way.

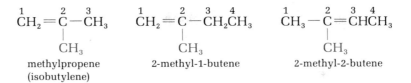

methylpropene   2-methyl-1-butene   2-methyl-2-butene
(isobutylene)

Note how the rules are applied in the following examples:

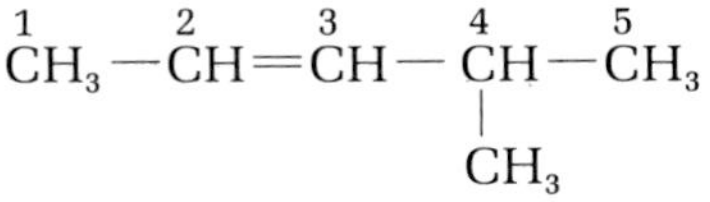

$$\overset{1}{CH_3}-\overset{2}{CH}=\overset{3}{CH}-\overset{4}{CH}-\overset{5}{CH_3} \qquad \overset{1}{CH_2}=\overset{2}{C}-\overset{3}{CH_2}\overset{4}{CH_3} \qquad \overset{1}{CH_2}=\overset{2}{CH}-\overset{3}{CH}=\overset{4}{CH_2}$$

         | |
       CH₃   CH₂CH₃

4-methyl-2-pentene   2-ethyl-1-butene   1,3-butadiene
(Not 2-methyl-3-pentene; the chain   (Named this way, even   (Note the *a* added to the
is numbered so that the double bond though there is a five-   name, to help in
gets the lowest number.)   carbon chain present,   pronunciation.)
                                because that chain does
                                not include both carbons
                                of the double bond.)

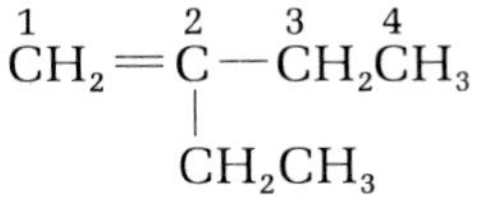

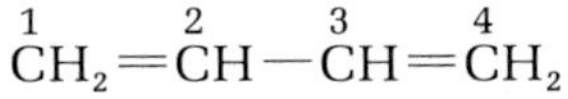

With cyclic hydrocarbons, start numbering the ring with the carbons of the multiple bonds.

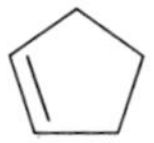

cyclopentene   3-methylcyclopentene   1,3-cyclohexadiene   1,4-cyclohexadiene
(No number is neces- (start numbering at,
sary, because there   and number through,
is only one possible the double bond to
structure.)   give the substituent
                      the lowest number;
                      5-methylcyclopentene
                      and 1-methyl-2-cyclo-
                      pentene are incorrect.

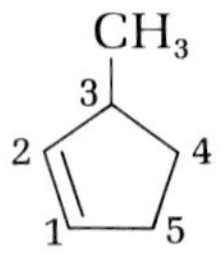

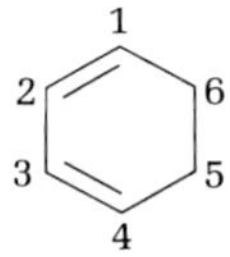

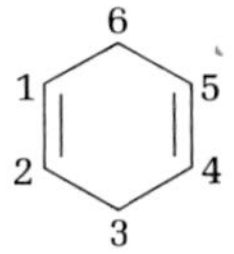

Problem 3.3   **Name each of the following structures by the IUPAC system.**
a. $ClCH=CHCH_3$ b. $(CH_3)_2C=C(CH_3)_2$ c. $CH_2=C(CH_3)CH=CH_2$
d.   CH₃   e. $CH_2=C(Cl)CH_3$ f. $HC≡C(CH_2)_3CH_3$

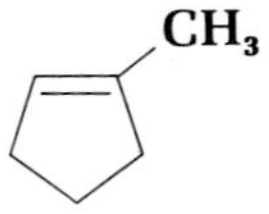

Problem 3.4   **Write the structural formula for:**
a. **2,4-dimethyl-2-pentene**   b. **2-hexyne**
c. **1,2-dibromocyclobutene**   d. **2-chloro-1,3-butadiene**

In addition to the IUPAC rules, it is important to learn a few widely used common names. For example, although the simplest members of the alkene and alkyne series are named ethene and ethyne in the IUPAC system, they are usually called by their common names **ethylene** and **acetylene,** as is the three-carbon alkene **propylene.**

$$CH_2=CH_2 \qquad HC\equiv CH \qquad CH_3CH=CH_2$$

ethylene   acetylene   propylene
(ethene)   (ethyne)   (propene)

   Three important groups derived from ethylene, propylene, and propyne have common names. They are vinyl, allyl, and propargyl groups.

$$CH_2=CH- \qquad CH_2=CH-CH_2- \qquad HC\equiv C-CH_2-$$

vinyl group   allyl group   propargyl group

These groups are used in common names.

$$CH_2=CHCl \qquad CH_2=CH-CH_2Cl \qquad HC\equiv C-CH_2Br$$

vinyl chloride   allyl chloride   propargyl bromide
(chloroethene)   (3-chloropropene)   (3-bromopropyne)

Problem 3.5 **Write the structural formula for:**
a. vinylcyclohexane   **b. allyl alcohol**   **c. propargyl alcohol**

3.3
SPECIAL FEATURES
OF DOUBLE
BONDS

Carbon–carbon double bonds have some special features that distinguish them from single bonds. For example, each carbon atom that is part of a double bond is connected to only *three* other atoms (instead of four atoms, as with tetrahedral carbon). We speak of such a carbon as being **trigonal.** Furthermore, the two carbon atoms of a double bond and the four atoms attached to them lie in a single plane. This planarity is shown in Figure 3.1 for ethylene. The H—C—H and H—C=C angles in ethylene are approximately 120°. Unlike the situation with single (σ) bonds, *rotation around double bonds is restricted.* Ethylene remains planar and does not adopt any other conformation; one carbon with its two attached hydrogens is not free to rotate with respect to the other carbon of the double bond. Finally, carbon–carbon double bonds are ordinarily shorter than carbon–carbon single bonds.

FIGURE 3.1

Three models of ethylene, each showing that the four atoms attached to a carbon–carbon double bond lie in a single plane.

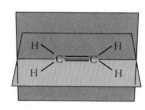

space-filling
model

Newman-type
projection

	Property	C—C	C=C
TABLE 3.1 Comparison of C—C and C=C bonds	1. Number of atoms attached to a carbon	4 (tetrahedral)	3 (trigonal)
	2. Rotation	relatively free	restricted
	3. Geometry	many conformations are possible; staggered is preferred	planar
	4. Bond angles (common)	109.5°	120°
	5. Bond length (common)	1.54 Å	1.34 Å

These differences between single and double bonds are summarized in Table 3.1. Let us see how the orbital model for bonding can explain the structure and properties of double bonds.

3.4
THE ORBITAL MODEL OF A DOUBLE BOND; THE PI BOND

Figure 3.2 shows what must be done with the atomic orbitals of carbon to accommodate trigonal bonding, bonding to only three other atoms. The first two parts of this figure are exactly the same as Figure 1.6. But, after one of the 2s electrons is promoted to a 2p orbital, we combine *only three* of the orbitals, to make *three equivalent sp²-hybridized orbitals* (called sp² because they are formed by combining one s and two p orbitals). These orbitals lie in a plane and are directed to the corners of an equilateral triangle. The angle between them is 120°. At this angle, repulsion between the electrons is minimized. The remaining valence electron is placed in the remaining 2p orbital, whose axis is perpendicular to the plane formed by the three sp² hybrid orbitals. The result of sp² hybridization is illustrated in Figure 3.3.

Now let us see what happens when two sp²-hybridized carbons are brought together to form a double bond. The process can be imagined as

FIGURE 3.2

The formation of three sp² hybrid orbitals (compare with Figure 1.6).

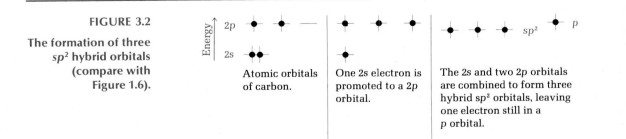

Atomic orbitals of carbon.

One 2s electron is promoted to a 2p orbital.

The 2s and two 2p orbitals are combined to form three hybrid sp² orbitals, leaving one electron still in a p orbital.

FIGURE 3.3

A trigonal carbon showing three sp^2 orbitals in a plane with a 120° angle between them. The remaining p orbital is perpendicular to the hybrid sp^2 orbitals. There is a small back lobe to each sp^2 orbital, which has been omitted for ease of representation.

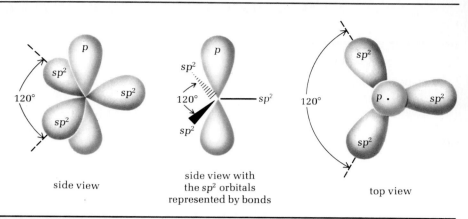

side view

side view with the sp^2 orbitals represented by bonds

top view

FIGURE 3.4

Schematic formation of a carbon–carbon double bond. Two sp^2 carbons form a sigma (σ) bond (overlap of two sp^2 orbitals) and a pi (π) bond (overlap of properly aligned p orbitals).

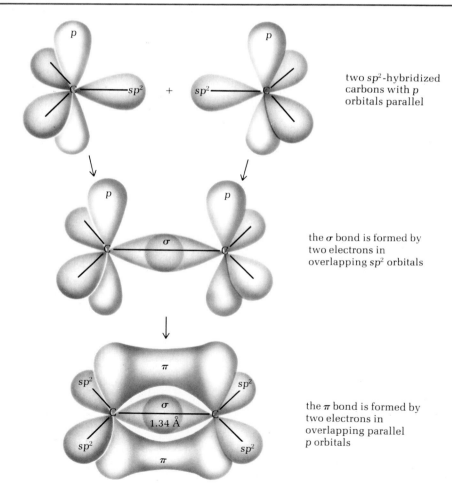

two sp^2-hybridized carbons with p orbitals parallel

the σ bond is formed by two electrons in overlapping sp^2 orbitals

the π bond is formed by two electrons in overlapping parallel p orbitals

FIGURE 3.5

The bonding in ethylene consists of one sp^2–sp^2 carbon–carbon σ bond, four sp^2–s carbon–hydrogen σ bonds, and one p–p π bond.

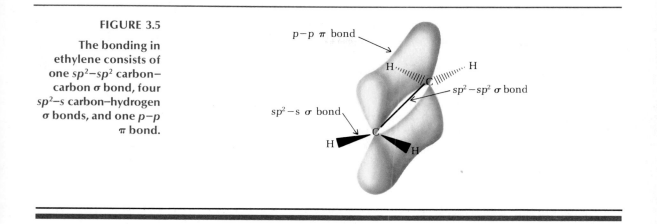

occurring stepwise, as illustrated in Figure 3.4. One of the two bonds, formed by overlap of two sp^2 orbitals, is a sigma (σ) bond. The second bond of the double bond is formed somewhat differently. If the two carbons are properly aligned so that the p orbitals on each carbon are parallel, lateral overlap can occur, as shown at the bottom of Figure 3.4. The bond formed by this type of p-orbital overlap is called a **pi (π) bond.**

To summarize, the bonding in ethylene is shown in Figure 3.5.

Note how this model explains the facts about double bonds that are listed in Table 3.1. Rotation about the double bond is restricted because, for rotation to occur, we would have to "break" the π-bond, as seen in Figure 3.6. For ethylene, it would cost about 62 kcal/mol (259 kJ/mol) to break the π bond, much more energy than is thermally available at room temperature. With the π bond intact, the remaining sp^2 orbitals on each

FIGURE 3.6
Rotation of one sp^2 carbon 90° with respect to another orients the p orbitals so that no overlap (and therefore no π bond) is possible.

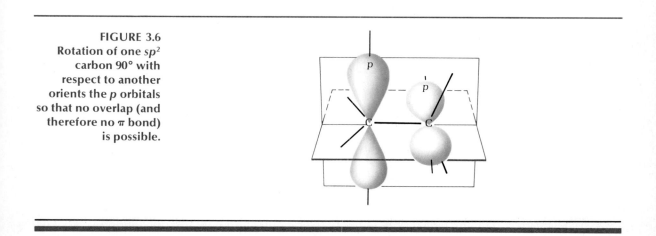

carbon lie in a single plane. The 120° angle between those bonds mini-mizes repulsion between the electrons in the sp^2 orbitals. Finally, the carbon–carbon double bond is shorter than the carbon–carbon single bond, because the two electron pairs draw the nuclei closer together than a single electron pair does.

To recap, this model of the carbon–carbon double bond consists of one σ bond and one π bond. The two electrons in the σ bond lie along the internuclear axis, whereas the two electrons in the π bond lie in a region of space above and below the plane formed by the two carbons and the four atoms that are attached to them. The π electrons are more exposed than the σ electrons and, as we shall see, are subject to attack by various electron-seeking reagents.

But before we consider reactions at the double bond, let us examine an important result of the restricted rotation around double bonds.

3.5
CIS-TRANS
ISOMERISM IN
ALKENES

Because rotation at carbon–carbon double bonds is restricted, *cis-trans* isomerism becomes possible in appropriately substituted alkenes (review Sec. 2.10). For example, 1,2-dichloroethene exists in two different forms:

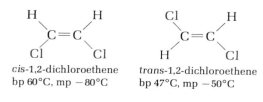

cis-1,2-dichloroethene   trans-1,2-dichloroethene
bp 60°C, mp −80°C   bp 47°C, mp −50°C

Example 3.2   Are *cis-trans* isomers possible for 1-butene and 2-butene?

Solution   Only for 2-butene:

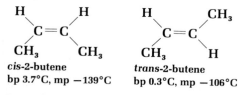

cis-2-butene   trans-2-butene
bp 3.7°C, mp −139°C   bp 0.3°C, mp −106°C

For 1-butene, carbon-1 has two identical groups attached (H's); therefore, only one structure is possible.

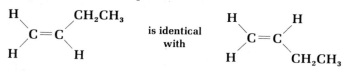

Problem 3.6   Which of the following compounds can exist as *cis-trans* isomers?
a. propene   b. 3-hexene   c. 2-hexene   d. 2-methyl-2-butene

Geometric double-bond isomers can be interconverted if they acquire sufficient energy to break the π bond and allow rotation about the remaining, somewhat stronger, σ bond (eq. 3.2). The energy may be supplied by light or heat.

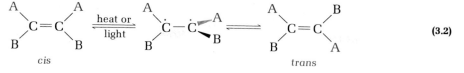

(3.2)

a word about

2. How We See: The Chemistry of Vision

Cis-trans isomerism is important in several biological processes, one of which is vision. The rod cells in the retina of the eye contain a red, light-sensitive pigment called rhodopsin. This pigment is a complex of a protein, opsin, with the highly unsaturated aldehyde 11-cis-retinal.

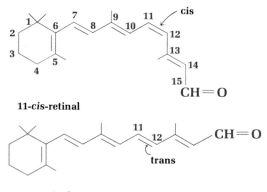

11-*cis*-retinal

trans-retinal

When visible light with the appropriate energy is absorbed by the rhodopsin, cis-retinal is isomerized to the trans isomer. As you can see from the structures, the shapes of the cis and the trans isomers are quite different. The trans-retinal complex with opsin (called lumi-rhodopsin) is less stable than the cis-complex, and it dissociates into opsin and trans-retinal. This change in geometry also triggers a response in the rod nerve cells that is transmitted to the brain and perceived as vision.

Were this all that happened, we would be able to see for only a few moments, because all of the 11-cis-retinal present in the rod cells would be quickly consumed. Fortunately there is an enzyme, retinal isomerase, that in the presence of light converts the trans-retinal back to the 11-cis isomer, so that the cycle can be repeated. Figure 3.7 illustrates the various steps we have just described for the chemistry of vision.

Vitamin A is important for vision. It has nearly the same structure as retinal, except that the aldehyde group is replaced by an alcohol group ($-CH_2OH$ in place of $-CH=O$ at the end of the chain). The vitamin A that we eat is oxidized (by an enzyme) to retinal.

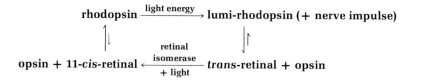

FIGURE 3.7 The vision cycle. This representation is somewhat simplified, because there are actually several intermediate complexes between rhodopsin and the fully dissociated *trans*-retinal and opsin.

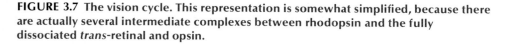

**3.6
PHYSICAL
PROPERTIES OF
ALKENES**

Unsaturated hydrocarbons have physical properties similar to those of alkanes. They are less dense than water and not very soluble in it. As with the alkanes, compounds with four or fewer carbons are colorless gases, whereas the more common five-carbon and higher homologs are volatile liquids.

**3.7
ADDITION AND
SUBSTITUTION
REACTIONS**

We saw in Chapter 2 that the most common reaction of alkanes was **substitution** (halogenation, for example). This reaction type can be expressed by the general equation

$$R-H + A-B \rightarrow R-A + H-B \qquad \text{substitution} \qquad (3.3)$$

With alkenes, the most common reaction is **addition:**

$$\diagup\!\!\!\!\!C\!=\!\!C\diagdown\!\!\!\! + A-B \rightarrow \ -\underset{A}{\overset{|}{C}}-\underset{B}{\overset{|}{C}}- \qquad \text{addition} \qquad (3.4)$$

In addition, group A of the reagent A—B becomes attached to one carbon of the double bond; group B becomes attached to the other. Here are specific examples of the two reaction types:

$$CH_3-CH_3 + Br_2 \xrightarrow[\text{(light)}]{h\nu} \underset{\text{bromoethane}}{CH_3CH_2Br} + HBr \qquad \text{substitution} \qquad (3.5)$$

$$CH_2\!=\!CH_2 + Br_2 \rightarrow \underset{Br}{\overset{|}{CH_2}}-\underset{Br}{\overset{|}{CH_2}} \qquad \text{addition} \qquad (3.6)$$
$$\text{1,2-dibromoethane}$$

In an addition reaction, the π bond of the alkene is broken but the σ bond remains intact. The σ bond of the reagent, however, is broken. Two new σ bonds are formed. In other words, we break a π and a σ bond, and we make two σ bonds. Because σ bonds are usually stronger than π bonds, the net reaction is generally very favorable.

Problem 3.7   **Why, in general, is a σ bond between atoms stronger than a π bond between the same two atoms?**

In the next sections we will describe a few typical alkene addition reactions. Afterwards, we will consider how these reactions occur.

**3.8
ADDITION OF
HALOGENS**

Alkenes readily add chlorine or bromine.

$$CH_3CH\!=\!CHCH_3 + Cl_2 \rightarrow CH_3CH-CHCH_3$$
$$\underset{\text{Cl}\qquad\text{Cl}}{|\qquad\ |} \qquad (3.7)$$

2-butene
bp 1–4°C

2,3-dichlorobutane
bp 117–119°C

$$CH_2{=}CH{-}CH_2{-}CH{=}CH_2 + 2\,Br_2 \rightarrow \underset{\underset{Br}{|}}{CH_2}{-}\underset{\underset{Br}{|}}{CH}{-}CH_2{-}\underset{\underset{Br}{|}}{CH}{-}\underset{\underset{Br}{|}}{CH_2}$$ (3.8)

1,4-pentadiene
bp 26.0°C

1,2,4,5-tetrabromopentane
mp 85–86°C

Usually the halogen is dissolved in some inert solvent such as carbon tetrachloride or chloroform, and then this solution is added dropwise to the alkene. Reaction is usually instantaneous, even at room temperature. These reaction conditions are much milder than those needed for substitution reactions.

Problem 3.8 **Write an equation for the reaction of bromine at room temperature with:**
a. 1-butene   b. 2-methyl-2-butene

The addition of bromine is frequently used as a qualitative test for unsaturation in an organic compound. Bromine solutions in carbon tetrachloride are dark reddish-brown, whereas the unsaturated compound and its bromine adduct are usually colorless. As the bromine solution is added to the unsaturated compound, the bromine color disappears. If the compound being tested were saturated, it would not react with bromine under these conditions and the color would persist.

3.9
ADDITION OF
HYDROGEN

Hydrogen adds to alkenes in the presence of an appropriate catalyst. The process is called **hydrogenation.**

$$\overset{\diagdown}{\underset{\diagup}{C}}{=}\overset{\diagup}{\underset{\diagdown}{C}} + H_2 \xrightarrow{\text{catalyst}} {-}\overset{|}{\underset{\underset{H}{|}}{C}}{-}\overset{|}{\underset{\underset{H}{|}}{C}}{-}$$ (3.9)

The catalyst is usually a finely divided metal such as nickel, platinum, or palladium. We know that such metals absorb hydrogen gas on the metal surface and activate the hydrogen–hydrogen bond. Both hydrogen atoms usually add to the same face of the double bond. For example, 1,2-dimethylcyclopentene gives cis-1,2-dimethylcyclopentane:

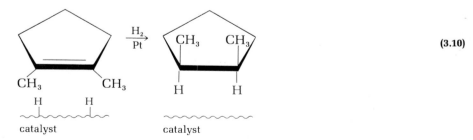

(3.10)

Catalytic hydrogenation of double bonds is used commercially to convert vegetable oils to margarine and other cooking fats (Sec. 11.11).

Problem 3.9   **Write an equation for the catalytic hydrogenation of:**
a. 2-methylpropene   b. 1,2-dimethylcyclobutene

3.10
ADDITION OF
WATER
(HYDRATION)

Water adds to alkenes in the presence of an acid as a catalyst. It adds as H—OH, and the products are alcohols.

$$CH_2{=}CH_2 + H{-}OH \xrightarrow{H^+} \underset{\underset{H \quad\ OH}{|\qquad |}}{CH_2{-}CH_2} \qquad (or\ CH_3CH_2OH) \tag{3.11}$$

ethyl alcohol, or ethanol

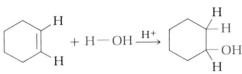 (3.12)

cyclohexene
bp 83.0°C

cyclohexanol
bp 161.1°C

This reaction is used to synthesize alcohols, both commercially and in the laboratory. The necessary role of the acid catalyst will be explained when we discuss the mechanism of these addition reactions.

Problem 3.10   **Write an equation for the acid-catalyzed addition of water to:**
a. 2-butene   b. cyclopentene

3.11
ADDITION OF
ACIDS

A variety of acids add to the double bond of alkenes. The hydrogen ion (or proton) adds to one carbon and the remainder of the acid adds to the other carbon.

$$\underset{}{\overset{}{C}}{=}\underset{}{\overset{}{C} } + \overset{\delta+\ \ \ \delta-}{H{-}A} \rightarrow \underset{\underset{H \quad A}{|\quad\ |}}{-C{-}C{-}} \tag{3.13}$$

Acids that add in this way are the hydrogen halides (HF, HCl, HBr, HI), sulfuric acid (H—OSO$_3$H), and organic carboxylic acids (H—O$\overset{\overset{\displaystyle O}{\|}}{C}$R). Here are two typical examples:

$$CH_2{=}CH_2 + H{-}Cl \rightarrow \underset{\underset{H \quad\ Cl}{|\qquad |}}{CH_2{-}CH_2} \qquad (or\ CH_3CH_2Cl) \tag{3.14}$$

ethylene   hydrogen   ethyl chloride
                         chloride   (chloroethane)

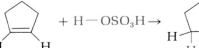 (3.15)

H   H

cyclopentene   sulfuric
                      acid

cyclopentyl
hydrogen sulfate

Problem 3.11 **Write an equation for each of the following reactions:**
a. 2-butene + HI
b. cyclopentene + HBr

Before we discuss the mechanism of addition reactions, we had better face up to a complication that all the examples we have chosen so far were carefully selected to avoid.

**3.12
ADDITION OF
UNSYMMETRIC
REAGENTS TO
UNSYMMETRIC
ALKENES;
MARKOVNIKOV'S
RULE**

Reagents and alkenes can be classified as either symmetric or unsymmetric with respect to addition reactions. Table 3.2 illustrates what we mean. If a reagent and/or an alkene is symmetric, only one addition product is possible. If you check back through all the equations and problems in Sec. 3.7 through Sec. 3.11, you will see that either the alkene or the reagent (or both) was symmetric. But if *both* the reagent and the double bond are unsymmetric, two products are, in principle, possible:

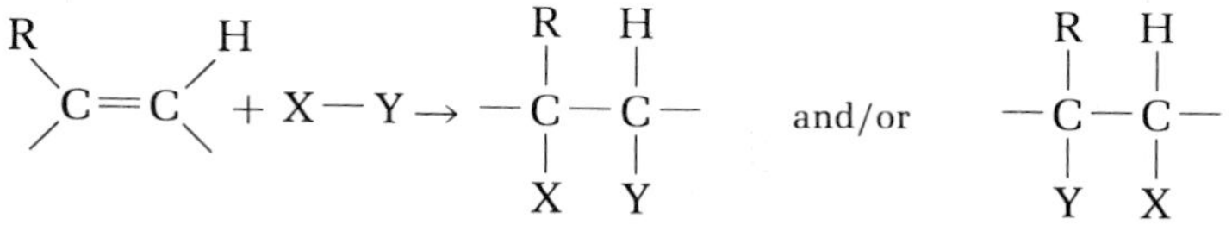

$$
\underset{R}{\overset{H}{\diagup}}C=C\underset{\diagdown}{\overset{H}{\diagup}} + X-Y \rightarrow \quad -\underset{\underset{X}{|}}{\overset{\overset{R}{|}}{C}}-\underset{\underset{Y}{|}}{\overset{\overset{H}{|}}{C}}- \quad \text{and/or} \quad -\underset{\underset{Y}{|}}{\overset{\overset{R}{|}}{C}}-\underset{\underset{X}{|}}{\overset{\overset{H}{|}}{C}}- \qquad \textbf{(3.16)}
$$

The products of eq. 3.16 are sometimes referred to as **regioisomers.** If a reaction of this type gives *only one* of the two possible regioisomers, it is said to be a **regiospecific** reaction. If it gives *mainly* one product, it is said to be **regioselective.**

Let us consider, as a specific example, the acid-catalyzed addition of water to propylene. In principle, two products could be formed, 1-propanol or 2-propanol.

TABLE 3.2		Symmetric		Unsymmetric	
Symmetry classification of reagents and alkenes with regard to addition reactions	**Reagents**	$Br-Br$ $Cl-Cl$ $H-H$		$H-Br$ $H-OH$ $H-OSO_3H$	
	Alkenes	$CH_2=CH_2$ $CH_3CH=CHCH_3$		$CH_3CH=CH_2$ $CH_3CH_2CH=CHCH_3$ CH_3	

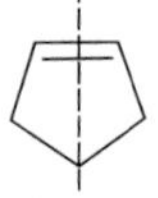

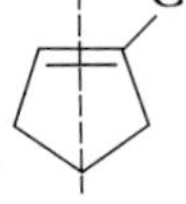

$$\overset{3}{C}H_3\overset{2}{C}H_2\overset{1}{C}H_2OH \xleftarrow[\text{H}^+]{\text{H--OH}} \overset{3}{C}H_3\overset{2}{C}H=\overset{1}{C}H_2 \xrightarrow[\text{H}^+]{\text{H--OH}} \overset{3}{C}H_3\overset{2}{C}H\overset{1}{C}H_3 \qquad \text{(3.17)}$$

$$\underset{\text{OH}}{}$$

1-propanol   2-propanol

That is, the H of the water could add to C-1 and the OH to C-2 of propylene, or vice versa. In fact, when the experiment is carried out, *only one product is formed. The addition is regiospecific, and the only product is 2-propanol.*

Most addition reactions to alkenes show a similar preference for the formation of only (or mainly) one of the two possible addition products. Here are some examples:

$$CH_3CH=CH_2 + H-Cl \rightarrow CH_3\underset{|}{\overset{}{C}}HCH_3 \qquad (\underline{not}\ CH_3CH_2CH_2Cl) \qquad \text{(3.18)}$$
$$\underset{Cl}{}$$

$$CH_3\underset{|}{\overset{}{C}}=CH_2 + H-OH \xrightarrow{H^+} CH_3\underset{|}{\overset{OH}{C}}CH_3 \qquad (\underline{not}\ CH_3\underset{|}{\overset{}{C}}HCH_2OH) \qquad \text{(3.19)}$$
$$\underset{CH_3}{} \qquad\qquad \underset{CH_3}{} \qquad\qquad \underset{CH_3}{}$$

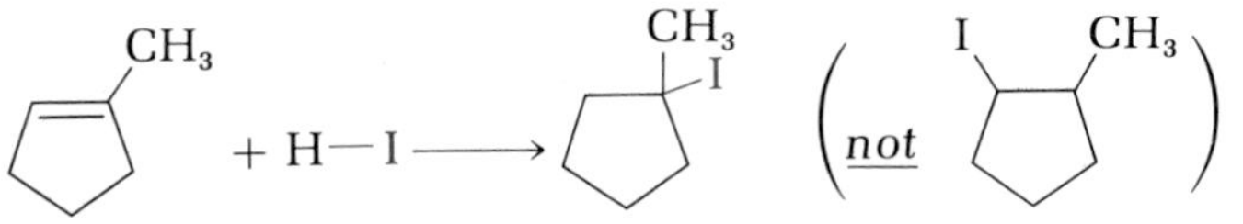

$$\text{(3.20)}$$

After studying a large number of such addition reactions, the Russian chemist Vladimir Markovnikov formulated the following rule over 100 years ago: *When an unsymmetric reagent adds to an unsymmetric alkene, the electropositive part of the reagent adds to the carbon of the double bond that is bound to the greater number of hydrogen atoms.*

Problem 3.12 **Use Markovnikov's rule to predict which regioisomer predominates in each of the following reactions:**
a. 1-butene + HCl   b. 2-methyl-2-butene + H₂O   (H⁺ catalyst)

Problem 3.13 **What two products are possible in principle from the addition of HCl to 2-pentene? Would you expect the reaction to be regiospecific?**

Let us now see whether we can develop a rational explanation for Markovnikov's rule in terms of modern chemical theory.

3.13
MECHANISM OF
ELECTROPHILIC
ADDITION TO
ALKENES

The π electrons of a double bond form a weaker bond and are more exposed to reagents than the σ electrons. It is these π electrons, then, that are involved in additions to alkenes. The double bond acts as a supplier of electrons to electron-seeking reagents. We call these electron-seeking reagents **electrophiles.** They may be positive ions (cations) or otherwise electron-deficient species.

As an example, consider the addition of acids to alkenes. The proton H^+ is the attacking electrophile. As it approaches the π bond, the two π electrons are used to form a σ bond between the proton and one of the two carbon atoms. Because this bond uses *both* π electrons, the *other* carbon acquires a positive charge, producing a **carbocation** (also called a *carbonium ion*).

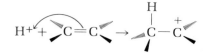

a carbocation
(carbonium ion)

$$(3.21)$$

Carbocations are, in general, extremely reactive because they have only six (instead of the usual eight) electrons around the positive carbon. The carbocation rapidly combines with some species that can supply it with two electrons. We call such electron-supplying reagents **nucleophiles;** they are usually negative ions or species with an unshared electron pair.

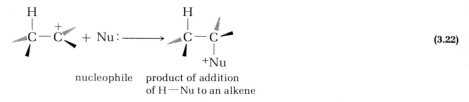

nucleophile   product of addition
of H—Nu to an alkene

$$(3.22)$$

Eq. 3.23 shows the steps involved in the addition of $H-Cl, H-OSO_3H$, or $H-OH$ to a double bond. The electrophile H^+ adds to give a carbocation that, in the second step, reacts with a nucleophile.

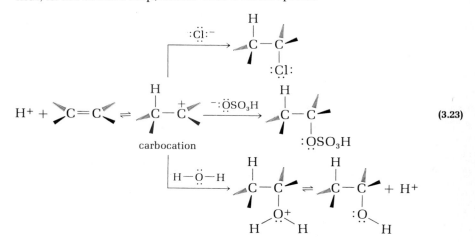

$$(3.23)$$

carbocation

With most alkenes the first step in this process, the formation of the carbocation, is the slower of the two steps. The resulting carbocations are usually so reactive that combination with the nucleophile is extremely rapid. Because the slow step in the addition is attack by the electrophile, the whole process is called **electrophilic addition.**

Example 3.3   **Describe the bonding of a carbocation in orbital terms.**

Solution   **The carbon atom is positive, and hence has only three valence electrons. Each of these electrons is located in an *sp²* orbital. The three *sp²* orbitals are used for bonding; they lie in one plane with 120° angles between them. The remaining *p* orbital is perpendicular to that plane and vacant.**

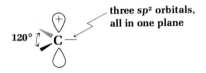

three *sp²* orbitals, all in one plane

120°

3.14 MARKOVNIKOV'S RULE EXPLAINED

To explain Markovnikov's rule, let us consider the addition of H—Cl to propylene. The first step of the reaction is addition of a proton. This can occur in two ways, to give an isopropyl or an *n*-propyl cation.

$$\overset{3}{CH_3}-\overset{2}{CH}=\overset{1}{CH_2} \xrightarrow{H^+}$$

adds to C$_1$ → CH$_3\overset{+}{C}$HCH$_3$
isopropyl cation

adds to C$_2$ → CH$_3$CH$_2$CH$_2^+$
n-propyl cation

(3.24)

At this stage the structure of the product is already determined; in combining with chloride ion, the isopropyl cation can give only 2-chloropropane and the *n*-propyl cation will give 1-chloropropane. The only observed product is 2-chloropropane, so we must conclude that *the proton adds to C-1 to form only the isopropyl cation.* Why?

From the study of many reactions that proceed by way of carbocation intermediates, and also from the direct study of carbocations in very strong acid solutions wherein they are stable and long-lived, we know that alkyl groups on the positive carbon stabilize carbocations.

Carbocations can be classified as tertiary, secondary, or primary, depending on whether the positive carbon atom bears three groups, two groups, or only one group. Their observed stability decreases in the following order:

$$\underset{\substack{| \\ R}}{\overset{\substack{R \\ |}}{R-C^+}} > \underset{\substack{| \\ R}}{R-\overset{+}{C}H} \gg R-CH_2^+ > CH_3^+$$

tertiary (3°)   secondary (2°)   primary (1°)   methyl (unique)

most stable ——————————————————————→ least stable

A carbocation is more stable the more the positive charge can be delocalized over the other atoms in the ion. In alkyl cations this delocalization is accomplished by some drift of electron density toward the positive carbon from the other σ bonds in the ion. If the positive carbon is surrounded by other carbon atoms (alkyl or R groups) instead of hydrogen atoms, there are more σ bonds to help delocalize the charge. This is the main reason for the observed stability order of carbocations.

Markovnikov's rule may now be restated: *The addition of an unsymmetric reagent to an unsymmetric double bond proceeds in such a direction as to involve the most stable intermediate carbocation.*

Problem 3.14   Write out the steps in the electrophilic additions in eq. 3.19 and eq. 3.20, and verify that, in each case, reaction occurs via the most stable carbocation.

3.15
THE MECHANISM
OF HALOGEN
ADDITION

Addition of halogens to alkenes occurs by a slightly different mechanism from the addition of acids. In the first step of the addition of bromine to a double bond, the π electrons displace bromide ion from a bromine molecule. The product is a **cyclic bromonium ion.**

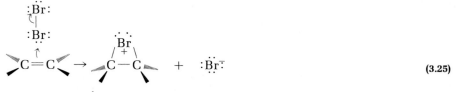

a bromonium ion

(3.25)

Reaction of bromide ion with the bromonium ion then gives the product.

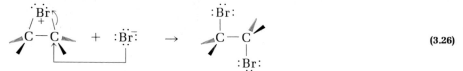

(3.26)

Problem 3.15   By calculating the formal charge at each atom of the three-membered ring, show that, in a bromonium ion, the bromine carries the + charge.

Example 3.4   Explain why the reaction of ethylene with bromine *and water* gives a mixture of $BrCH_2CH_2Br$ and $BrCH_2CH_2OH$ as the products. Under the same conditions, $BrCH_2CH_2Br$ does *not* react with water to give $BrCH_2CH_2OH$.

Solution   The first step in the mechanism is as shown in eq. 3.25. The resulting bromonium ion will then react rapidly with *any* available nucleophile. It can react either with bromide ion or with water molecules.

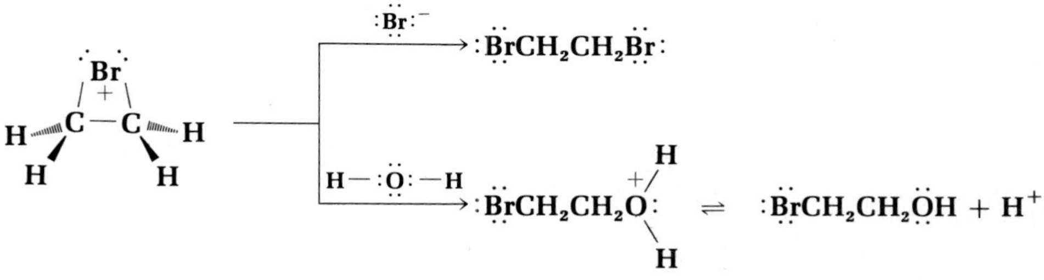

The observation of both products provides experimental evidence that the addition of bromine occurs stepwise.

Problem 3.16   The addition of bromine to cyclopentene gives *trans*-1,2-dibromocyclopentane. Almost none of the *cis* isomer is formed. Show that this observation is nicely explained by the bromonium ion intermediate.

**3.16
1,4-ADDITION TO
CONJUGATED
DIENES**

When one mole of hydrogen bromide is added to one mole of 1,3-butadiene, a rather surprising result is obtained. Two products are isolated:

$$\overset{1}{CH_2}=\overset{2}{CH}-\overset{3}{CH}=\overset{4}{CH_2} \xrightarrow{HBr}$$

1,3-butadiene

$$\xrightarrow{80\%} \underset{\underset{H}{|}}{CH_2}-\underset{\underset{Br}{|}}{CH}-CH=CH_2 \quad \text{(1,2-addition)}$$

3-bromo-1-butene

$$\xrightarrow{20\%} \underset{\underset{H}{|}}{CH_2}-CH=CH-\underset{\underset{Br}{|}}{CH_2} \quad \text{(1,4-addition)}$$

1-bromo-2-butene

(3.27)

In one product, HBr has added to one of the two double bonds, and the other double bond is still present in its original position. We call this the product of **1,2-addition.** The other product may at first seem unexpected. The hydrogen and bromine have added to carbon-1 and carbon-4 of the original diene, and a new double bond has appeared between carbon-2 and carbon-3. This process is called **1,4-addition** and is quite general for conjugated systems. How can we explain it?

In the first step, the proton adds to the terminal carbon atom.

$$H^+ + CH_2=CH-CH=CH_2 \rightarrow CH_3-\overset{+}{CH}-CH=CH_2 \quad (3.28)$$

The resulting carbocation can be stabilized by resonance (Sec. 1.12) and in fact is a hybrid of two contributing resonance structures.

$$CH_3-\overset{+}{CH}-CH=CH_2 \leftrightarrow CH_3-CH=CH-\overset{+}{CH_2}$$

The positive charge is delocalized over carbon-2 and carbon-4. When the carbocation reacts with bromide ion, it can react either at carbon-2, to give the product of 1,2-addition, or at carbon-4, to give the product of 1,4-addition.

$$\left. \begin{array}{c} CH_3-\overset{+}{CH}-CH=CH_2 \\ \\ \updownarrow \\ \\ CH_3-CH=CH-\overset{+}{CH_2} \end{array} \right\} \xrightarrow{Br^-} \begin{array}{c} CH_3CH-CH=CH_2 \\ \underset{Br}{|} \\ + \\ CH_3-CH=CH-CH_2 \\ \underset{Br}{|} \end{array} \quad (3.29)$$

Problem 3.17 **Explain why, in the first step in the addition of HBr to 1,3-butadiene, the proton adds to C-1 (eq. 3.28), and not to C-2.**

The carbocation intermediate in these reactions is a single species, a resonance hybrid, even though we need two contributing structures to describe it. This type of carbocation, with a carbon–carbon double bond adjacent to the positive carbon, is called an **allylic cation.** Such ions are more stable than saturated alkyl cations with similar carbon skeletons.

Problem 3.18 **Write an equation for the expected products of 1,2-addition and 1,4-addition of bromine to 1,3-butadiene.**

3.17 FREE-RADICAL POLYMERIZATION; VINYL POLYMERS

Some reagents add to alkenes by a free-radical mechanism instead of by an ionic mechanism. From a commercial standpoint, the most important of these free-radical additions are those that lead to polymers.

A **polymer** is a large molecule, usually with a high molecular weight, built up from small repeating units. The simple molecule from which these repeating units are derived is called a **monomer,** and the process of converting a monomer to a polymer is called **polymerization.**

Some polymers, such as starch, cellulose, and silk, are **natural polymers;** they are produced in nature by plants or animals. Others are **synthetic polymers,** because they are made in the laboratory. The term **plastic** is sometimes used interchangeably with *synthetic polymer.* The word comes from the Latin (*plasticus,* fit for molding) and usually refers to synthetic polymers that can be melted and molded, cast, or extruded into various shapes, but it also includes films or filaments used as textile fibers.

Vinyl polymers constitute one of the main classes of synthetic polymers. They are made by linking vinyl monomers together.

$$CH_2 = CHX \xrightarrow{\text{catalyst}} \left(CH_2 - \underset{\underset{X}{|}}{CH} \right)_n \tag{3.30}$$

vinyl monomer   vinyl polymer

The structure in parentheses in the product of eq. 3.30 is called the **repeating unit.** The number of repeating units n in any one molecule of a polymer can vary from just a few up to the thousands. The bonds at either end of the polymer chain may be connected to a catalyst molecule, or the chain may be terminated in some other way. When n is a large number, the end groups comprise a relatively insignificant part of the whole molecule. The group X in eq. 3.30 may be a hydrogen, an alkyl group, a halogen, or any of a rather large number of groups. Table 3.3 lists some of the more common vinyl monomers, together with the names and uses of the polymers derived from them. Note that, in some examples, the monomer

TABLE 3.3	Monomer	Polymer	Uses
Common industrial vinyl monomers and polymers	$CH_2{=}CH_2$	polyethylene	sheets and films, blow-molded bottles, injection-molded toys and housewares, wire and cable coverings, shipping containers
	$CH_2{=}CHCH_3$	polypropylene	fiber products such as indoor–outdoor carpeting, car and truck parts, packaging, toys and housewares
	$CH_2{=}C(CH_3)_2$	polyisobutene	adhesives such as those used on plastic bandages
	$CH_2{=}CHCl$	polyvinyl chloride (PVC)	plastic pipe and pipe fittings, films and sheets, floor tile, records, coatings
	$CH_2{=}CHCN$	polyacrylonitrile (Orlon, Acrilan)	sweaters and other clothing
	$CH_2{=}CH{-}\langle\bigcirc\rangle$	polystyrene	packaging and containers (Styrofoam), toys, recreational equipment, appliance parts, disposable food containers and utensils, insulation
	$CH_2{=}CH{-}\overset{\overset{\displaystyle O}{\|}}{O}CCH_3$	polyvinyl acetate	adhesives, and latex paints
	$CH_2{=}C(CH_3){-}\overset{\overset{\displaystyle O}{\|}}{O}CCH_3$	polymethyl methacrylate (Plexiglas, Lucite)	objects that must be clear, transparent, and tough
	$CH_2{=}CCl_2$	polyvinylidene chloride (Saran)	food packaging
	$CF_2{=}CF_2$	polytetrafluoroethylene (Teflon)	coatings for cooking utensils, electric insulators, lenses for high-intensity discharge lamps

corresponds to the general formula $CH_2 = CX_2$ or $CH_2 = CXY$ or even $CX_2 = CX_2$.

How is a vinyl monomer converted to a vinyl polymer? There are several types of catalysts, but the most common are **free-radical chain initiators.** These catalysts contain a weak bond of some sort, one that can easily be broken by heat to give free radicals. One of the most common types of catalyst is an organic peroxide. The $O—O$ single bond is quite weak, and, on heating, this bond breaks, with one electron going to each of the oxygens.*

$$R—O \frown O—R \xrightarrow{\text{heat}} 2\,R—O\cdot \qquad\qquad \textbf{(3.31)}$$

organic peroxide   two radicals

A radical from the catalyst adds to the carbon–carbon double bond of the vinyl monomer. Let us use ethylene as an example.

$$RO\cdot \quad CH_2 = CH_2 \rightarrow RO—CH_2—CH_2\cdot \qquad\qquad \textbf{(3.32)}$$

catalyst   a carbon
radical   free radical

The result is a carbon free radical, which adds to another ethylene molecule, and another, and another, and so on:

$$ROCH_2CH_2\cdot \xrightarrow{CH_2 = CH_2} ROCH_2CH_2CH_2CH_2\cdot \xrightarrow{CH_2 = CH_2}$$

$$ROCH_2CH_2CH_2CH_2CH_2CH_2\cdot, \text{and so on} \qquad \textbf{(3.33)}$$

The carbon chain continues to grow in length until some chain-termination reaction occurs (perhaps combination of two radicals).

We might think that only a single long chain of carbons would be formed, but this is not always the case. A "growing" polymer chain may abstract a hydrogen atom from its own back, so to speak, to cause chain branching:

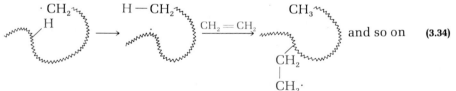

and so on   **(3.34)**

In the end, a giant molecule with long and short chains is formed:

branched polyethylene

---

*Note that we use a curved arrow, $\frown$, to describe the movement of an electron *pair*, whereas we use a "fishhook" or half-headed arrow, $\frown$, to show the movement of only *one* electron.

The degree of chain-branching, the molecular weight, and other features of the polymer structure often can be controlled by choice of catalyst and reaction conditions. The free-radical polymerization of ethylene is usually carried out at high pressures.

The production of vinyl polymers is not a small-scale operation. For example, approximately 10 billion pounds of polyethylene are produced in the United States alone each year by the process we have just described, and overall production of vinyl polymers runs well over 30 billion pounds annually.

Example 3.5   **Write out the structure of three or four repeating units of polypropylene.**

Solution   **The general formula is** $\left(\begin{array}{c}\\\end{array}\right)$ **; a segment of four units is**

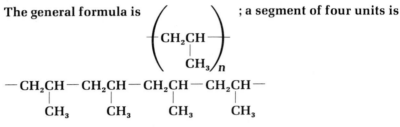

$$ -CH_2CH-CH_2CH-CH_2CH-CH_2CH- $$
$$ \quad\quad\quad CH_3 \quad\quad CH_3 \quad\quad CH_3 \quad\quad CH_3 $$

Example 3.6   **Explain why the methyl substituents appear on alternate carbon atoms and why units of the following type do not ordinarily appear in polypropylene.**

$$ -CH_2-CH \,+\, CH-CH_2- \quad\quad or \quad\quad -CH-CH_2 \,+\, CH_2-CH- $$
$$ \quad\quad CH_3 \quad CH_3 \quad\quad\quad\quad\quad CH_3 \quad\quad\quad\quad CH_3 $$

Solution   The catalyst radical can add to propylene in one of two ways:

$$ RO\cdot \,+\, CH_2{=}CH \rightarrow RO-CH_2-\overset{\cdot}{C}H \quad or \quad \overset{\cdot}{C}H_2-CH-OR $$
$$ \quad\quad\quad\quad CH_3 \quad\quad\quad\quad\quad CH_3 \quad\quad\quad\quad\quad CH_3 $$

Either a secondary or a primary free carbon radical is formed. The order of free-radical stability is the same as that of carbocation stability: $3° > 2° > 1°$. Therefore only the secondary radical is produced. When it adds to the next propylene unit, another secondary radical is formed, and so on.

$$ RO-CH_2-\overset{\cdot}{C}H \,+\, CH_2{=}CH \rightarrow RO-CH_2-CH-CH_2-\overset{\cdot}{C}H \rightarrow polymer $$
$$ \quad\quad\quad CH_3 \quad\quad CH_3 \quad\quad\quad\quad CH_3 \quad\quad CH_3 $$

Consequently, the methyl groups end up on alternate carbons of the polymer chain.

3.18
NATURAL AND
SYNTHETIC RUBBER

When Christopher Columbus arrived in the New World, he saw Native Americans play a game with balls made from the gum of rubber trees. Undoubtedly the natives also coated outer garments with this material for protection against rain, and they knew how to prepare footwear and bottles by coating clay molds with rubber and allowing them to dry. Of course it was not until much later that the chemical nature of rubber was recognized. The name *rubber* was given to the substance by Joseph Priestley (discoverer of oxygen), who used it to rub out pencil marks.

Natural rubber is an unsaturated hydrocarbon polymer. It is obtained commercially from the milky sap (latex) of the rubber tree. Its chemical structure was deduced in part from the observation during the nineteenth century that, when latex is heated *in the absence of air,* it gives mainly a single unsaturated hydrocarbon product called **isoprene.**

$$\text{natural rubber} \xrightarrow{\text{heat}} \underset{\underset{CH_3}{|}}{CH_2=C-CH=CH_2} \tag{3.35}$$

isoprene
(2-methyl-1,3-butadiene)

It seemed natural to assume that isoprene molecules were somehow linked in long chains to form rubber.

It is now possible to synthesize a material that is essentially identical to natural rubber by treating isoprene with a special catalyst (such as triethylaluminum, $(CH_3CH_2)_3Al$, and titanium trichloride, $TiCl_3$). The isoprene molecules add to one another by a head-to-tail 1,4-addition:

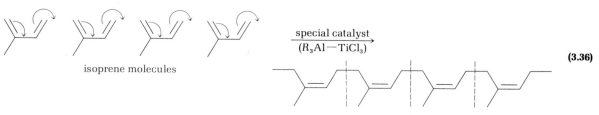

isoprene molecules

$$\xrightarrow[\text{(R}_3\text{Al}-\text{TiCl}_3)]{\text{special catalyst}} \tag{3.36}$$

natural rubber

The double bonds in natural rubber are "isolated"; that is, they are separated from one another by more than one single bond. The double bonds in natural rubber have a *cis* geometry. By this we mean that, to proceed down the length of the chain, we must enter and leave each double bond from the same "side." Most rubber has a molecular weight in excess of 1,000,000, though the value varies with the source and method of processing. Crude plantation rubber contains, in addition to polyisoprene, about 2.5–3.5% protein, 2.5–3.2% fats, 0.1–1.2% water, and traces of inorganic matter.

The five-carbon repeating unit that makes up the natural rubber molecule (it is marked off, in the structure, by dashed lines) is called an **isoprene unit.** It consists of a four-carbon chain with a one-carbon branch at carbon-2:

$$\overset{1}{C}-\underset{\underset{C}{|}}{\overset{2}{C}}-\overset{3}{C}-\overset{4}{C}$$

isoprene unit

This five-carbon unit is extremely common in many natural products.

Problem 3.19   **Draw the structures of the following natural products, and mark off with dashed lines the isoprene units in each.**
a. geraniol (Figure 1.11)   b. limonene (Figure 1.12)
c. α-pinene (Figure 1.12)

Although natural rubber has many useful properties, it also has some undesirable ones. Early manufactured rubber goods were often sticky and smelly, and they softened in warm weather and hardened with cold. Some of these weaknesses were overcome when Charles Goodyear invented **vulcanization,** a process of cross-linking polymer chains by heating rubber with sulfur. The cross-links add strength to the rubber and act as a kind of "memory" that helps the polymer recover its original shape after stretching.

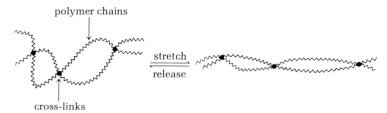

In spite of such improvements, there were still problems. For example, it was not uncommon years ago to have to check the air pressure in tires almost every time the automobile gas tank was filled. Therefore, there was a need to develop **synthetic rubber,** a name given to polymers with rubber-like properties but superior to and somewhat different chemically from natural rubber.

Many monomers or mixtures of monomers form **elastomers** (rubberlike substances) when they are polymerized. The largest-scale commercial synthetic rubber is a **copolymer** of 25% styrene and 75% 1,3-butadiene, called SBR (styrene-butadiene rubber):

$$n\text{CH}_2\!\!=\!\!\text{CHC}_6\text{H}_5 + 3n\text{CH}_2\!\!=\!\!\text{CH}-\text{CH}\!\!=\!\!\text{CH}_2 \xrightarrow[\text{initiator}]{\text{free-radical}}$$

styrene   butadiene

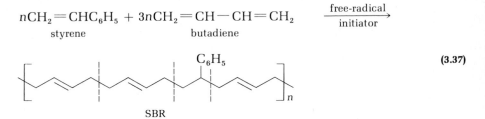

(3.37)

SBR

The structure is approximately as shown in eq. 3.37, although about 20% of the butadiene adds 1,2- instead of 1,4- as shown. The double bonds in the polymer have a *trans* geometry. The dashes in the structure show the units from which the polymer is constructed. About two-thirds of SBR goes into tire manufacture. Its annual production exceeds that of natural rubber by a factor of 2.

Hydroboration was discovered by Professor Herbert C. Brown (Purdue University). The reaction is so useful in synthesis that Brown's work in this field earned him a Nobel Prize (1979). We will describe here only one major use of hydroboration, a two-step alcohol synthesis from alkenes.

Hydroboration involves addition of the hydrogen–boron bond to an alkene. The $H-B\diagdown$ bond is polarized with the hydrogen as $\delta-$ and the boron as $\delta+$. Addition occurs so that the boron adds to the less substituted carbon.

$$R-CH{=}CH_2 + \overset{\delta-}{H}{-}\overset{\delta+}{B}\diagup \rightarrow R-CH_2-CH_2-B\diagup \qquad (3.38)$$

For example, 3 moles of propylene react with 1 mole of borane, BH_3, to give tri-n-propylborane.

$$3CH_3CH{=}CH_2 + BH_3 \rightarrow CH_3CH_2CH_2{-}B\underset{\diagdown CH_2CH_2CH_3}{\diagup CH_2CH_2CH_3} \qquad (3.39)$$

propylene   borane   tri-n-propylborane

The trialkylboranes made in this way are usually not isolated as such but are treated with some other reagent to obtain the desired final product. For example, trialkylboranes are readily oxidized by hydrogen peroxide and base to give alcohols.

$$(CH_3CH_2CH_2)_3\!B + 3\,H_2O_2 + 3\,NaOH \rightarrow 3\ CH_3CH_2CH_2OH + Na_3BO_3 + 3\,H_2O \qquad (3.40)$$

tri-n-propylborane n-propyl alcohol   sodium borate

One great advantage of the hydroboration–oxidation sequence is that it allows us to make alcohols that are *not* obtainable by the acid-catalyzed hydration of alkenes (review eq. 3.17).

$$R-CH{=}CH_2 \begin{cases} \xrightarrow[\;H^+\;]{H-OH} & R-\underset{\underset{OH}{|}}{CH}-CH_3 \quad \text{Markovnikov addition} \\ \\ \xrightarrow[2.\,H_2O_2,OH^-]{1.\,BH_3} & R-CH_2-CH_2OH \quad \text{anti-Markovnikov addition} \end{cases} \qquad (3.41)$$

Example 3.7   Identify the alcohol obtained from the sequence

$$CH_3-\overset{\overset{\displaystyle CH_3}{|}}{C}{=}CH_2 \xrightarrow{BH_3} \xrightarrow[OH^-]{H_2O_2}$$

Solution   The boron adds to the less substituted carbon; oxidation gives the corresponding alcohol.

$$CH_3-\underset{\underset{CH_3}{|}}{C}=CH_2 \xrightarrow{BH_3} (CH_3-\underset{\underset{CH_3}{|}}{CH}-CH_2)_3 B \xrightarrow[OH^-]{H_2O_2} CH_3-\underset{\underset{CH_3}{|}}{CH}-CH_2OH$$

Problem 3.20   What alcohol is obtained by applying the hydroboration–oxidation sequence to 2-methyl-2-butene?

Problem 3.21   What alkene will give ⬠—CH_2CH_2OH via the hydroboration–oxidation sequence?

3.20
OXIDATION OF
ALKENES WITH
PERMANGANATE

Alkenes are in general more easily oxidized than alkanes by chemical oxidizing agents. These reagents attack the π electrons of the double bond.
   Alkenes react with alkaline potassium permanganate to form **glycols** (compounds with two adjacent hydroxyl groups).

$$3 \underset{/}{\overset{\backslash}{C}}=\overset{/}{\underset{\backslash}{C}} + 2\,K^+MnO_4^- + 4\,H_2O \rightarrow 3\,-\underset{\underset{OH}{|}}{C}-\underset{\underset{OH}{|}}{C}- + 2\,MnO_2 + 2\,K^+OH^- \quad \textbf{(3.42)}$$

| alkene | potassium permanganate (purple) | a glycol | manganese dioxide (brown-black) |

As the reaction occurs, the purple color of the permanganate ion is replaced by the brown precipitate of manganese dioxide. The reaction, via its color change, is used as a test to distinguish alkenes from alkanes. Unfortunately, the test is not foolproof because certain other functional groups may also be oxidized by the reagent. However, if the unknown is a hydrocarbon, this test for unsaturation is quite reliable.

3.21
OZONOLYSIS OF
ALKENES

Alkenes react rapidly and quantitatively with ozone, O_3. The products, called **ozonides,*** are usually not isolated because many are explosive. Instead, they are immediately treated with a reducing agent (usually zinc dust and aqueous acid), to give aldehydes or ketones.

$$\underset{/}{\overset{\backslash}{C}}=\overset{/}{\underset{\backslash}{C}} \xrightarrow{O_3} \underset{/}{\overset{\backslash}{C}}\overset{\overset{\displaystyle O}{\diagup\;\diagdown}}{\underset{O-O}{}}\overset{/}{\underset{\backslash}{C}} \xrightarrow[H_3O^+]{Zn} \underset{/}{\overset{\backslash}{C}}=O + O=\overset{/}{\underset{\backslash}{C}} \quad \textbf{(3.43)}$$

| alkene | an ozonide | two carbonyl compounds (aldehyde or ketone) |

*Ozonides are not formed in one step. Ozone first adds to the double bond to give

$$-\underset{\underset{O}{\diagdown}}{\overset{\diagup}{C}}-\underset{\underset{O}{\diagup}}{\overset{\diagdown}{C}}-$$

, and this intermediate then rearranges to the ozonide.

The net result is to "break" the double bond of the alkene and form two carbon–oxygen double bonds (carbonyl groups), one at each carbon of the original double bond. The overall process is called **ozonolysis.**

Ozonolysis can be used to locate the position of a double bond. For example, on ozonolysis, 1-butene gives two different aldehydes, whereas 2-butene gives a single aldehyde.

$$CH_2\!=\!CHCH_2CH_3 \xrightarrow[\text{2. Zn, H}^+]{\text{1. O}_3} CH_2\!=\!O + O\!=\!CHCH_2CH_3 \qquad\qquad \textbf{(3.44)}$$
$$\text{1-butene} \text{formaldehyde} \text{propionaldehyde}$$

$$CH_3CH\!=\!CHCH_3 \xrightarrow[\text{2. Zn, H}^+]{\text{1. O}_3} 2\ CH_3CH\!=\!O \qquad\qquad \textbf{(3.45)}$$
$$\text{2-butene} \text{acetaldehyde}$$

In this way, one can easily tell which butene isomer is which. By working backwards from the structures of ozonolysis products, one can deduce the structure of an unknown alkene.

Example 3.8 **An alkene, on ozonolysis, gives equal amounts of acetone, $(CH_3)_2C\!=\!O$, and formaldehyde, $CH_2\!=\!O$. Deduce its structure.**

Solution **Connect to each other by a double bond the carbons that are bound to oxygen in the ozonolysis products. The alkene is $(CH_3)_2C\!=\!CH_2$.**

Problem 3.22 **What alkene would give only acetone, $(CH_3)_2C\!=\!O$, as the ozonolysis product?**

a word about ——————————————————————————————

3. Ozone

Ozone is prepared by passing an electric discharge through a stream of oxygen. It is formed during lightning, and its pleasant, characteristic "fresh" odor can often be smelled in the air after a thunder storm. Pure ozone is a bluish, explosive gas.

Though small concentrations of ozone in the earth's upper atmosphere shield us from dangerous ultraviolet radiation, ozone in the lower atmosphere is a hazard. It is a toxic, powerful oxidizing agent. Because of its rapid reaction with double bonds, it causes the breakdown and deterioration of unsaturated compounds (for example, the rubber of automobile tires). The normal ozone concentration in the atmosphere at sea level is about 0.05 ppm, but it can reach higher, dangerous levels as a result of sunlight-catalyzed reactions with automobile exhaust fumes.

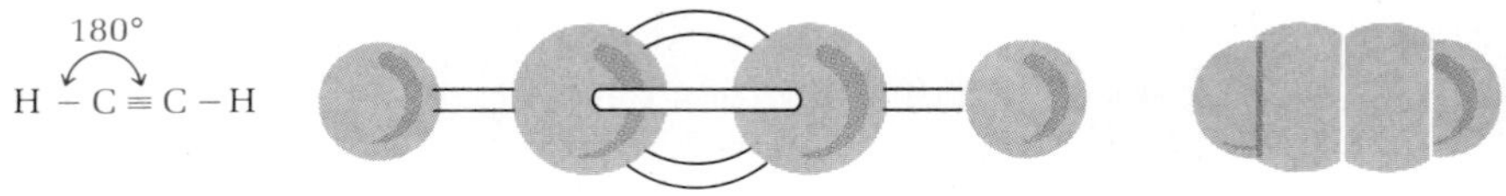

FIGURE 3.8

Models of acetylene,
showing its linearity.

$$H - C \equiv C - H$$

180°

3.22
SOME FACTS
ABOUT TRIPLE
BONDS

In the final sections of this chapter, we will describe some of the special features of triple bonds and alkynes.

A carbon that is part of a triple bond is attached to only *two* other atoms, and the bond angle is 180°. Thus acetylenes are linear, as shown in Figure 3.8. The carbon–carbon triple bond distance is about 1.21 Å, appreciably shorter than that of the average double bond (1.34 Å) or single bond (1.54 Å). Apparently three electron pairs between two carbons draw them even closer together than do two pairs. Because of their linear geometry, no *cis-trans* isomerism is possible for alkynes.

Now let us see how the orbital theory of bonding can be adapted to explain these facts.

3.23
THE ORBITAL
MODEL OF A
TRIPLE BOND

We begin, as with sp^3 and sp^2 bonding, by promoting a carbon 2s electron to the vacant 2p orbital (Figure 3.9). But, because the carbon of an acetylene is connected to only two other atoms, we need combine the 2s with only one 2p orbital, to make two sp hybrid orbitals. These orbitals extend in opposite directions from the carbon atom. The angle between the orbitals is 180°, so as to minimize repulsion between any electrons placed in these two hybrid orbitals. The remaining two valence electrons occupy two p orbitals that are mutually perpendicular and also perpendicular to the hybrid sp orbitals.

The formation of a triple bond from two sp-hybridized carbons is shown in Figure 3.10. The end-on overlap of two sp orbitals forms the σ bond between the two carbons, and the overlap of the properly aligned p orbitals forms two π bonds (designated π_1 and π_2 in the figure). This model nicely explains the linearity of acetylenes.

FIGURE 3.9

The formation of two
sp hybrid orbitals.

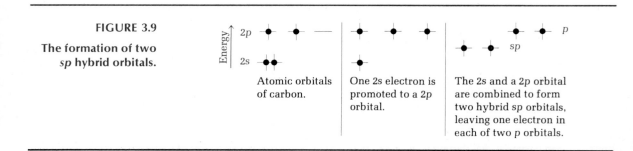

Atomic orbitals
of carbon.

One 2s electron is
promoted to a 2p
orbital.

The 2s and a 2p orbital
are combined to form
two hybrid sp orbitals,
leaving one electron in
each of two p orbitals.

FIGURE 3.10

The triple bond consists of the end-on overlap of two *sp* hybrid orbitals to form a σ bond plus the edgewise overlap of two sets of parallel-oriented *p* orbitals to form two mutually perpendicular π bonds.

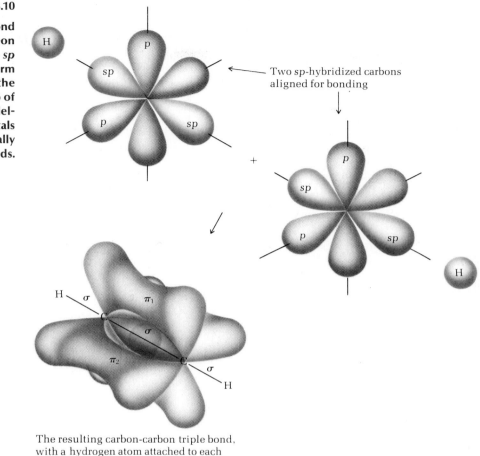

Two *sp*-hybridized carbons aligned for bonding

The resulting carbon-carbon triple bond, with a hydrogen atom attached to each remaining *sp* bond. (The orbitals involved in the C—H bonds are omitted for clarity.)

3.24 ADDITION REACTIONS OF ALKYNES

Many addition reactions described for alkenes also occur with alkynes. For example, bromine adds as follows:

$$H-C\equiv C-H \xrightarrow{Br_2} \underset{Br}{\overset{H}{}}C=C\underset{H}{\overset{Br}{}} \xrightarrow{Br_2} H-\underset{\underset{Br}{|}}{\overset{\overset{Br}{|}}{C}}-\underset{\underset{Br}{|}}{\overset{\overset{Br}{|}}{C}}-H \qquad (3.46)$$

trans-1,2-dibromoethene   1,1,2,2-tetrabromoethane

Note that the addition occurs mainly in a *trans* manner, as shown.

A palladium catalyst (called Lindlar's catalyst) can control hydrogen addition so that only *one* mole of hydrogen adds. In this case, the product is a cis-alkene, because both hydrogens add from the catalyst surface.

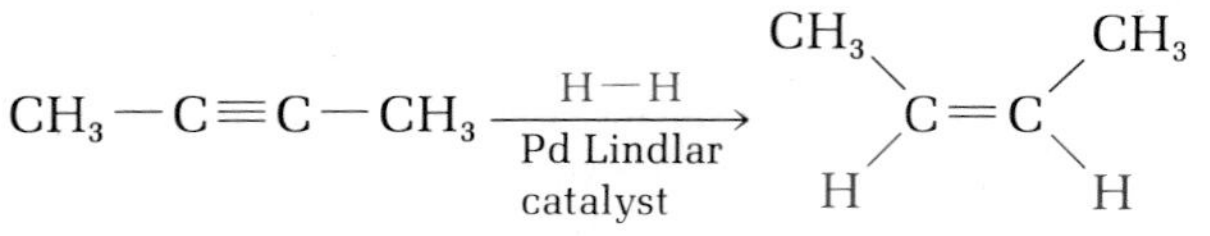

$$CH_3-C\equiv C-CH_3 \xrightarrow[\substack{Pd\ Lindlar\\catalyst}]{H-H} \underset{\substack{H \qquad\qquad H}}{\overset{\substack{CH_3 \qquad\qquad CH_3}}{C=C}} \qquad (3.47)$$

2-butyne   cis-2-butene
bp 27°C   bp 3.7°C

With unsymmetric triple bonds and unsymmetric reagents, Markovnikov's rule is followed in each step, as shown in the following example:

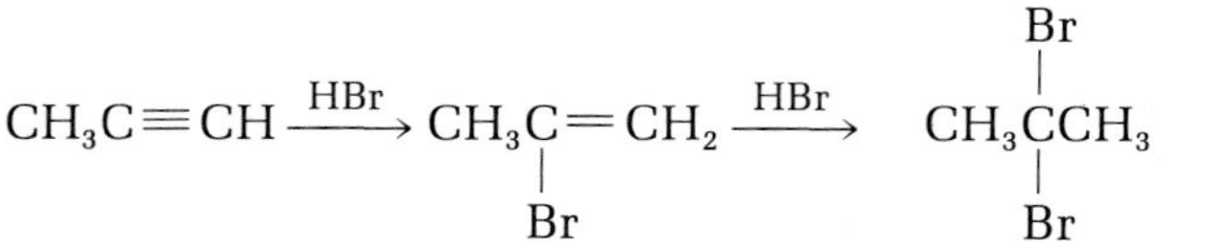

$$CH_3C\equiv CH \xrightarrow{HBr} \underset{Br}{\overset{}{CH_3C=CH_2}} \xrightarrow{HBr} \underset{Br}{\overset{Br}{CH_3CCH_3}} \qquad (3.48)$$

propyne   2-bromopropene   2,2-dibromopropane

Addition of water to alkynes requires not only an acid catalyst but mercuric ion as well. The mercuric ion forms a complex with the triple bond and activates it for addition. Although the reaction is similar to that of alkenes, the initial product—a vinyl alcohol or enol—is not stable and rearranges.

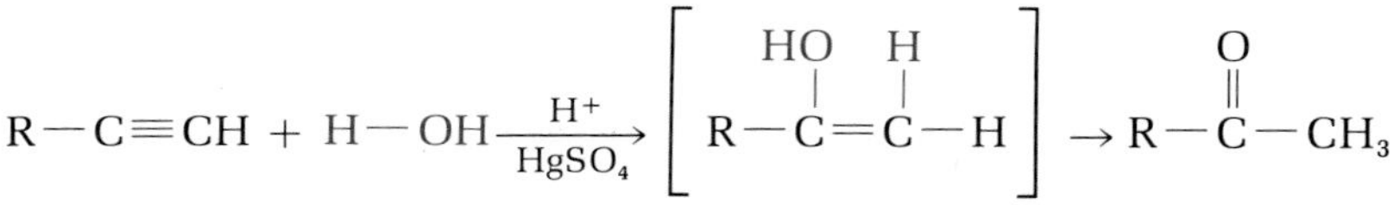

$$R-C\equiv CH + H-OH \xrightarrow[HgSO_4]{H^+} \left[\underset{\substack{}}{\overset{\substack{HO \quad H}}{R-C=C-H}} \right] \to \underset{}{\overset{O}{R-C-CH_3}} \qquad (3.49)$$

a vinyl alcohol
or enol

The product is a methyl ketone or, in the case of acetylene itself (R=H), acetaldehyde. We will discuss the chemistry of enols and the mechanism of the second step in Chapter 9.

Problem 3.23   **Write equations for the following reactions.**
   a. $CH_3C\equiv CH + Cl_2$ (1 mol)
   b. $CH_3C\equiv CH + Cl_2$ (2 mol)
   c. 1-butyne + HBr (1 and 2 mol)
   d. 1-butyne + H_2O (Hg^{2+}, H^+)

**3.25
ACIDITY OF
ALKYNES**   A hydrogen on a triple bond is weakly acidic and can be removed by a very strong base. Sodium amide, for example, converts acetylenes to acetylides.

$$R-C\equiv C-H + Na^+NH_2^- \xrightarrow{liquid\ NH_3} R-C\equiv C:{}^-Na^+ + NH_3 \qquad (3.50)$$

this hydrogen is   sodium amide   a sodium acetylide
weakly acidic

This type of reaction occurs easily with a hydrogen attached to a triple bond, but not with hydrogens attached to a double or single bond. Why? Consider the hybridization of the carbon atom in C—H bonds:

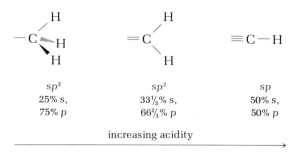

$$sp^3 \qquad\qquad sp^2 \qquad\qquad sp$$
$$25\% \ s, \qquad\qquad 33\tfrac{1}{3}\% \ s, \qquad\qquad 50\% \ s,$$
$$75\% \ p \qquad\qquad 66\tfrac{2}{3}\% \ p \qquad\qquad 50\% \ p$$

increasing acidity

As the hybridization at carbon becomes more s-like and less p-like, the acidity of the attached hydrogen increases. Recall that s orbitals are closer to the nucleus than are p orbitals. Consequently, the bonding electrons are closest to the carbon atom in the ≡C—H bond, making it easiest for a base to remove that type of proton. Sodium amide is a sufficiently strong base for this purpose.

Problem 3.24 **Write an equation for the reaction of 1-butyne with sodium amide in liquid ammonia.**

Problem 3.25 **Will 2-butyne react with sodium amide? Explain.**

In Sec. 6.6, we will see how the acidity of alkynes can be used to develop a general method for their synthesis.

a word about

4. Petroleum, Petroleum Refining, and Octane Numbers

Petroleum is at present the most important of the fossil fuels. The need for petroleum to keep our industrial society going sometimes seems second only to our need for food, air, water, and shelter. What is this "black gold," and how do we use it?

Petroleum is a complex mixture of hydrocarbons that formed over eons of time through the gradual decay of buried animal and vegetable matter. Crude oil is a viscous black liquid that collected in vast underground pockets in sedimentary rock (the word

petroleum literally means "rock oil," from the Latin petra, rock, and oleum, oil). It must be brought to the surface via drilling and pumping. To be most useful, the crude oil must subsequently be refined.

The first step in petroleum refining is usually distillation. The crude oil is heated to about 400°C and the vapors rise through a tall fractionating column. The lower-boiling fractions rise faster and higher in the column before condensing to liquids, whereas higher-boiling fractions do not rise so high. By drawing off liquid at various column levels, technicians separate crude oil roughly into the fractions shown in Table 3.4.

The gasoline fraction comprises only about 25% of crude oil. It is the most valuable fraction, both as a

lower boiling points. The carbon chain can break at many points:

$$
\begin{array}{ccc}
 & alkane & alkene \\
 & C_5H_{12} & + \quad C_5H_{10} \\
C_{10}H_{22} \longrightarrow & C_8H_{18} & + \quad C_2H_4 \\
\text{alkane} & C_2H_6 & + \quad C_8H_{16} \\
 & C_4H_{10} & + \quad (C_4H_8 + C_2H_4)
\end{array}
$$

To balance the number of hydrogens, any particular alkane must give at least one alkane and one alkene as products. Thus catalytic cracking converts larger alkanes into a mixture of smaller alkanes and alkenes and increases the yield of gasoline from petroleum.

During cracking, large amounts of the lower gaseous hydrocarbons are formed—ethylene, propylene, butanes, and butenes. Some of these, especially ethylene, are used as petrochemical raw materials. In order to obtain more gasoline, investigators sought methods to convert these low-molecular-weight hydrocarbons to larger hydrocarbons that boil in the gasoline range. One such process is alkylation, which is the combination of an alkane with an alkene to form a higher-boiling alkane:

$$C_2H_6 + C_4H_8 \xrightarrow{\text{catalyst}} C_6H_{14}$$
$$C_4H_{10} + C_4H_8 \xrightarrow{\text{catalyst}} C_8H_{18}$$

These processes, which were developed in the late 1930s, were of prime importance for producing aviation fuel during World War II and are still used extensively for making high-octane gasoline.

This brings us to octane number and why it is important. Some hydrocarbons, especially those with highly branched structures, burn smoothly in an engine and drive the piston forward evenly. Other hydrocarbons, especially those with straight (or unbranched) carbon chains, tend to explode in the cylinder and drive the piston forward violently. These undesirable explosions produce knocks. An arbitrary scale was set up to evaluate the knock properties of gasolines. Iso-octane (2,2,4-trimethylpentane), an excellent fuel with a highly branched structure, was given a rating of 100, and heptane, a very poor fuel in this respect, was given a rating of 0. A "regular" gasoline with an octane number of 87 has the same "knock" properties as a mixture that is 87% iso-octane and 13% heptane.

The addition of small amounts of tetraethyllead, $(CH_3CH_2)_4Pb$, to gasoline improves its octane rating

TABLE 3.4   Common petroleum fractions

Boiling range, °C	Name	Range of carbon atoms per molecule	Use
below 20	gases	C_1 to C_4	heating, cooking, and chemical raw material
20–200	naphtha; straight run gasoline	C_5 to C_{12}	fuel; lighter fractions (such as petroleum ether, bp 30–60°C) are also used as laboratory solvents.
200–300	kerosene	C_{12} to C_{15}	fuel
300–400	fuel oil	C_{15} to C_{18}	heating homes, diesel fuel
over 400		over C_{18}	lubricating oil, greases, paraffin waxes, asphalt

fuel and as a source material for the petrochemical industry. Thus, many processes have been developed for converting other fractions into gasoline.

Higher-boiling fractions can be "cracked" by heat and catalysts (mainly silica and alumina), to give products with shorter carbon chains and therefore but is undesirable for environmental reasons. However, unleaded gasoline must contain a very high percentage of hydrocarbons with a high octane rating. The object, then, is to convert straight-chain hydrocarbons to branched-chain hydrocarbons, which have higher octane ratings, or to develop less toxic additives than tetraethyllead.

Certain catalysts can produce branched-chain alkanes from straight-chain alkanes.

$$CH_3CH_2CH_2CH_3 \xrightarrow[\text{alumina}]{\text{AlCl}_3,\ \text{HCl}} CH_3\overset{\overset{\displaystyle CH_3}{|}}{C}HCH_3$$

n-butane   isobutane

This process, called isomerization, is carried out on a large scale commercially.

Aromatic hydrocarbons such as benzene and toluene have a high octane rating. A platinum catalyst used in a process called platforming cyclizes and dehydrogenates alkanes to cycloalkanes and to aromatics:

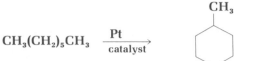

$$CH_3(CH_2)_5CH_3 \xrightarrow[\text{catalyst}]{\text{Pt}}$$

methylcyclohexane

$$\xrightarrow[\text{catalyst}]{\text{Pt}}$$

toluene

Of course, large amounts of hydrogen gas are also formed during platforming. Millions of gallons of aromatic hydrocarbons are produced daily by such processes, not only to add to unleaded gasoline to improve its octane rating, but also to supply raw materials for many other petrochemically based products, as we shall see in the next chapter.

ADDITIONAL PROBLEMS

3.26. Write structural formulas and IUPAC names for all possible isomers of the following compounds, with the indicated number of multiple bonds.
  a. C_4H_8 (one double bond)   **b.** C_5H_{10} (one double bond)
  c. C_5H_8 (two double bonds)   **d.** C_5H_8 (one triple bond)

3.27. Name the following compounds by the IUPAC system.
  a. $CH_3CH_2CH{=}CHCH_3$ **b.** $(CH_3)_2C{=}CHCH_3$
  c.   **d.** $CH_3C{\equiv}CCH_2CH_3$
                                    f.

  e. $CH_2{=}CCl{-}CH{=}CH_2$ **g.**

3.28. Write a structural formula for each of the following compounds.
  a. 3-hexene   **b.** cyclobutene
  c. 1,3-dibromo-2-butene   **d.** 3-methyl-1-pentyne
  e. 1,4-hexadiene   **f.** vinyl bromide
  g. allyl chloride   **h.** vinylcyclopentane
  i. 4-methylcyclohexene   **j.** 2,3-dibromo-1,3-cyclopentadiene

3.29. Explain why the following names are incorrect, and give a correct name in each case.
  a. 3-butene   **b.** 3-pentyne
  c. 2-ethyl-1-propene   **d.** 2-methylcyclopentene
  e. 3-methyl-1,3-butadiene   **f.** 1-methyl-2-butene

3.30. **a.** What are the usual lengths for the single (sp^3–sp^3), double (sp^2–sp^2), and triple (sp–sp) carbon–carbon bonds?
b. The *single* bond in the following compounds has the length shown. Suggest a possible expanation for the observed shortening.

$$CH_2{=}CH{-}CH{=}CH_2 \qquad CH_2{=}CH{-}C{\equiv}CH \qquad CH{\equiv}C{-}C{\equiv}CH$$
$$\qquad\quad \uparrow \qquad\qquad\qquad\qquad \uparrow \qquad\qquad\qquad\qquad \uparrow$$
$$\qquad\quad 1.47\,\text{Å} \qquad\qquad\qquad\quad 1.43\,\text{Å} \qquad\qquad\qquad 1.37\,\text{Å}$$

3.31. Which of the following compounds can exist as *cis-trans* isomers? If such isomerism is possible, draw the structures in a way that clearly illustrates the geometry.
a. 1-pentene   **b.** 2-pentene   **c.** 1-chloropropene
d. 3-chloropropene   **e.** 1,3,5-hexatriene   **f.** 1,2-dibromocyclodecene

3.32. The mold metabolite and antibiotic *mycomycin* has the formula

$$HC{\equiv}C{-}C{\equiv}C{-}CH{=}C{=}CH{-}CH{=}CH{-}CH{=}CH{-}CH_2\overset{\displaystyle O}{\overset{\|}{C}}{-}OH$$

Number the carbon chain, starting with the carbonyl carbon.
a. Which multiple bonds are conjugated?
b. Which multiple bonds are cumulative?
c. Which multiple bonds are isolated?

3.33. Write the structural formula and name of the product when each of the following reacts with 1 mole of bromine.
a. 2-butene   **b.** vinyl bromide   **c.** 1-methylcyclopentene
d. 1,3-cyclohexadiene   **e.** 1,4-cyclohexadiene   **f.** 2,3-dimethyl-2-butene

3.34. Write an equation for the reaction of 1-butene with each of the following reagents.
a. chlorine   **b.** hydrogen chloride
c. hydrogen (Pt catalyst)   **d.** ozone, followed by Zn,H$^+$
e. H$_2$O,H$^+$ **f.** B$_2$H$_6$ followed by H$_2$O$_2$,OH$^-$

3.35. Which unsaturated hydrocarbon would react with what reagent to form each of the following compounds?
a. CH$_3$CHBrCHBrCH$_3$ **b.** (CH$_3$)$_3$COH   **c.** (CH$_3$)$_2$CHOSO$_3$H
d.   **e.** CH$_3$CH{=}CHCH$_2$Br   **f.** CH$_3$CCl$_2$CCl$_2$CH$_3$

g.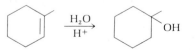

3.36. *β*-Carotene, a yellow pigment present in carrots and many other plants, is a polyene with the molecular formula C$_{40}$H$_{56}$. Complete hydrogenation of *β*-carotene gives a saturated hydrocarbon with the formula C$_{40}$H$_{78}$. How many double bonds and how many rings are present in *β*-carotene?

3.37. The acid-catalyzed hydration of 1-methylcyclohexene gives 1-methyl-cyclohexanol:

Write every step in the mechanism of this reaction.

3.38. Predict the structures of the two possible monohydration products of limonene (Figure 1.12). These alcohols are called terpineols. Predict the structure of the diol (di-alcohol) obtained by hydrating both double bonds in limonene. These alcohols are used in the cough medicine "elixir of terpin hydrate" as an expectorant.

3.39. When propylene is treated with bromine in methanol (CH_3OH), two products are formed, with the molecular formulas $C_3H_6Br_2$ and C_4H_9BrO. Give their structures and explain mechanistically (with equations) how each is formed.

3.40. Adding 1 mol of hydrogen bromide to 2,4-hexadiene gives two products. Give their structures and write all the steps in a reaction mechanism that explains how each product is formed.

3.41. Using Table 3.3 to obtain the monomer structures, write the structure of the repeating unit for each of the following polymers.
a. Orlon   **b.** polyvinyl chloride   **c.** polystyrene   **d.** Saran

3.42. Write out the steps in a mechanism that explains the free-radical copolymerization of 1,3-butadiene and styrene to give the synthetic rubber SBR (eq. 3.37).

3.43. Write an equation that shows clearly the structure of the alcohol product obtained from the sequential hydroboration and H_2O_2/OH^- oxidation of:
a. 2-methyl-1-butene   **b.** 1,2-dimethylcyclopentene

3.44. Write equations to show how  $=CH_2$ could be converted to:

a.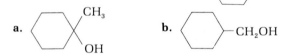

b. $-CH_2OH$

3.45. Describe two simple chemical tests that could be used to distinguish cyclohexane from cyclohexene.

3.46. Give the formulas of the alkenes that, on ozonolysis, give:
a. only $CH_3CH_2CH=O$ **b.** $(CH_3)_2C=O$ and $CH_3CH=O$
c. $CH_2=O$ and $(CH_3)_2CHCH=O$ **d.** $O=CHCH_2CH_2CH=O$

3.47. The ozonolysis of natural rubber gives levulinic aldehyde,

$$CH_3CCH_2CH_2CH=O.$$
$$\underset{O}{\overset{\|}{}}$$

Explain how this result is consistent with its formula, shown in eq. 3.36.

3.48. Write equations for the following reactions.
a. 2-pentyne + Cl_2 (2 mol)
b. 3-hexyne + H_2 (1 mol, Lindlar's catalyst)
c. propyne + H_2O (H^+,$HgSO_4$ catalyst)
d. propyne + sodium amide in liquid ammonia

3.49. Determine what alkyne and what reagent would give:
a. 2,2-dibromobutane   **b.** 2,2,3,3-tetrabromobutane

four

Aromatic Compounds

Spices and herbs have played a long and romantic role in shaping the course of history. They bring to mind mysterious commodities such as frankincense and myrrh and the great explorers of past centuries—Vasco da Gama, Christopher Columbus, Ferdinand Magellan, Sir Francis Drake—who, in their quest for spices, opened up the Western world. Trade in these substances was immensely profitable. It is not surprising, therefore, that spices and herbs were among the first types of natural products studied by organic chemists. If one could isolate from plants pure substances responsible for these desirable fragrances and flavors and determine their structures, these early chemists knew, one could synthesize them in large quantity and at a low cost.

As natural products go, many of these fragrant or **aromatic** substances have relatively simple structures. Many of them contain a common six-carbon unit that passes unscathed through various chemical reactions that involve the rest of the structure. This group, C_6H_5-, is common to many substances, including **benzaldehyde** (isolated from the oil of bitter almonds), **benzyl alcohol** (isolated from gum benzoin, a balsam resin obtained from certain Southeast Asian trees), and **toluene** (a hydrocarbon isolated from tolu balsam). When any of these three compounds is oxidized, the C_6H_5-, group remains intact, the product in each case being **benzoic acid** (another constituent of gum benzoin). On heating, the calcium salt of this acid gives the parent hydrocarbon, C_6H_6, to which all of these substances are structurally related (eq. 4.1).

$$C_6H_5CH{=}O \xrightarrow{\text{oxidize}}$$
benzaldehyde

$$C_6H_5CH_2OH \xrightarrow{\text{oxidize}} C_6H_5CO_2H \xrightarrow[\text{2. heat}]{\text{1. CaO}} C_6H_6 \qquad (4.1)$$
benzyl alcohol $\qquad$ benzoic acid $\qquad$ benzene

$$C_6H_5CH_3 \xrightarrow{\text{oxidize}} \uparrow$$
toluene

This same hydrocarbon had first been isolated by Michael Faraday in 1825 from compressed illuminating gas. The compound is now called **benzene,** and it is the parent hydrocarbon of a class of substances that we call **aromatic compounds** *not* because of their aroma, but because of their special chemical properties. Let us see what some of those special properties are.

4.2
SOME FACTS
ABOUT BENZENE

The carbon-to-hydrogen ratio in the molecular formula for benzene, C_6H_6 or $(CH)_6$, suggests a highly unsaturated structure. Compare it, for example, with the saturated alkane hexane, C_6H_{14}, or the saturated cycloalkane cyclohexane, C_6H_{12}.

Problem 4.1 **Draw at least five isomeric structures, all of which fit the molecular formula C_6H_6. Note that they are highly unsaturated and/or contain small, strained rings.**

Despite its molecular formula, benzene for the most part does not behave as though it were unsaturated. For instance, it does not decolorize bromine solutions the way alkenes and alkynes do, nor is it easily oxidized by potassium permanganate. It does not undergo the typical addition reactions of alkenes or alkynes, for example, with hydrogen chloride or sulfuric acid. Instead, *benzene reacts mainly by substitution.* For example, when treated with bromine in the presence of ferric bromide as a catalyst, benzene gives bromobenzene and hydrogen bromide.

$$C_6H_6 + Br_2 \xrightarrow[\text{catalyst}]{FeBr_3} C_6H_5Br + HBr \qquad (4.2)$$

benzene   bromobenzene

Chlorine, with a ferric chloride catalyst, reacts similarly.

$$C_6H_6 + Cl_2 \xrightarrow[\text{catalyst}]{FeCl_3} C_6H_5Cl + HCl \qquad (4.3)$$

chlorobenzene

Only one monobromobenzene or monochlorobenzene has ever been isolated; that is, there are no isomers obtained in either of these reactions. This observation implies that *all six hydrogens in benzene are chemically equivalent.* It does not matter which hydrogen is replaced by bromine; we get the same monobromobenzene. This fact will have to be explained by any structure proposed for benzene.

When bromobenzene is treated with a second equivalent of bromine and the same ferric bromide catalyst, *three* dibromobenzenes are obtained.

$$C_6H_5Br + Br_2 \xrightarrow[\text{catalyst}]{FeBr_3} C_6H_4Br_2 + HBr \qquad (4.4)$$

dibromobenzenes
(*three* isomers)

The isomeric dibromobenzenes are not formed in equal amounts. Two of them predominate and only a small amount of the third isomer is formed, but the important point is that there are three isomers—no more and no less. Similar results are obtained when chlorobenzene is further chlorinated to give dichlorobenzenes. These facts will also have to be explained by any structure proposed for benzene.

The problem of benzene's structure does not sound overwhelming, yet it took many decades to solve. Let us examine the main proposals that led to our modern view of its structure.

**4.3
THE KEKULÉ
STRUCTURE OF
BENZENE**

In 1865 Kekulé proposed the first reasonable structure for benzene.* He suggested that the six carbon atoms are located at the corners of a regular hexagon, with one hydrogen atom attached to each carbon atom. To give each carbon atom a valence of 4, he suggested alternating single and double bonds around the ring (what we now call a conjugated system of double bonds). Of course, this structure is highly unsaturated. To explain benzene's negative tests for unsaturation, Kekulé suggested that the single and double bonds exchange positions around the ring so rapidly that the typical reactions of alkenes cannot take place.

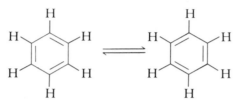

the Kekulé structures for benzene

Problem 4.2   Write out eq. 4.2 and eq. 4.4 using the Kekulé structure for benzene. Does the Kekulé proposal for benzene explain the existence of only one monobromobenzene and only three dibromobenzenes?

Problem 4.3   How would Kekulé explain the fact that there is only one dibromobenzene with the bromines on adjacent carbon atoms, although we can draw two different structures, with either a double or a single bond between the bromine-bearing carbons?

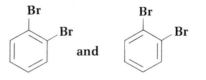

**4.4
THE RESONANCE
MODEL FOR
BENZENE**

Kekulé's model for the structure of benzene is nearly, but not quite, correct. We now recognize that *Kekulé's two structures for benzene differ only in the arrangement of the electrons;* all the atoms occupy the same positions in both structures. This is precisely the requirement for **resonance** (review Sec. 1.12). That is, Kekulé's formulas represent two identical contributing structures to the single resonance hybrid structure of benzene. Instead of writing an equilibrium symbol between them, as Kekulé did, we write the double-headed arrow symbol for a resonance hybrid:

*Friedrich August Kekulé (1829–1896) was a pioneer in the development of structural formulas in organic chemistry. He was among the first to recognize the tetracovalence of carbon and the importance of carbon atom chains in organic structures. He is best known for his proposal regarding the structure of benzene and other aromatic compounds. It is interesting that Kekulé initially studied architecture and only later switched to chemistry. Judging from his contributions, he apparently viewed chemistry as molecular architecture.

FIGURE 4.1
Scale model of
benzene.

Properties
colorless liquid
bp 80°C
mp 5.5°C

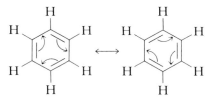

Benzene is a resonance hybrid of
these two contributing structures.

To express this model of benzene another way, all benzene molecules are identical and their structure is not adequately represented by either contributing structure. Being a resonance hybrid, benzene is more stable than either of the contributing structures. There are no single or double bonds in the resonance hybrid structure for benzene, only one type of carbon–carbon bond, which is of an intermediate type. Consequently, it is not surprising that benzene does not react exactly like alkenes.

Modern physical measurements support this model for the benzene structure. Benzene molecules are planar, and each carbon atom is at the corner of a regular hexagon. All the carbon–carbon bond lengths are identical. The bonds do not alternate in length, as the Kekulé structure might suggest. This length is 1.39 Å, intermediate between typical single (1.54 Å) and double (1.34 Å) carbon–carbon bond lengths. Figure 4.1 shows a scale model of benzene.

4.5
ORBITAL MODEL
FOR BENZENE

Orbital theory, which was so useful in rationalizing the geometries of alkanes, alkenes, and alkynes, is also useful in explaining the structure of benzene. Each carbon atom in benzene is connected to only three other atoms (two carbons and a hydrogen). Hence each carbon should be sp^2-hybridized, as in ethylene. Two sp^2 orbitals of each carbon atom overlap with similar orbitals of adjacent carbon atoms to form the σ bonds of the hexagonal ring. The third sp^2 orbital of each carbon overlaps the hydrogen 1s orbital to form the C—H σ bonds. Perpendicular to the plane of the

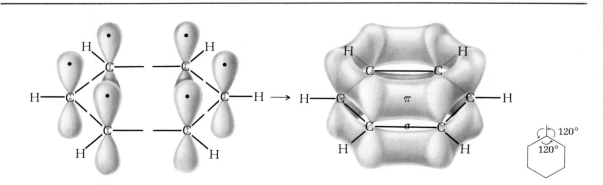

FIGURE 4.2   An orbital representation of the bonding in benzene. Sigma bonds are formed by the overlap of *sp²* orbitals. In addition, each carbon contributes one electron to the π system by overlap of its *p* orbital with the *p* orbital of its two neighbors.

three *sp²* orbitals is a *p* orbital containing one electron. The *p* orbitals on all six carbon atoms overlap to form π orbitals that create a cloud of electrons above and below the plane of the ring. The construction of a benzene ring from six *sp²*-hybridized carbons is shown schematically in Figure 4.2. Note that this model nicely explains the planarity of benzene. It also explains the hexagonal shape, with H—C—C and C—C—C angles of 120°.

4.6
SYMBOLS FOR
BENZENE

Two symbols are used to represent benzene. One is the standard Kekulé structure, and the other is a hexagon with an inscribed circle, to represent the delocalized π electron cloud:

Kekulé   delocalized π

Regardless of which symbol is used, the hydrogens are usually not written explicitly, but we must remember that there is one hydrogen attached to the carbon at each corner of the hexagon.

The delocalized π symbol emphasizes the fact that the electrons are distributed evenly around the ring, and in this sense it is perhaps the more accurate symbol for benzene. However, the Kekulé symbol reminds us very clearly that there are six π electrons in benzene. For this reason, it is particularly useful in allowing us to keep track of the valence electrons during chemical reactions of benzene. In this book we will use the Kekulé symbol. As we use this symbol, however, we must keep in mind that the "double bonds" are not fixed in the positions shown, nor are they really double bonds at all.

Example 4.1 **Write the structural formula for benzaldehyde (eq. 4.1).**

Solution

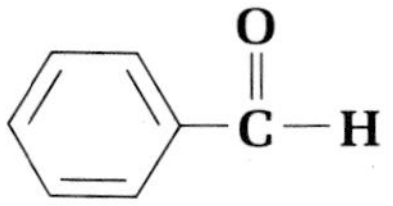

Problem 4.4 **Write the formulas for benzyl alcohol, toluene, and benzoic acid (eq. 4.1).**

4.7
NOMENCLATURE
OF AROMATIC
COMPOUNDS

Because aromatic chemistry developed in a haphazard fashion many years before systematic methods of nomenclature were developed, common names acquired historic respectability and are still frequently used. Examples include

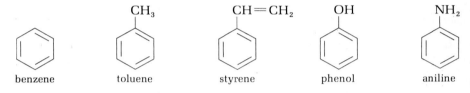

benzene   toluene   styrene   phenol   aniline

There are, however, some systematic features. Aromatic hydrocarbons, as a class, are called **arenes.** Monosubstituted benzenes are named as derivatives of benzene.

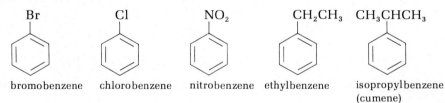

bromobenzene   chlorobenzene   nitrobenzene   ethylbenzene   isopropylbenzene (cumene)

When two substituents are present, three isomeric structures are possible. They are designated by the prefixes *ortho-*, *meta-*, and *para-*, which are frequently abbreviated as *o-*, *m-*, and *p-*, respectively.

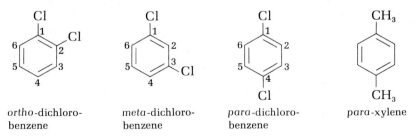

ortho-dichloro-   *meta*-dichloro-   *para*-dichloro-   *para*-xylene
benzene   benzene   benzene

Problem 4.5 **Draw the structures for *ortho*-xylene and *meta*-xylene.**

The *ortho-*, *meta-*, and *para-* prefixes may be used even when the substituents differ from one another.

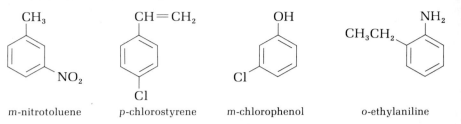

o-bromochlorobenzene   m-nitrotoluene   p-chlorostyrene   m-chlorophenol   o-ethylaniline

The positions of substituents can also be designated by numbering the ring, a method especially important when more than two substituents are present.

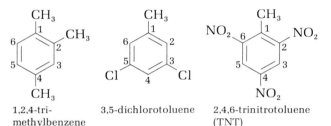

1,2,4-tri-   3,5-dichlorotoluene   2,4,6-trinitrotoluene
methylbenzene   (TNT)

Problem 4.6   **Draw the structures of:**
a. *para*-nitrotoluene   **b. *ortho*-bromophenol**
c. *meta*-dinitrobenzene   **d. *para*-divinylbenzene**

Problem 4.7   **Draw the structures of:**
a. 1,3,5-trimethylbenzene   **b. 2,6-dibromo-4-chlorotoluene**

Two groups with special names occur frequently in aromatic compounds. They are the **phenyl group** and the **benzyl group.**

C_6H_5- or $C_6H_5CH_2-$ or

phenyl group   benzyl group

The symbol Ph (or sometimes ϕ, phi) is used as an abbreviation for the phenyl group. The symbol **Ar** is the general symbol for an **aryl** (aromatic) group, just as R is the general symbol for an alkyl group. The use of these group names is illustrated in the following examples:

$CH_3CHCH_2CH_2CH_3$

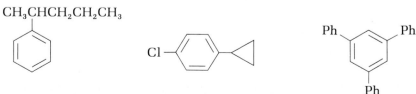

2-phenylpentane   *para*-chlorophenylcyclopropane   1,3,5-triphenylbenzene

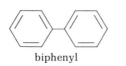

biphenyl

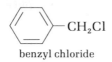

benzyl chloride

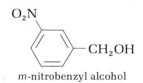

m-nitrobenzyl alcohol

Problem 4.8   **Draw the structures for:**
a. phenylcyclohexane   **b. benzyl alcohol**
c. para-phenylstyrene   **d. bibenzyl**

Problem 4.9   **Name the following structures:**

4.8
THE RESONANCE
ENERGY OF
BENZENE

We have asserted in Sec. 1.12 and Sec. 4.4 that a resonance hybrid is always more stable than any of its contributing structures. Fortunately, in the case of benzene, we can prove this assertion quite easily experimentally, and we can even measure how much more stable benzene is than the hypothetical molecule 1,3,5-cyclohexatriene.

The hydrogenation of a carbon–carbon double bond is an exothermic reaction. The amount of energy released is about 26–30 kcal/mol of alkene.

$$\text{C=C} + \text{H—H} \rightarrow \text{—C—C—} + 26\text{–}30 \text{ kcal/mol} \tag{4.5}$$

(The exact value depends on the number of alkyl groups attached to the double bond.) When two double bonds are hydrogenated, twice as much heat is evolved. Thus the heat of hydrogenation of 1,3-cyclohexadiene is approximately twice that of cyclohexene.

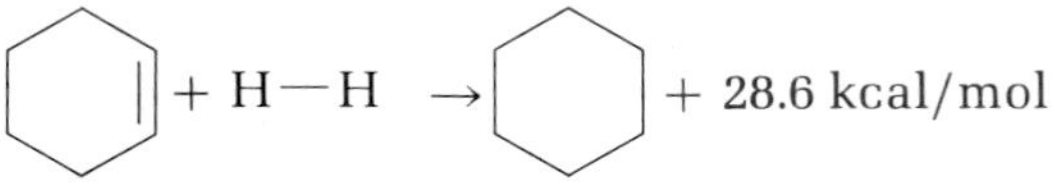

$$\bigcirc + \text{H—H} \rightarrow \bigcirc + 28.6 \text{ kcal/mol} \tag{4.6}$$

$$\bigcirc + 2\text{ H—H} \rightarrow \bigcirc + 55.4 \text{ kcal/mol} \tag{4.7}$$

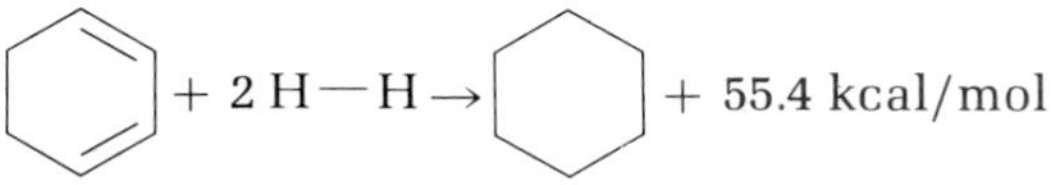

We infer that the heat of hydrogenation of the *hypothetical* 1,3,5-cyclohexatriene would correspond to the value for three double bonds, or about 84–86 kcal/mol. However, when we measure the heat of hydrogenation of benzene to cyclohexane, we find that the *experimental* value for the heat released is *much lower*: only 49.8 kcal/mol.

$$\text{benzene} + 3\,\text{H}-\text{H} \rightarrow \text{cyclohexane} + 49.8\,\text{kcal/mol} \qquad (4.8)$$

benzene   cyclohexane

We can only conclude that *real benzene molecules are more stable than one of the contributing structures,* the hypothetical molecule 1,3,5-cyclo-hexatriene, *by about 36 kcal/mol* (86 − 50 = 36).

We define the **stabilization energy** or **resonance energy** of a substance as the difference between the actual energy of the real molecule (the resonance hybrid) and the calculated energy of the most stable contributing structure. For benzene this value is about 36 kcal/mol. This is a substantial amount of energy. Consequently, as we shall see, benzene and other aromatic compounds react in such a way as to retain their aromatic structure and therefore retain their resonance energy.

**4.9
ELECTROPHILIC
AROMATIC
SUBSTITUTION**

The most common reactions of aromatic compounds involve substitution of other atoms or groups for the ring hydrogen, as we saw in eq. 4.2–eq. 4.4. Here are some typical substitution reactions of benzene:

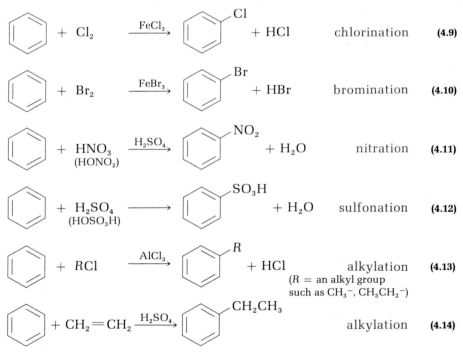

$$\text{benzene} + Cl_2 \xrightarrow{FeCl_3} C_6H_5Cl + HCl \qquad \text{chlorination} \qquad (4.9)$$

$$\text{benzene} + Br_2 \xrightarrow{FeBr_3} C_6H_5Br + HBr \qquad \text{bromination} \qquad (4.10)$$

$$\text{benzene} + HNO_3\ (HONO_2) \xrightarrow{H_2SO_4} C_6H_5NO_2 + H_2O \qquad \text{nitration} \qquad (4.11)$$

$$\text{benzene} + H_2SO_4\ (HOSO_3H) \longrightarrow C_6H_5SO_3H + H_2O \qquad \text{sulfonation} \qquad (4.12)$$

$$\text{benzene} + RCl \xrightarrow{AlCl_3} C_6H_5R + HCl \qquad \text{alkylation} \qquad (4.13)$$
$$(R = \text{an alkyl group such as } CH_3{}^-, CH_3CH_2{}^-)$$

$$\text{benzene} + CH_2{=}CH_2 \xrightarrow{H_2SO_4} C_6H_5CH_2CH_3 \qquad \text{alkylation} \qquad (4.14)$$

With benzene, most of these reactions are carried out at temperatures between about 0°C and 50°C, but these conditions may have to be milder or more severe if other substituents are already present on the benzene ring. Ordinarily the conditions can also be adjusted to introduce more than one substituent, if desired.

In the following sections we will consider how these reactions occur, why substitution occurs instead of addition, and what effect a substituent already present on the aromatic ring has on further substitution reactions.

4.10
THE MECHANISM
OF ELECTROPHILIC
AROMATIC
SUBSTITUTION

Much evidence indicates that all the reactions mentioned in the previous section involve initial attack on the benzene ring by an electrophilic re-agent. Consider chlorination (eq. 4.9) as a specific example. The reaction of benzene with chlorine is exceedingly slow without a catalyst, but it occurs quite briskly with the catalyst. What does the catalyst do? It acts as a Lewis acid and converts chlorine from a weak to a strong electrophile by polarizing the Cl—Cl bond and creating a positive chloronium ion.

$$:\ddot{C}l-\ddot{C}l: + Fe-Cl \rightleftharpoons \overset{\delta+}{Cl}\cdots\overset{\delta-}{Cl}\cdots Fe-Cl \rightleftharpoons :\ddot{C}l^+ + Cl-Fe^--Cl \qquad (4.15)$$

chloronium ion

Reaction of the electrophile with the benzene ring begins in exactly the same way as electrophilic addition to a carbon–carbon double bond (Sec. 3.13). The electrophile *adds* to the aromatic ring by using two of the π electrons from the aromatic π cloud to form a σ bond with one of the ring carbon atoms.

$$(4.16)$$

a benzenonium ion

The resulting carbocation is called a **benzenonium ion.** The positive charge in this carbocation can be delocalized by resonance to the carbon atoms that are *ortho*- and *para*- to the carbon to which the chlorine atom became attached.

resonance forms of a benzenonium ion   a composite representation
                                                    of a benzenonium ion

However, this delocalization only partially makes up for the aromatic res-onance energy that is lost in forming the ion. The reaction is completed by loss of a proton (from the ring carbon to which the electrophile became attached) to regenerate the aromatic ring:

$$\rightarrow \qquad -Cl + H^+ \qquad (4.17)$$

FIGURE 4.3	$E^+ = Cl^+$ or Br^+	NO_2^+	SO_3 or SO_3H^+	R^+
The electrophiles in common aromatic substitutions.	halonium ion	nitronium ion	sulfur trioxide or its protonated form	carbocation
	(halogenation)	(nitration)	(sulfonation)	(alkylation)

To summarize, electrophilic aromatic substitution occurs in two steps. In the first step (eq. 4.16), the electrophile adds to the aromatic ring; in the second step (eq. 4.17), a proton is lost from the intermediate benzenonium ion. We can express the general mechanism for all electrophilic aromatic substitutions by the equation

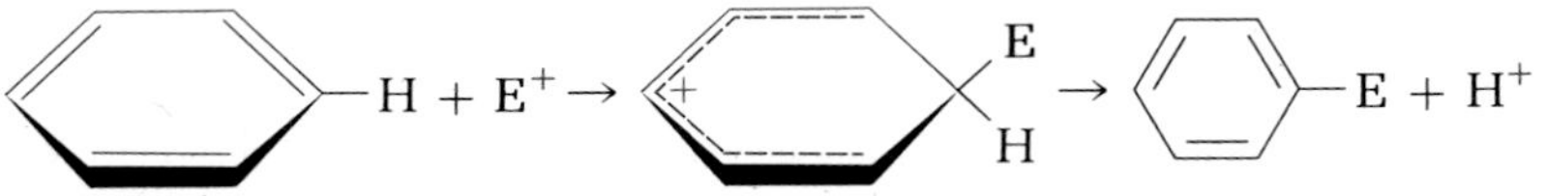

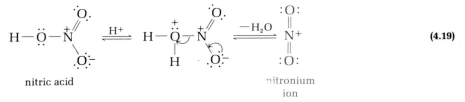

$$(4.18)$$

In Figure 4.3 we show the various electrophiles that are involved in typical aromatic substitution reactions (eq. 4.9–eq. 4.14).

In nitrations (eq. 4.11), the sulfuric acid catalyst protonates the nitric acid to generate the nitronium ion.

$$(4.19)$$

nitric acid   nitronium
                                                    ion

The nitronium ion then attacks the aromatic ring, as shown in eq. 4.18.

Example 4.2 **Write out the steps in the mechanism for nitration of benzene.**

Solution **The first step, formation of the electrophile NO_2^+, is shown in eq. 4.19. Then**

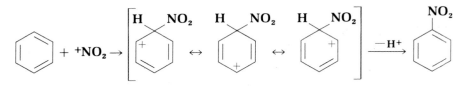

In sulfonation (eq. 4.12), we use either concentrated or fuming sulfuric acid, and the electrophile may be sulfur trioxide or protonated sulfur trioxide, $^+SO_3H$.

Problem 4.10   Write out the steps in the mechanism for sulfonation of benzene.

The alkylation of aromatic compounds (eq. 4.13 and eq. 4.14) is often referred to as the **Friedel–Crafts reaction,** after Charles Friedel (French) and James Mason Crafts (American), who first discovered the reaction in 1877. The electrophile is a carbocation, which may be formed either by removing halide ion from an alkyl halide with a Lewis acid catalyst (for example, $AlCl_3$) or by adding a proton to an alkene (as shown in eq. 4.14). Eq. 4.20 and eq. 4.21 illustrate these steps for the synthesis of ethylbenzene.

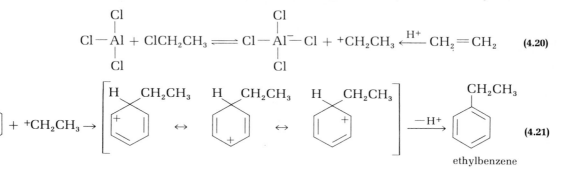

$$(4.20)$$

$$(4.21)$$

ethylbenzene

Problem 4.11   **Which product would you expect if propylene were used in place of ethylene in eq. 4.14: *n*-propylbenzene or isopropylbenzene? Explain.**

The Friedel–Crafts alkylation reaction has some limitations. In general, it cannot be applied to an aromatic ring that already has on it a nitro or sulfonic acid group, because these groups form complexes with and deactivate the aluminum chloride catalyst.

**4.11
RING-ACTIVATING
AND RING-
DEACTIVATING
SUBSTITUENTS**

In this section and the next we will present some of the experimental evidence that supports the electrophilic aromatic substitution mechanism just described. We will do this by examining how substituents already present on an aromatic ring affect further substitution reactions.

For example, consider the relative rates of the following compounds toward nitration, all under the same reaction conditions:

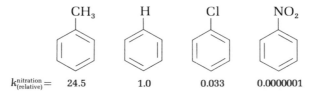

	CH_3	H	Cl	NO_2
$k^{nitration}_{(relative)} =$	24.5	1.0	0.033	0.0000001

We see that some substituents (for example, CH$_3$) make the ring more reactive than benzene itself, whereas other substituents (Cl, NO$_2$) make the ring less reactive. We know from other evidence that the methyl group is electron-donating compared to hydrogen, whereas the chloro or nitro groups are electron-withdrawing compared to hydrogen.

These observations are entirely consistent with the electrophilic mechanism for substitution. If the reaction rate depends on attack of an electrophile on the ring, then substituents that release electrons to the ring will speed up the reaction, whereas substituents that withdraw electrons from the ring will decrease the availability of the π electrons to the attacking electrophile and therefore slow down the reaction. This reactivity pattern is exactly what is observed, not only with nitration, but with all of the aromatic substitution reactions we have considered.

**4.12
ORTHO,PARA-
DIRECTING AND
META-DIRECTING
GROUPS**

Substituents already present on an aromatic ring determine the position taken by a new substituent. For example, we find that the nitration of toluene gives mainly a mixture of the *ortho-* and *para-*nitrotoluenes.

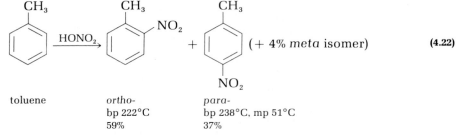

(+ 4% *meta* isomer)   **(4.22)**

toluene   ortho-   para-
                       bp 222°C   bp 238°C, mp 51°C
                       59%   37%

On the other hand, nitration of nitrobenzene gives mainly the *meta*-isomer.

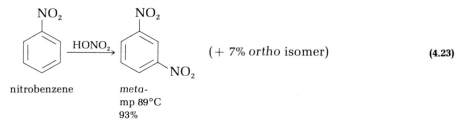

(+ 7% *ortho* isomer)   **(4.23)**

nitrobenzene   meta-
                      mp 89°C
                      93%

This pattern is followed for other electrophilic aromatic substitutions as well—chlorination, bromination, sulfonation, and so on. Toluene undergoes mainly *ortho,para*-substitution, whereas nitrobenzene undergoes *meta*-substitution.

In general, we find that certain groups are *ortho,para*-directing and others are *meta*-directing. Table 4.1 lists some of the common groups in each category.

TABLE 4.1 Directing effects of common functional groups

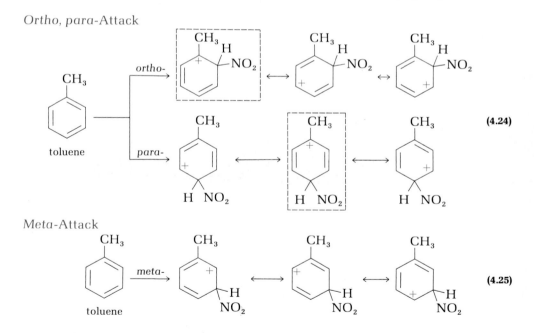

Ortho,Para-Directing	*Meta*-Directing
$-CH_3$, $-CH_2CH_3$ (alkyl, –R)	$-\overset{+}{N}(=O)-O^-$, $-S(=O)_2-OH$
$-F$, $-Cl$, $-Br$, $-I$ $-OH$, $-OCH_3$, $-OR$ $-NH_2$, $-NHR$, $-NR_2$	$-C(=O)-R$, $-C(=O)-OH$, $-C(=O)-OR$ $-C\equiv N$

Let us see how consistent our mechanism is with these facts. Consider first the nitration of toluene. In the first step of the mechanism, the nitronium ion can attack *ortho-*, *meta-*, or *para-* to the methyl group:

Ortho, para-Attack

(4.24)

Meta-Attack

(4.25)

Note that, in one of the three contributors to the resonance hybrid benzenonium ion intermediate for *ortho-* or for *para*-substitution (the ones shown in the boxes), the positive charge is on the methyl-bearing carbon. No such arrangement is present in the intermediate for *meta*-substitution (eq. 4.25).

Because alkyl groups such as CH_3 stabilize carbocations, *ortho,para*-substitution is preferred.

Now let us examine the nitration of nitrobenzene in the same way to see whether we can explain the *meta*-directing effect of the nitro group. The corresponding equations are

Ortho, para-Attack

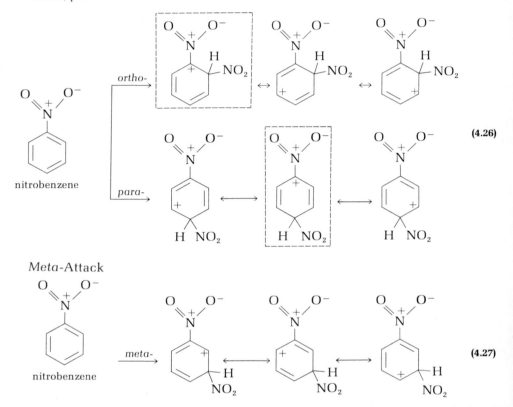

(4.26)

Meta-Attack

(4.27)

Note in eq. 4.26 that one of the three contributors to the resonance hybrid intermediate for *ortho*- or for *para*-substitution (the ones shown in the boxes) has two adjacent positive charges, a highly undesirable arrangement. This is not true for the intermediate in *meta*-substitution (eq. 4.27). Consequently, *meta*-substitution is preferred.

Can we generalize these explanations to the other groups in Table 4.1? First, note that all the *ortho,para*-directing groups in the table (except for the alkyl groups) have an atom with an unshared electron pair (that is, an O, N, or halogen) directly attached to the aromatic ring. This electron pair can help delocalize (and therefore stabilize) an adjacent positive charge. This stabilization is illustrated for *ortho*-substitution and *para*-substitution in the following contributors to the resonance structure of the immediate carbocation:

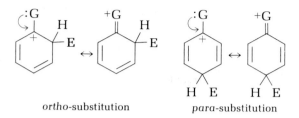

ortho-substitution   para-substitution

Groups with unshared electrons on the atom attached to the ring are therefore ortho,para-*directing.*

Note that all the *meta*-directing groups in Table 4.1 are connected to the ring by an atom that is part of a double or triple bond and that the atom at the other end of that double or triple bond is more electronegative than carbon (that is, O or N). In all such cases the atom directly attached to the benzene ring will carry a partial positive charge (like the nitrogen in the nitro group). This is because of resonance contributors such as

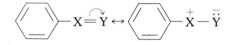

Y is an electron-withdrawing atom such as O or N; atom X carries a partial (+) charge.

Consequently, all such groups will be *meta*-directing for the same reason that the nitro group is *meta*-directing.

Problem 4.12 **Draw the important resonance contributors for the intermediate in the bromination of anisole, and explain why *o,p*-substitution predominates.**

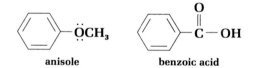

anisole   benzoic acid

Problem 4.13 **Repeat Problem 4.12 but for benzoic acid, and explain why the main product is *m*-bromobenzoic acid.**

Finally, note that in all *meta*-directing groups the atom connected to the ring will tend to withdraw electrons from the ring because of the atom's partial positive charge. *All* meta-*directing groups are therefore also ring-deactivating groups.* On the other hand, *ortho,para-directing groups in general can supply electrons to the ring and are therefore ring-activating.* With the halogens (F, Cl, Br, and I) two opposing effects bring about the only important exception to these rules. Because of their electronegativity, *the halogens are ring-deactivating, although their unshared electron pairs make them ortho,para-directing.*

**4.13
THE IMPORTANCE
OF DIRECTING
EFFECTS IN
SYNTHESIS**

When designing a multistep synthesis involving electrophilic aromatic substitution, we must keep in mind the directing and activating effects of the groups involved. Consider, for example, the bromination and nitration of benzene to make bromonitrobenzene. If we brominate first and then nitrate, we will get a mixture of the *ortho*- and *para*-isomers.

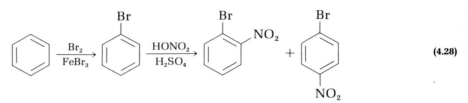

(4.28)

This is because the Br atom in bromobenzene is *o,p*-directing. On the other hand, if we nitrate first and then brominate, we will get mainly *meta*-product because of the *m*-directing effect of the nitro group.

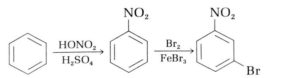

(4.29)

The *sequence* in which we carry out the same reactions, bromination and nitration, is therefore very important. It determines which type of product is formed.

Problem 4.14   **Devise a synthesis for each of the following, starting with benzene.**
a. m-chlorobenzenesulfonic acid
b. p-nitrotoluene

Problem 4.15   **Would it be possible to prepare m-bromochlorobenzene or p-nitroben-zenesulfonic acid by carrying out two successive electrophilic aromatic substitutions? Explain.**

If two groups are already present on a ring, they may reinforce or they may oppose each other when we try to introduce a third group. Thus electrophilic substitution in *m*-dinitrobenzene or *p*-nitrotoluene goes mainly in the position indicated by the arrows, because *both* groups direct to that position.

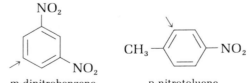

m-dinitrobenzene   *p*-nitrotoluene

With *meta*-nitrotoluene the *o,p*- and *m*-directing groups conflict. In such cases, the *o,p*-directing group usually wins out, because it is also ring-activating:

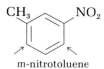

m-nitrotoluene

substitution occurs
o- and *p*- to the
methyl group

Problem 4.16   **Predict the major bromination product(s) of:**
a. *p*-nitrophenol   b. *m*-ethylbenzoic acid

Problem 4.17   **Which disubstituted benzene would be the best starting material for a one-step synthesis of pure 3-bromo-5-nitrobenzoic acid?**

**4.14
POLYCYCLIC
AROMATIC
HYDROCARBONS**

The concept of **aromaticity**—*the unusual stability of certain fully conjugated cyclic systems*—can be extended well beyond benzene itself or simple substituted benzenes. In the remaining sections of this chapter we will describe a few of these extensions.

Coke, which is required in huge quantities for the manufacture of steel, is obtained by heating coal in the absence of air. A by-product of this conversion of coal to coke is a distillate called **coal tar**, a complex mixture containing many aromatic hydrocarbons (including benzene, toluene, xylenes) and heterocyclic compounds. **Naphthalene,** $C_{10}H_8$, was the first pure compound to be obtained from the higher-boiling fractions of coal tar. It was easily isolated because it sublimes readily from the tar as a beautiful colorless crystalline solid, mp 80°C. Naphthalene is a planar molecule with two *fused* benzene rings. The two rings share two carbon atoms.

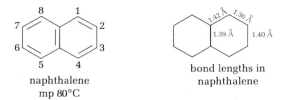

naphthalene
mp 80°C

bond lengths in
naphthalene

The bond lengths in naphthalene are not all identical, but they all approximate the bond length in benzene (1.39Å). Naphthalene has a resonance energy of about 60 kcal/mol, somewhat less than twice that of benzene. It undergoes electrophilic substitution reactions (halogenation, nitration, and so on), usually under somewhat milder conditions than benzene.

Problem 4.18   **Draw the three Kekulé contributors to the naphthalene structure. Are all three structures likely to be equally important? [*Hint*: Examine each structure to see whether *both* rings have a Kekulé-type structure.]**

Problem 4.19   **Can you suggest, from your answer to Problem 4.18, why the carbon-1–carbon-2 bond in naphthalene is somewhat shorter than the carbon-2–carbon-3 bond?**

Problem 4.20   **How many monobromonaphthalenes are possible? Draw their structures.**

Naphthalene is the parent compound of a whole series of fused polycyclic hydrocarbons, a few examples of which are

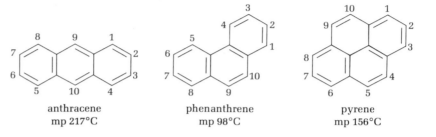

anthracene
mp 217°C

phenanthrene
mp 98°C

pyrene
mp 156°C

a word about

5. Polycyclic Aromatic Hydrocarbons and Cancer

Certain polycyclic aromatic hydrocarbons are carcinogenic (that is, they produce cancers). The most potent will produce a skin tumor on mice in a short time when only trace amounts are painted on the skin. These carcinogenic hydrocarbons are present not only in coal tar but also in soot and tobacco smoke. Indeed, their biological effect was noted as long ago as 1775, when soot was identified as the cause of a high incidence of scrotal cancer in chimney sweeps. A similar occurrence of lung and lip cancer is known in habitual smokers.

The way in which these carcinogens produce cancer is now rather well understood. In an effort to eliminate hydrocarbons, the body usually oxidizes them to render them more water-soluble so that they can be excreted. These metabolic oxidation products appear to be the real culprits in causing cancer. For example, one of the most potent carcinogens of this type is benzo[a]pyrene. In an effort to excrete benzo[a]pyrene, enzymatic oxidation converts it to the diol-epoxide shown:

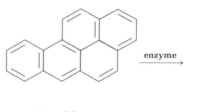

enzyme →

benzo[a]pyrene

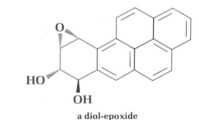

a diol-epoxide

The diol-epoxide reacts with cellular DNA, causing mutations that eventually prevent the cells from reproducing normally.

Benzene itself is quite toxic to humans and can cause severe liver damage. Toluene is very much less toxic. To eliminate benzene the aromatic ring must be oxidized, and intermediates in this oxidation are damaging. However, the methyl side chain of toluene can be oxidized to give benzoic acid, which can be eliminated. None of the intermediates in this process causes problems.

a word about———————————————————

6. Benzene, Chemical of Commerce

When one speaks of benzene, the prototype of aromatic compounds, one is speaking of big business indeed. Approximately 13 billion pounds are produced annually in the United States alone! As large-scale organic chemicals go, it comes in third, just behind propylene and at a little less than half the production of the leader, ethylene.

Most benzene is produced by catalytic reforming of alkanes and cycloalkanes (page 93) or by cracking of certain gasoline fractions. Only about 8% of it comes from coal tar, though this was its major source at one time.

What is all this benzene used for? About half is used to make styrene, according to the sequence

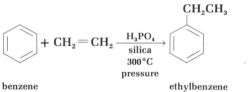

benzene   ethylbenzene

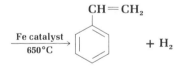

styrene

The first step is a Friedel–Crafts alkylation (eq. 4.14) and the second is a catalytic dehydrogenation. The styrene is converted to polymers (Table 3.3) and to synthetic rubber (eq. 3.37).

The second largest use of benzene is to make phenol and acetone:

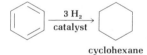

acetone   phenol   cumene hydroperoxide

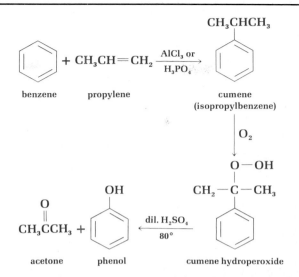

benzene   propylene   cumene
                                        (isopropylbenzene)

Again the first step is a Friedel–Crafts reaction, this time with propylene. Air oxidation of the resulting cumene gives a hydroperoxide, which is decomposed by acid to acetone and phenol. Both of these products have many commercial uses.

Large quantities of benzene are also hydrogenated catalytically to cyclohexane.

cyclohexane

Cyclohexane is a major raw material for the manufacture of nylon.

Together, these three commercial processes—production of styrene, phenol, and cyclohexane—consume over 80% of commercial benzene. Substantial amounts of benzene are used to raise the octane rating of unleaded gasoline. The rest is used to make aniline, chlorobenzene, and a variety of other industrial chemicals. At present all of these important materials come almost entirely from petroleum, further evidence of how dependent we are on that natural resource.

4.15 AROMATICITY: THE HÜCKEL RULE   Benzene and its derivatives have six π electrons in a conjugated, cyclic array. We might well ask whether only systems with six π electrons are aromatic, or whether we might do with fewer or more π electrons. For example, we can imagine four π or eight π conjugated cyclic systems, such as cyclobutadiene or cyclooctatetraene:

cyclobutadiene   cyclooctatetraene

Are these compounds aromatic? In fact, neither of them has aromatic properties, even though we can write resonance contributors for each that would delocalize the electrons.

   Without going into detail, one of the triumphs of orbital theory (over resonance theory) is that it explains these observations. Monocyclic conjugated π systems with $(4n + 2)$ π electrons (where n is an integer—0, 1, 2 . . .) are aromatic (**the Hückel Rule**). Thus monocyclic systems with 2, 6, 10, 14 . . . conjugated π electrons are aromatic. Systems of this type, but with $4n$ electrons (4, 8, 12 . . .) are not aromatic. Of the aromatic systems, by far the most common are those with six π electrons. In the final section of this chapter we will see that not all of these six π electrons must be supplied by carbon atoms, however.

**4.16
HETEROCYCLIC
AROMATIC
COMPOUNDS**

One can replace the carbon atoms of an aromatic ring with heteroatoms—most commonly nitrogen, oxygen or sulfur—and still maintain the properties of an aromatic system. Heterocyclic aromatic compounds fall in two general categories, those in which the heteroatom supplies *one* electron to the aromatic π system, and those in which the heteroatom supplies *two* electrons to the π system. These types are typified by **pyridine** and **pyrrole,** respectively:

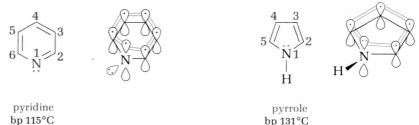

pyridine
bp 115°C
soluble in water

pyrrole
bp 131°C
insoluble in water

In each compound the nitrogen is sp^2-hybridized. Pyridine has five carbon atoms in the ring, each of which contributes one electron to the aromatic π system, so that the nitrogen atom need contribute only one electron to bring the total to six π electrons. On the other hand, in pyrrole there are only four sp^2 carbon atoms, so the nitrogen must contribute a pair of electrons to make up the six π electrons of the aromatic system. As a consequence of this difference in electron demand from the nitrogen, pyridine and pyrrole have some rather different properties.

In pyridine, the unshared electron pair on the nitrogen is *not* part of the aromatic π electron system. This electron pair is therefore available for accepting a proton. Consequently, pyridine can form hydrogen bonds with (and hence is soluble in) water and can react with acids to form salts.

pyridinium chloride
mp 82°C

(4.30)

In pyrrole, on the other hand, the unshared electron pair is an essential part of the aromatic π system. Protonation of the nitrogen in pyrrole would destroy the aromatic system (the nitrogen would become sp^3-hybridized). Consequently, pyrrole does not readily form hydrogen bonds and is insoluble in water; it is also a very weak base and does not form salts with aqueous acids.

Other important aromatic rings of the pyridine type, wherein a hetero-atom contributes only one electron to the aromatic π system, include

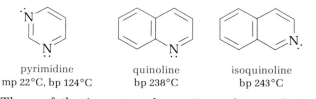

pyrimidine   quinoline   isoquinoline
mp 22°C, bp 124°C   bp 238°C   bp 243°C

Three of the important bases in nucleic acids (cytosine, thymine, and uracil; see Chapter 16) are pyrimidines. The quinoline and isoquinoline ring systems are present in many drugs obtained from plants (Sec. 12.13b).

Other aromatic rings of the pyrrole type, wherein a heteroatom contributes two electrons to the aromatic π system, include

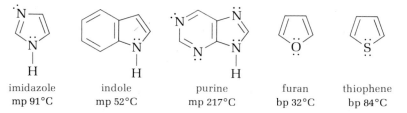

imidazole   indole   purine   furan   thiophene
mp 91°C   mp 52°C   mp 217°C   bp 32°C   bp 84°C

All of these rings occur in natural products. The purine ring system, for example, which consists of fused pyrimidine and imidazole rings, is present in the important DNA and RNA bases adenine and guanine (Chapter 16).

Problem 4.21   **Only one of the two nitrogens in imidazole is quite basic. Tell which one it is, and explain the reason for your choice.**

4.22 Write structural formulas for the following compounds.
a. 1,3,5-tribromobenzene   **b.** *m*-chlorotoluene
c. *o*-diethylbenzene   **d.** isopropylbenzene
e. benzyl bromide   **f.** *p*-chlorophenol
g. 2,3-diphenylbutane   **h.** *p*-bromostyrene
i. 2-chloro-4-ethyl-3,5-dinitrotoluene   **j.** *m*-chlorobenzenesulfonic acid
k. *p*-bromobenzoic acid   **l.** 2,4,6-trimethylaniline

4.23. Name the following compounds.

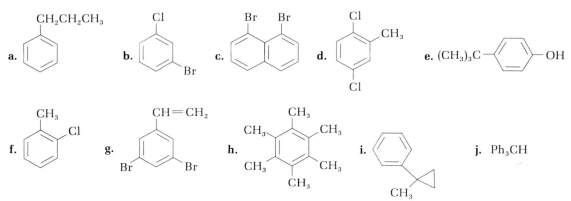

4.24. Give the structures and names for all the isomers of:
a. trimethylbenzenes   **b.** dichloronitrobenzenes

4.25. There are three dibromobenzenes (*o*-, *m*-, and *p*-). Suppose we have samples of each in separate bottles, but we don't know which is which. Let us call them A, B, and C. On nitration, compound A (mp 87°C) gives only *one* nitrodibromobenzene. What is the structure of A? B and C are both liquids. On nitration, B gives *two* nitrodibromobenzenes and C gives *three* nitrodibromobenzenes (of course, not in equal amounts). What are the structures of B and C and of their mononitration products? This method, known as Körner's method, was used years ago to assign structures to isomeric benzene derivatives.

4.26. Give the structure and name of each of the following aromatic hydrocarbons.
a. C_8H_{10}; has three possible ring-substituted monobromo derivatives
b. C_9H_{12}; can give only one mononitro product on nitration
c. C_9H_{12}; can give four mononitro derivatives on nitration

4.27. The observed amount of heat evolved when 1,3,5,7-cyclooctatetraene (Sec. 4.15) is hydrogenated is 110 kcal/mol. What does this fact tell you about the possible resonance energy of this compound?

4.28. The structure of the nitro group $-NO_2$ is usually shown as

Yet experiments show that the two nitrogen–oxygen bonds have the same length of 1.21 Å, intermediate between 1.36 Å for the N—O single bond and 1.18 Å for the N=O double bond. Draw structural formulas that explain this observation.

4.29. Draw all reasonable electron-dot formulas for the nitronium ion, $(NO_2)^+$, the electrophile in aromatic nitrations. Show any formal charges. Which structure would be favored, and why?

4.30. Write out all steps in the mechanism for the reaction of:
a. p-xylene + nitric acid ($+H_2SO_4$ catalyst)
b. benzene + t-butyl chloride + $AlCl_3$

4.31. Draw all possible contributing structures to the carbocation intermediate in the chlorination of chlorobenzene. Explain why the major products are *ortho-* and *para*-dichlorobenzene (*Note:* p-Dichlorobenzene is produced commercially this way, for use against clothes moths).

4.32. Repeat Problem 4.31 for the chlorination of benzenesulfonic acid, and explain why the product is m-chlorobenzenesulfonic acid.

4.33. Indicate the main monosubstitution products in each of the following reactions. Keep in mind that certain substituents are *meta*-directing and others are *ortho,para*-directing.
a. toluene + chlorine (Fe catalyst)
b. nitrobenzene + concentrated sulfuric acid (heat)
c. bromobenzene + chlorine (Fe catalyst)
d. chlorobenzene + bromine (Fe catalyst)
e. benzenesulfonic acid + concentrated nitric acid (heat)
f. ethylbenzene + bromine (Fe catalyst)
g. iodobenzene + bromine (Fe catalyst)

4.34. Suggest a reason why $FeCl_3$ is used as a catalyst for aromatic chlorinations, and $FeBr_3$ for brominations (that is, why the iron halide used has the same halogen as the halogenating agent).

4.35. Using benzene or toluene as the only aromatic organic starting materials, devise syntheses for each of the following.
a. m-bromonitrobenzene   **b.** p-toluenesulfonic acid
c. p-nitroethylbenzene   **d.** methylcyclohexane
e. 2,6-dibromo-4-nitrotoluene   **f.** p-bromonitrobenzene
g. 2-chloro-4-nitrotoluene   **h.** 3,5-dinitrochlorobenzene

4.36. Which of the following compounds can probably be prepared in a pure state by electrophilic substitution, starting with an already disubstituted benzene? Tell how.

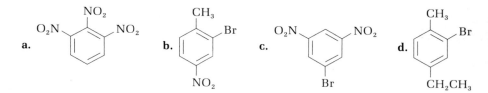

4.37. Predict whether the following substituents on a benzene ring are likely to be o,p- or m-directing and whether they are likely to be ring-activating or ring-deactivating:

a. $-SCH_3$ **b.** $-\overset{+}{N}(CH_3)_3$ **c.** $-O-\overset{\overset{\displaystyle O}{\|}}{C}-CH_3$ **d.** $-\overset{\overset{\displaystyle O}{\|}}{C}-NH_2$

4.38. The explosive TNT (2,4,6-trinitrotoluene) can be made by nitrating toluene with a mixture of nitric and sulfuric acids, but the reaction conditions must gradually be made more severe as the nitration proceeds. Explain why.

4.39. Bromination of naphthalene (by heating with a solution of bromine in carbon tetrachloride) gives mainly 1-bromonaphthalene. Very little of the 2- isomer is formed. Draw all possible contributors to the carbocation intermediate for bromination at each position and compare (a) the total number of contributors, and (b) the number of contributors that preserve the benzenoid character (three double bonds) of the ring that is not attacked by the electrophile. Can you now explain why substitution in the 1-position is favored?

4.40. Anthracene and phenanthrene mainly undergo electrophilic substitution at the 9- (or 10-) position. Use resonance structures, as in Problem 4.39, to explain why this position for substitution is preferred.

4.41. Draw the product that would be obtained if pyrrole were protonated on the nitrogen by acid. Compare with eq. 4.30, and explain why pyrrole is not as basic as pyridine.

4.42. Which of the nitrogens in purine contribute one electron to the aromatic π system, and which contribute two electrons? Which of the four nitrogens will be the *least* basic (the least likely to be protonated by acid)?

five

Stereoisomerism and
Optical Activity

**5.1
INTRODUCTION**

Stereoisomers are compounds with the same atom connectivities, or order of attachment of the atoms, but with different arrangements of the atoms in space. Two subsets of isomers that come under this general definition have already been described: conformational isomers (Sec. 2.8, 2.9) and *cis-trans* isomers (Sec. 2.10, 3.5). Let us now classify stereoisomers somewhat more systematically.

Stereoisomers can be categorized conveniently into two types according to their structures. Either they are mirror images (called **enantiomers**) or they are *not* mirror images (called **diastereomers**). The stereoisomers that we described before belong in the latter category. For example, *cis*- and *trans*-2-butene (Example 3.2, page 68) are *not* mirror images of each other and are therefore diastereomers. In this chapter, we will discuss other types of diastereomers, and we will discuss stereoisomers that are enantiomers (mirror images).

Stereoisomers can also be categorized into two types according to the ease with which they are interconverted. Either they can be interconverted by rotation about single bonds (**conformational isomers** or **conformers**), or they can be interconverted only by breaking and remaking covalent bonds (**configurational isomers**). The staggered and eclipsed stereoisomers of ethane fall into the first category, whereas *cis*- and *trans*-2-butene belong in the second. The interconversion of conformers usually requires only a small amount of energy, so conformers usually have a short lifetime and defy separation. On the other hand, interconversion of configurational isomers usually requires substantial energy, enough to break bonds. Hence configurational isomers usually have long lifetimes and can be separated into pure forms.

In this chapter we will focus our attention on configurational isomers of a type not previously described. Some of these are enantiomers and others are diastereomers. A thorough discussion of these configurational isomers will require some understanding of their interaction with plane-polarized light, which is where we will begin.

**5.2
PLANE-POLARIZED
LIGHT**

An ordinary light beam consists of waves vibrating in all possible planes perpendicular to its path. A cross section of such a beam, coming out of the page toward you, is shown in Figure 5.1. The waves are propagated at any angle from 0° to 360°. However, as shown in the right half of the figure, any wave vibrating in a particular plane can be resolved into a vertical and a horizontal component. If the light beam were to pass through some substance that permitted only one of these components to pass through, the resulting beam would have all of its waves vibrating along the same plane. Such a light beam is said to be **plane-polarized** and is illustrated in Figure 5.2.

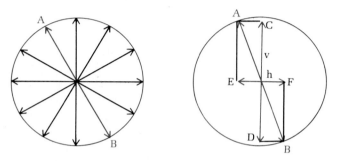

FIGURE 5.1   *Left:* An ordinary light beam vibrating in all possible planes, coming toward the reader. *Right:* Beam AB can be resolved into vertical (CD) and horizontal (EF) components. Any beam of light in the figure at the left can be similarly resolved into vertical and horizontal components.

FIGURE 5.2

A beam of light AB, initially vibrating in all directions, passes through a polarizing substance that "strains" the light so that only the vertical component emerges.

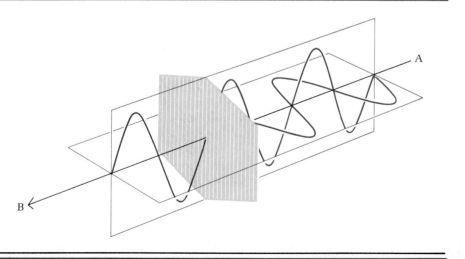

FIGURE 5.3

The two sheets of polarizing material shown have their axes aligned perpendicularly to one another. Although each disk alone is transparent, the area where they overlap is opaque. You can check this out for yourself by using two pairs of Polaroid sunglasses. (Courtesy of the Polaroid Corporation.)

In 1808 the French physicist Etienne Malus discovered that light can be polarized. One convenient way is to pass ordinary light through a device composed of Iceland spar (crystalline calcium carbonate) called a **Nicol prism,** invented in 1828 by the British physicist William Nicol. A more recently developed polarizing material is **Polaroid,** invented by the American E. H. Land. It contains a crystalline organic compound properly oriented and embedded in a transparent plastic. Sunglasses are often made from Polaroid.

An ordinary light beam will pass through two samples of polarizing material only if their polarizing axes are aligned. If the axes are perpendicular, no light will pass through (Figure 5.3).

5.3 THE POLARIMETER AND OPTICAL ACTIVITY

A **polarimeter** is an instrument used to study the effect of various substances on plane-polarized light. It uses the idea, illustrated in Figure 5.3, that two samples of a polarizing material fixed with their axes at right angles to one another do not allow a light beam to pass through.

The instrument is shown schematically in Figure 5.4. Here is how it works. With the light on and the sample tube empty, the analyzer prism is rotated so that the light beam is completely blocked and the field of view is dark. The prism axes of the polarizer prism and analyzer prism will

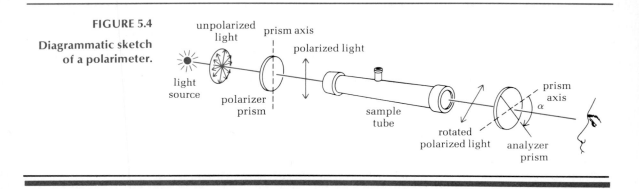

FIGURE 5.4

Diagrammatic sketch of a polarimeter.

be perpendicular to one another. Now the sample is placed in the sample tube. If the substance is **optically inactive,** nothing will happen. The field of view will remain dark. But if an **optically active** substance is placed in the tube, it will rotate the plane of polarization and some light will pass through the analyzer to the observer. By turning the analyzer prism clockwise or counterclockwise, the observer can block the light beam once again and restore the dark field. The angle through which the analyzer must be rotated is α, called the **observed rotation,** and is equal to the number of degrees that the optically active substance has rotated the beam of plane-polarized light. If the analyzer must be rotated to the *right* (clockwise), the optically active substance is said to be **dextrorotatory.** If the analyzer must be rotated to the *left* (counterclockwise), the substance is **levorotatory.** By convention, we designate a dextrorotatory substance by a $(+)$ sign and a levorotatory substance by a $(-)$ sign.

The observed rotation α of a given sample of an optically active substance depends on several factors. Of course it depends on the molecular structure, but we will defer that question until later in the chapter. The observed rotation also depends on the length of the sample tube, the concentration of the optically active substance in solution, the temperature, the wavelength of the light source, and the nature of the solvent. All these factors are usually standardized when we want to compare the optical activity of different substances. The **specific rotation** $[\alpha]$ of an optically active substance is defined as the observed rotation caused by a solution containing 1 gram of the substance per milliliter of solution placed in a sample tube 1 decimeter (10 centimeters) long. It is expressed as follows:

$$\text{specific rotation} = [\alpha]_\lambda^t = \frac{\alpha}{l \times c} \quad \text{(solvent)}$$

where l is the length of the sample tube in decimeters, c is the concentration in grams per milliliter, t is the temperature of the solution, and λ is the wavelength of light. The solvent used is indicated in parentheses. Measurements are usually made at room temperature, and the most common light source is the D-line of a sodium vapor lamp ($\lambda = 589.3$ nm).

The specific rotation of an optically active substance under carefully specified conditions is as definite a property of the substance as is its melting point, its boiling point, or its density.

Example 5.1 **The observed rotation for 100 mL of an aqueous solution containing 1 g of sucrose (ordinary sugar), placed in a 2-decimeter sample tube, is +1.33° at 25°C (using a sodium lamp). Calculate and express the specific rotation of sucrose.**

Solution

$$[\alpha]_D^{25} = \frac{+1.33}{2 \times 0.01} = +66.5 \quad (\text{H}_2\text{O})$$

Note that $c = 1 \text{ g}/100 \text{ mL} = 0.01 \text{ g/mL}$

Problem 5.1 **Camphor (1.5 g) dissolved in ethanol to a total volume of 50 mL, placed in a 5-cm sample tube, gives an observed rotation of +.66° at 20°C (using the sodium D-line). Calculate and express the specific rotation of camphor.**

In the early nineteenth century the French physicist Jean Baptiste Biot studied the behavior of a great many substances in a polarimeter. Some, such as turpentine, lemon oil, solutions of camphor in alcohol, and solutions of cane sugar in water, were optically active. Others, such as water, alcohol, and solutions of salt in water, were optically inactive. Later, many natural products (carbohydrates, proteins, and steroids, to name just a few) were added to the list of optically active compounds. What is it about the structure of molecules that causes some to be optically active and others inactive? It will be helpful, in finding an explanation, to examine certain symmetry properties of molecules.

**5.4
CHIRAL AND
ACHIRAL OBJECTS;
ENANTIOMERS**

We all know that it is quite easy for two right-handed (or two left-handed) persons to shake hands, but it is not possible to shake a right hand with a left hand in the usual way. As we shall see, certain molecules may also possess the property of "handedness," and this property affects their chemistry. Let us examine the idea of molecular "handedness."

A **chiral** object (or molecule) is one that exhibits the property of handedness. The word *chiral* (pronounced "kai-ral" to rhyme with *spiral*) comes from the Greek (*cheir*, hand). Typical chiral objects include hands, feet, gloves, shoes, screws and other threaded objects, and helices (for example, a spiral staircase or the double helix of DNA). *A chiral object can be recognized by the fact that its mirror image is not identical with or superimposable on the object itself.* This is illustrated in Figure 5.5, where we see that the mirror image of a left hand is not another left hand, but a right hand.

An object (or a molecule) and its nonsuperimposable mirror image are said to be **enantiomers.** This word also comes from the Greek (*enantio,* opposite, and *meros,* part). A left and right hand are said to be a *pair of*

FIGURE 5.5

The mirror-image relationship of the right and left hands The mirror image of a left hand is not a left hand, but a right hand.

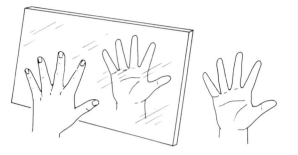

enantiomers (Figure 5.5). As we shall see, a similar relationship is possible with molecules.

Achiral objects (or molecules) do *not* have the property of handedness. Typical achiral objects are balls, spheres, cubes, squares, rectangles, and equilateral and isosceles triangles. *An achiral object can be recognized by the fact that its mirror image is identical with or superimposable on the object itself.* Such objects (or molecules) cannot exist in a right- or left-handed form. They cannot exist as a pair of enantiomers.

Problem 5.2   **Which of the following objects are chiral and which are achiral?**
a. golf club   b. tea cup
c. tennis racket   d. shoe

How can we tell quickly whether an object (or molecule) is chiral or achiral? One way, of course, is to compare the object directly with its mirror image, as we have just seen. Another way is to examine its symmetry properties.

**5.5
PLANES OF
SYMMETRY**

A **plane of symmetry** (sometimes called a mirror plane) is a plane that passes through an object (or molecule) in such a way that what is on one side of the plane is the exact reflection of what is on the other side. *Any object with a plane of symmetry is achiral. Chiral objects do not have a plane of symmetry.* Seeking a plane of symmetry is usually one quick way to tell whether an object (or molecule) is chiral or achiral.

Problem 5.3   **Describe the planes of symmetry in the typical achiral objects mentioned in Sec. 5.4 (ball, sphere, cube, square, rectangle, equilateral and isosceles triangles).**

Problem 5.4   **Search for a plane of symmetry in the typical chiral objects mentioned in Sec. 5.4 (hand, foot, glove, shoe, screw, helix). Can you find any?**

Problem 5.5 **Which of the objects in Problem 5.2 have a plane of symmetry? Compare this answer with your previous conclusion about whether each object was chiral or achiral.**

Problem 5.6 **Locate and describe all the planes of symmetry in:**
a. cyclobutane   b. the eclipsed conformation of ethane
Are these molecules chiral or achiral?

5.6
PASTEUR'S
EXPERIMENTS

The concepts of chirality, achirality, and planes of symmetry can be applied to molecules. The great French scientist Louis Pasteur was the first to recognize that optical activity is related to what we now call chirality (or handedness). He realized from the following experiments that molecules of the same substance that rotate plane-polarized light in opposite directions must be related to one another as an object and its nonsuperimposable mirror image (that is, as a pair of enantiomers). Here is how he came to that conclusion.

Working in the mid-nineteenth century in a country famous for its wine industry, Pasteur was aware of two isomeric acids that deposit in wine casks during fermentation: **tartaric acid** (dextrorotatory) and racemic acid, which is optically inactive.

Pasteur prepared various salts of these acids. He noticed that crystals of the sodium ammonium salt of *tartaric* acid were chiral; that is, the crystals had no plane of symmetry. Furthermore, all the crystals were of identical handedness or chirality. Let us say the crystals were all right-handed. When Pasteur examined crystals of the sodium ammonium salt of *racemic* acid, he found that they too were chiral but that some of the crystals were right-handed and others were left-handed. That is, the crystals were enantiomeric. With a magnifying lens and a pair of tweezers, Pasteur carefully separated the crystals into piles, the left-handed crystals and the right-handed crystals.

When Pasteur dissolved the two types of crystals separately in water and placed the solutions in a polarimeter, he found that each solution was optically active (remember, he obtained these crystals from racemic acid, which was optically inactive). One solution had a specific rotation identical to that of the sodium ammonium salt of tartaric acid! The other had an equal but opposite specific rotation. It must therefore be the mirror image, levorotatory tartaric acid. Pasteur correctly concluded that racemic acid was not a single substance but was a 50:50 mixture of (+) and (−) tartaric acids. The reason why racemic acid was optically inactive was that it contained equal amounts of two enantiomers. *We now define a* **racemic mixture** *as a 50:50 mixture of enantiomers,* and of course such a mixture is optically inactive.

Pasteur recognized that optical activity must be due to some property of the molecules of tartaric acid themselves (not just some property of the crystals), because the crystalline shape was lost when the crystals were dissolved in water. But the precise explanation eluded him.

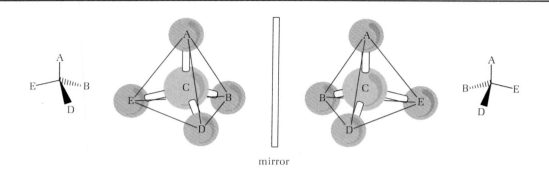

FIGURE 5.6   Four different groups may be arranged at the corners of a tetrahedron in two different ways.

5.7
THE VAN'T HOFF–
LeBEL
EXPLANATION*

Pasteur's experiments were performed at about the same time that Kekulé in Germany was developing his theories about organic structures. Kekulé recognized that carbon is tetravalent, and there is even a hint in some of his writings (about 1867) and also in the writings of the Russian chemist A. M. Butlerov (1862) and the Italian E. Paternò (1869) that the carbon atom might be tetrahedral. But it was not until 1874 that the Dutch physical chemist J. H. van't Hoff and his former fellow student, the Frenchman J. A. LeBel, simultaneously but quite independently made a bold hypothesis about carbon that would explain the optical activity of some organic molecules and the optical inactivity of others.

Van't Hoff and LeBel noticed that, when *four different groups* are attached to the corners of a tetrahedron, two arrangements are possible. These arrangements, shown in Figures 5.6 and 5.7, are related to one another as nonsuperimposable mirror images (that is, they are enantiomers). The chirality, or right- and left-handedness, of these models can readily be observed by sighting down one of the bonds, as shown in Figure 5.8.

The nonsuperimposability of these mirror images depends on all four groups being different. If two or more of the groups are identical, the structure has a plane of symmetry and will be achiral. That is, its mirror image will be identical with the original structure, as shown in Figure 5.9.

Van't Hoff and LeBel used these ideas to explain the optical activity of organic compounds. They proposed that the valences of carbon are directed from the nucleus to the corners of a regular tetrahedron. They defined an **asymmetric carbon atom** as *a carbon atom with four different groups attached to it.* Nowadays we refer to such a carbon atom as a **chiral**

---

*You may find it helpful to use molecular models to help you visualize the structures in the remaining sections of this chapter.

FIGURE 5.7

When the four different groups attached to an asymmetric carbon atom are arranged as mirror images, the resulting molecules are not superimposable. The models may be twisted or turned in any direction, but as long as no bonds are broken, only two of the four attached groups coincide.

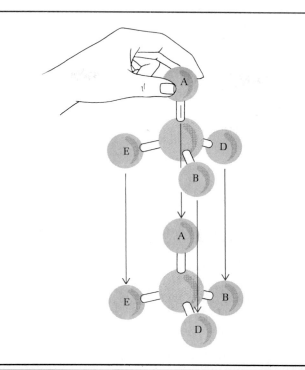

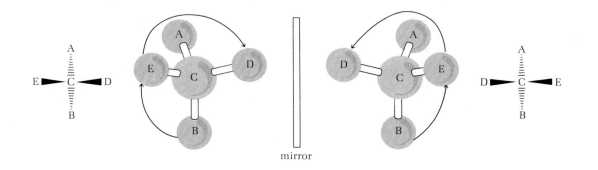

mirror

FIGURE 5.8 The chirality of enantiomers. Looking down the C—A bond, one must proceed in a clockwise direction to spell BED for the model on the left, but one must proceed in a counterclockwise direction for its mirror image.

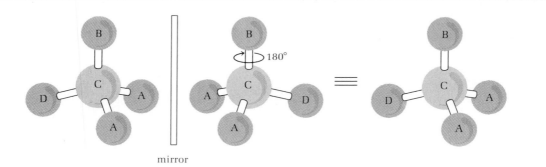

FIGURE 5.9   The tetrahedral model at the left has two corners occupied by identical groups A. It has a plane of symmetry that passes through atoms BCD and bisects angle ACA. Its mirror image is identical to itself as seen by a 180° rotation of the mirror image about the C—B bond. Hence the model is achiral.

center. A molecule with a chiral center may exist in two forms, as a pair of enantiomers. Being right- or left-handed, these enantiomers rotate plane-polarized light in opposite directions. At the time van't Hoff and LeBel proposed their theory, only 13 optically active compounds with established structures were known. Each of them did in fact contain at least one chiral center. Further experimentation has confirmed the correctness of their theory.

Example 5.2   Which of the following two compounds is chiral: 1-bromobutane or 2-bromobutane?

Solution $CH_3CH_2CH_2CH_2Br$ $CH_3\overset{*}{C}HCH_2CH_3$

                                            $|$
                                            Br

        1-bromobutane   2-bromobutane

The carbon marked with an asterisk in 2-bromobutane is a chiral center. That carbon atom has four different groups attached to it (H, Br, CH_3-, $-CH_2CH_3$). Therefore 2-bromobutane can exist as a pair of enantiomers, each of which will be optically active:

$$Br-\underset{CH_2CH_3}{\overset{CH_3}{C}}-H \quad or \quad \underset{CH_3CH_2 \quad Br}{\overset{CH_3}{C}\cdots H} \quad and \quad \underset{Br \quad CH_2CH_3}{H\cdots\overset{CH_3}{C}} \quad or \quad H-\underset{CH_2CH_3}{\overset{CH_3}{C}}-Br$$

All the other carbon atoms in *both* molecules have at least two identical groups attached. 1-Bromobutane is an achiral molecule, but 2-bromobutane is chiral.

Problem 5.7 **Pick out the chiral centers in:**
a. 3-methylhexane   **b. 2,3-dichlorobutane**
c. 3-methylcyclohexene   **d. 1-bromo-1-chloroethane**

Problem 5.8 **Which of the following compounds, if any, is chiral?**
a. 1-bromo-1-phenylethane
b. 1-bromo-2-phenylethane

5.8
PROPERTIES OF
ENANTIOMERS;
LACTIC ACID

Enantiomers differ from one another only with respect to chirality. In all other respects they are identical. For this reason, they can be expected to differ from one another only in properties that are also chiral. Let us illustrate this idea first with familiar objects.

A left-handed baseball player (chiral) can use the same ball (achiral) as a right-handed player. But of course a left-handed player (chiral) can use only a left-handed baseball glove (chiral). A bolt with a right-handed thread (chiral) can use the same washer (achiral) as a bolt with a left-handed thread, but it can fit only into a nut (chiral) with a right-handed thread. To generalize, chirality—left- or right-handedness—becomes significant only when it interacts with chirality; chirality is not important when it interacts with achirality.

Thus enantiomers have identical *achiral* properties, such as melting point, boiling point, density, and various types of spectra. Their solubilities in an ordinary, achiral solvent are also identical. However, enantiomers have different *chiral* properties, one of which is the *direction* in which they rotate plane-polarized light (clockwise or counterclockwise). Though enantiomers rotate plane-polarized light in opposite directions, they have identical specific rotations (except for sign), because the *number of degrees* is not a chiral property. Only the *direction* of rotation is a chiral property. Here is a specific example.

Lactic acid is an optically active hydroxy-acid that is important in several biological processes. It has one chiral center. Its structure and

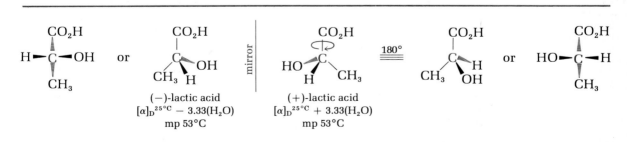

(−)-lactic acid
$[\alpha]_D^{25°C}$ − 3.33(H_2O)
mp 53°C

(+)-lactic acid
$[\alpha]_D^{25°C}$ + 3.33(H_2O)
mp 53°C

FIGURE 5.10   **The structures and properties of the lactic acid enantiomers.**

some of its properties are shown in Figure 5.10. Note that both enantiomers have identical melting points and, except for sign, identical specific rotations.

Enantiomers often have different biological properties. The reason is that the biological property usually involves a reaction with another chiral molecule. For example, the enzyme lactic acid dehydrogenase will oxidize (+)-lactic acid to pyruvic acid, but it will *not* oxidize the (−)-enantiomer:

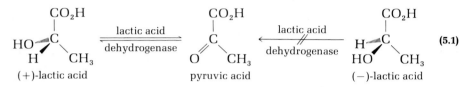

$$(+)\text{-lactic acid} \qquad \text{pyruvic acid} \qquad (-)\text{-lactic acid} \qquad (5.1)$$

The reason is that the enzyme itself is chiral and distinguishes between right- and left-handed lactic acid molecules.

Enantiomers differ in many types of biological activity. One enantiomer may be a drug, whereas the other enantiomer may be ineffective. For example, only (−)-adrenalin is a cardiac stimulant; the (+)-isomer is ineffective. One enantiomer may be toxic, another harmless. One may be an antibiotic, the other useless. One may be an insect sex attractant, the other without effect or actually a repellant. The importance of chirality as a factor in the biological world is paramount.

a word about

7. Odor and Chirality

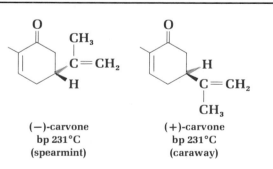

(−)-carvone
bp 231°C
(spearmint)

(+)-carvone
bp 231°C
(caraway)

Not a great deal is known about the relationship between chemical structure and odor, nor do we yet know very much about the physical and chemical processes that take place when an odoriferous substance interacts with human olfactory receptors embedded in the lining of the nose. We do know that the receptors are exceedingly sensitive, and we also know that some odors are chemically very complex. The aroma of coffee, for example, results from over 100 volatile organic compounds.

One thing we do know about the odor receptors in the human nose is that they are chiral. We know this because they register different odors for certain pairs of enantiomers. A good example is carvone.

Carvone has only one chiral center. The (−)-enantiomer is found in spearmint and has that odor, whereas the (+)-enantiomer is responsible for the odor of caraway seeds. Note that, although the enantiomers have different smells, they have identical boiling points. Any complete theory of odor that is eventually developed will have to take into account the chirality of the receptor sites.

**5.9
CONFIGURATION
AND THE *R–S*
CONVENTION**

Enantiomers differ in the arrangement of the groups attached to the chiral center. This arrangement of groups is called the **configuration** of the chiral center. *Enantiomers, such as those of lactic acid shown in Figure 5.10, are said to have opposite configurations.*

When referring to a particular enantiomer, we would like to be able to specify which configuration we mean without having to draw the structure. A system for doing this was developed by two English chemists, R. S. Cahn and C. K. Ingold, and a Swiss Nobel Prize winner, Vladimir Prelog. It is known as the Cahn–Ingold–Prelog system or the *R–S* system. Here is how it works.

The four groups attached to the chiral center are placed in a priority order (by a system we will describe in the next paragraph), $a \rightarrow b \rightarrow c \rightarrow d$. The chiral center is then *observed from the side opposite the lowest priority group, d*. If the remaining three groups ($a \rightarrow b \rightarrow c$) form a clockwise array, the configuration is designated *R* (from the Latin *rectus*, right). If they form a *counterclockwise* array, the configuration is designated as *S* (from the Latin *sinister*, left).

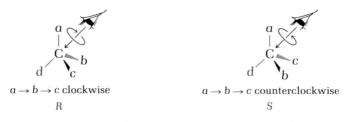

$a \rightarrow b \rightarrow c$ clockwise $a \rightarrow b \rightarrow c$ counterclockwise

R   *S*

The priority order is set in the following way. Atoms attached directly to the chiral center are arranged according to *decreasing atomic number*: the higher the atomic number, the higher the priority. For example,

$$I > Br > Cl > F > OH > NH_2 > CH_3 > H$$

high priority ——————————————→ low priority

If the chiral center has a hydrogen atom directly attached to it, the hydrogen will obviously be the lowest-priority group (atomic number 1).

If two or more of the atoms attached to the chiral center are identical, we must compare the next closest atoms. For example,

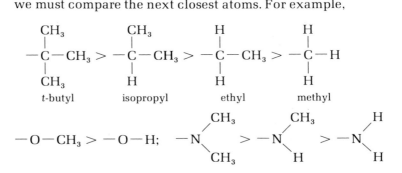

The rules are slightly complicated when applied to groups with multiple bonds. Double and triple bonds are treated as though each such bonded atom were duplicated or triplicated. For example,

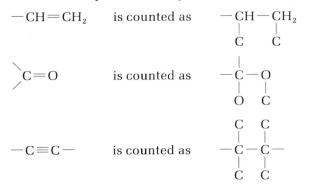

$-CH=CH_2$ is counted as

$>C=O$ is counted as

$-C\equiv C-$ is counted as

This rule leads to the following priority orders:

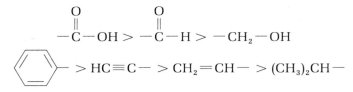

Let us apply these rules to the enantiomers of lactic acid (Figure 5.10). The priority order of the four groups attached to the chiral center is $OH > CO_2H > CH_3 > H$. The configuration of $(-)$-lactic acid is determined as follows:

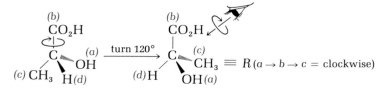

The priority orders are shown in parentheses. The molecule is turned so that it can best be viewed from the side opposite the lowest-priority group (H). This can be done in a variety of ways, only one of which is shown. The three remaining groups are arranged in a clockwise array, so the configuration is R, and the name can be written as (R)-$(-)$-lactic acid. Its enantiomer is (S)-$(+)$-lactic acid.

Example 5.3   **Draw the structure of (R)-2-bromobutane.**

Solution   **First write out the structure and prioritize the groups attached to the chiral center.**

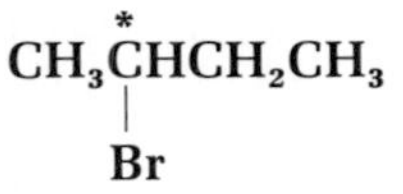

$$CH_3\overset{*}{C}HCH_2CH_3$$
$$|$$
$$Br$$

$$Br > CH_3CH_2 - > CH_3 - > H$$

Now make the drawing with the H (lowest-priority group) "away" from you, and place the three remaining groups in a clockwise (R) array.

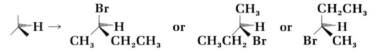

Problem 5.9   Determine the configuration (R or S) at the chiral center in:

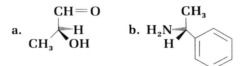

Problem 5.10   Draw the structure of:

a. (R)-3-methylhexane   b. (S)-3-methyl-1-pentene

Problem 5.11   Determine the configuration of (+)-carvone (see p. 132 for its structure).

There is no obvious relationship between configuration (R or S) and sign of rotation (+ or −). For example, (R)-lactic acid is levorotatory, as we have just seen. When (R)-lactic acid is converted to its methyl ester (eq. 5.2), the configuration is unchanged because none of the bonds to the chiral carbon is involved in the reaction. Yet the sign of rotation, a physical property of the product, is (+).

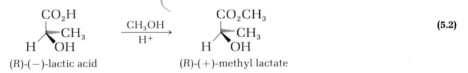

(5.2)

(R)-(−)-lactic acid   (R)-(+)-methyl lactate

5.10
THE *E–Z*
CONVENTION FOR
CIS-TRANS
ISOMERS

Before we go on to the more complex situation of molecules with two or more chiral centers, let us digress briefly to describe a useful extension of the Cahn–Ingold–Prelog system of nomenclature to *cis-trans* isomers (Sec. 3.5). Sometimes *cis-trans* nomenclature is ambiguous, as in the following examples:

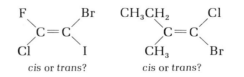

The system we have just discussed for chiral centers has been extended to double-bond isomers. We use exactly the same priority rules. The two groups attached to each carbon of the double bond are assigned priorities. If the higher-priority groups are on *opposite* sides of the double bond, the prefix E (from the German *entgegen*, opposite) is used. If the higher-priority groups are on the *same* side of the double bond, the prefix is Z (from the German *zusammen*, together). The highest-priority groups for the foregoing examples are shown here in color, and the correct names are shown below the structure.

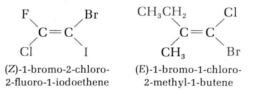

(Z)-1-bromo-2-chloro-   (E)-1-bromo-1-chloro-
2-fluoro-1-iodoethene   2-methyl-1-butene

Problem 5.12   **Name by the E-Z system:**

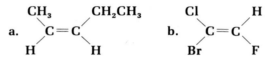

Problem 5.13   **Write the structure for:**
a. (E)-2-pentene   b. (Z)-1,3-pentadiene

Example 5.4   **How many stereoisomers of 4-bromo-2-pentene are possible?**

Solution   **First draw the structure:**

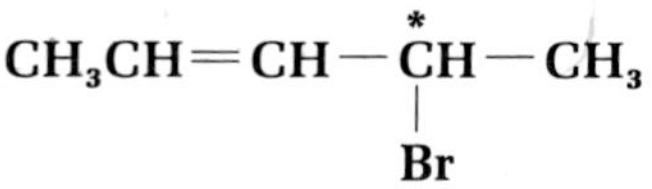

$$CH_3CH=CH-\overset{*}{CH}-CH_3$$
$$|$$
$$Br$$

We see that there are two possible arrangements for the double bond (E or Z) and that there is one chiral center (marked with an asterisk), which can be R or S. Consequently 2 × 2 or 4 structures in all are possible. They are

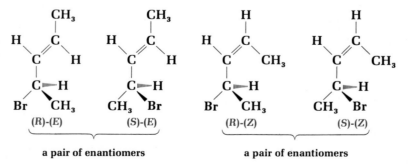

a pair of enantiomers   a pair of enantiomers

Problem 5.14   **How many stereoisomers of 3-methyl-2,4-hexadiene are possible? Name them according to the E-Z system.**

**5.11
COMPOUNDS
WITH MORE THAN
ONE CHIRAL
CENTER;
DIASTEREOMERS**

Many compounds in nature have more than one chiral center, so it is important to be able to determine how many isomers exist in such cases and how they are related to one another. Let us take an example *not* from nature, just to keep the structures simple. Consider the molecule 2-bromo-3-chlorobutane.

$$CH_3 - \overset{*}{C}H - \overset{*}{C}H - CH_3$$
$$\underset{Br}{|} \underset{Cl}{|}$$

2-bromo-3-chlorobutane

As indicated by the asterisks, the molecule has two chiral centers. Each of these could have the configuration *R* or *S*. Thus four isomers in all are possible: *RR, SS, RS,* and *SR,* where the first letter refers to the configuration at carbon-2 and the second letter refers to the configuration at carbon-3. We can draw these four isomers in many ways, only one of which is shown in Figure 5.11. Note that there are two pairs of enantiomers. The (*R,R*) and (*S,S*) forms are nonsuperimposable mirror images, and the (*R,S*) and (*S,R*) forms constitute another such pair.

But what is the relationship between, say, the (*R,R*) and the (*R,S*) forms? They are *not* mirror images because they have the *same* configuration at carbon-2 but opposite configurations at carbon-3. They are certainly stereoisomers, but they are not enantiomers. For such pairs of stereoisomers we use the term **diastereomers.** *Diastereomers are stereoisomers that are not mirror images of one another.*

There is a very important difference between enantiomers and diastereomers. Because they are mirror images, enantiomers differ *only* in mirror-image (chiral) properties. Because they have the same achiral properties, such as melting point, boiling point, and solubility in ordinary solvents, enantiomers cannot be separated from one another by methods that de-

---

FIGURE 5.11

The four stereoisomers of 2-bromo-3-chlorobutane, a compound with two different chiral centers.

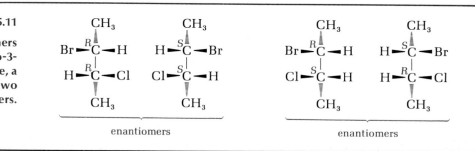

pend on these properties, such as recrystallization or distillation. On the other hand, diastereomers are not mirror images. They differ in all kinds of properties, whether the properties are chiral or achiral. As a consequence, diastereomers can differ in melting point, boiling point, solubility, and the number of degrees that they rotate plane-polarized light—in short, they behave as two different chemical substances. They can be separated from one another by ordinary means, such as distillation or recrystallization.

Note that *cis-trans* isomers are in fact diastereomers. That is, they are stereoisomers but not mirror images. We have already seen (in Sec. 2.10 and 3.5) that such pairs of isomers do indeed have different achiral properties, such as melting and boiling points.

Problem 5.15   **How do you expect the specific rotations of the (R,R)- and (S,S)-isomers of 2-bromo-3-chlorobutane to be related? Answer the same question for the (R,R)- and (S,R)-isomers.**

Can we generalize about the number of stereoisomers possible when a larger number of chiral centers is present? Suppose we add a third chiral center to the compounds shown in Figure 5.11 (consider, for example, 2-bromo-3-chloro-4-iodopentane). The new chiral center added to each of the four structures can once again have either an R or an S configuration, so the total possible number of isomers with three different chiral centers is eight. The situation is summed up in a simple rule: *If a molecule has* **n** *different chiral centers, it may exist in a maximum of* 2^n *stereoisomeric forms.* There will be a maximum of $2^n/2$ pairs of enantiomers.

Problem 5.16   **One formula for glucose (blood sugar) is:**

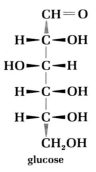

glucose

How many other stereoisomers of this sugar are possible?

Actually, the number of isomers predicted by this rule is the *maximum* number possible. Sometimes certain features of the structure reduce the actual number of isomers. In the next section we examine a case of this type.

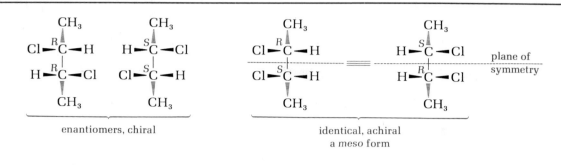

plane of symmetry

enantiomers, chiral

identical, achiral
a meso form

FIGURE 5.12   The *three* stereoisomers of 2,3-dichlorobutane.

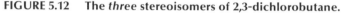

**5.12
MESO
COMPOUNDS; THE
STEREOISOMERS
OF TARTARIC ACID**

Consider the stereoisomers of 2,3-dichlorobutane. As with 2-bromo-3-chlorobutane, which was discussed in the previous section, there are two chiral centers. Each is marked with an asterisk:

$$CH_3 - \overset{*}{C}H - \overset{*}{C}H - CH_3$$
$$| |$$
$$Cl Cl$$

2,3-dichlorobutane

We can write out the stereoisomers just as we did in Figure 5.11. They are shown in Figure 5.12. Once again, the (*R,R*) and (*S,S*) isomers constitute a pair of nonsuperimposable mirror images, or enantiomers. However, the other "two structures," (*R,S*) and (*S,R*), in fact now represents a single compound! You can easily see this by rotating one of the structures 180° in the plane of the paper and noting that it is now superimposable on the other structure. This structure has a plane of symmetry that is perpendicular to the plane of the paper and bisects the central C—C bond. The structure therefore has a mirror image superimposable on itself, as we have just seen, and is achiral. We call this structure a **meso compound.** *A meso compound is an optically inactive achiral diastereomer of a compound with chiral centers.* The meso structure arises with 2,3-dichlorobutane because both chiral centers have the *same* four different groups attached:

$$Cl-, CH_3CH-, CH_3- \text{ and } H$$
$$|$$
$$Cl$$

Tartaric acid, the compound whose optical activity was so carefully studied by Louis Pasteur (Sec. 5.6), also has two such chiral centers.

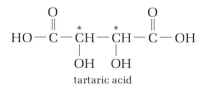

tartaric acid

FIGURE 5.13

The stereoisomers of
tartaric acid.

CO_2H	CO_2H	CO_2H
H—C—OH	HO—C—H	H—C—OH
HO—C—H	H—C—OH	H—C—OH
CO_2H	CO_2H	CO_2H

Configuration	(R, R)	(S, S)	meso
$[\alpha]_D^{20°}$ (H_2O)	+12	−12	0
Melting point, °C	170	170	140

(plane of symmetry — labeled at right of third structure)

The structures of its three stereoisomers and some of their properties are shown in Figure 5.13. Note that the enantiomers have identical properties except for the *sign* of the specific rotation, whereas the *meso* form, being a diastereomer of each enantiomer, differs from them in all properties.

For about 100 years after Pasteur's research it was still not possible to determine the particular configuration associated with a particular enantiomer. More specifically, it was not known whether (+)-tartaric acid had the (R,R) or the (S,S) configuration. It was known that (+)-tartaric acid had to have one of these two configurations and that (−)-tartaric acid had to have the opposite configuration from the (+) isomer, but which was which?

In 1951, the Dutch scientist J. M. Bijvoet developed a special x-ray technique that solved the problem. Using this technique on crystals of the sodium rubidium salt of (+)-tartaric acid, Bijvoet showed that it had the (R,R) configuration. Previously, the **relative configurations** of many chiral molecules had been established. For example, (+)-tartaric acid had been converted chemically in several steps to (−)-lactic acid.

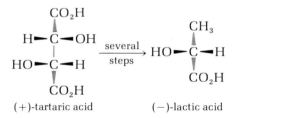

(5.3)

(+)-tartaric acid   (−)-lactic acid

Therefore these substances had the same configurations, whatever they were. With the configuration of (+)-tartaric acid established as (R,R), (−)-lactic acid had to have the (R) configuration. In this way, it became possible to assign **absolute configurations** to many pairs of enantiomers.

Example 5.5   Describe all the stereoisomers of 1,2-dimethylcyclobutane.

Solution   The carbons to which the methyl groups are attached are chiral. The methyls can be *cis* or *trans*. The *cis*-isomer has a plane of symmetry that is perpendicular to the plane of the cyclobutane ring and bisects the bond between C_1 and C_2. The *cis*-isomer is therefore an achiral *meso* form. The *trans*-isomer, on the other hand, has no plane of symmetry and exists as a pair of chiral enantiomers.

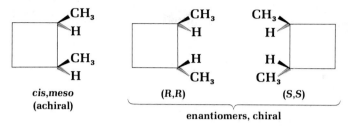

cis,meso   (R,R)   (S,S)
(achiral)

enantiomers, chiral

Problem 5.17 Describe the stereoisomers of 1,3-dimethylcyclobutane.

Problem 5.18 Describe the stereoisomers of 1,3-dimethylcyclopentane. (Note that a molecule can have a *meso* form even if the chiral centers are not adjacent.)

5.13 RESOLUTION OF RACEMIC MIXTURES

In most chemical reactions carried out in the laboratory or commercially, products with a chiral center are produced as a 50:50 mixture of enantiomers—that is, as a racemic mixture. Consider, for example, the addition of HBr to 1-butene to give 2-bromobutane in accordance with Markovnikov's rule.

$$CH_3CH_2CH\!=\!CH_2 + HBr \rightarrow CH_3CH_2\overset{*}{C}HCH_3 \qquad (5.4)$$
$$\underset{\text{Br}}{|}$$

  1-butene   2-bromobutane

The product has one chiral center, marked with an asterisk. Both enantiomers are formed in *exactly equal amounts*. Consider the reaction mechanism (Sec. 3.13):

$$CH_3CH_2CH\!=\!CH_2 + H^+ \rightarrow CH_3CH_2\overset{+}{C}HCH_3 \xrightarrow{\;Br^-\;} CH_3CH_2CHCH_3 \qquad (5.5)$$
$$\underset{\text{Br}}{|}$$

2-butyl cation

The intermediate 2-butyl cation is planar, and bromide ion reacts with it from the "top" or the "bottom" side with exactly equal probability:

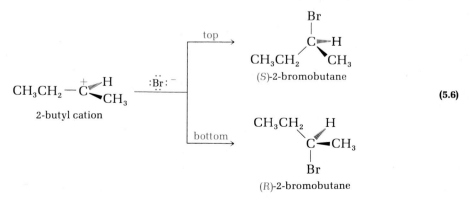

(5.6)

The product is therefore a racemic mixture, an optically inactive mixture of the two enantiomers.

Problem 5.19   **Show that the free-radical chlorination of butane at C_2 will give a 50:50 mixture of the 2-chlorobutane enantiomers.**

We frequently obtain racemic mixtures as reaction products. *The process of separating a racemic mixture into its (+) and (−) enantiomers is called* **resolution.** We have already seen, however, that enantiomers have identical achiral properties. How, then, can we resolve a racemic mixture into its components?

To differentiate between two enantiomers, we let them react with a chiral reagent. The product will be a pair of *diastereomers* and these, as we have seen, differ in all types of properties, chiral or achiral, and can be separated by ordinary methods. This principle is illustrated in the following equation:

$$\begin{Bmatrix} R \\ S \end{Bmatrix} \quad + \quad R \quad \rightarrow \quad \begin{Bmatrix} R - R \\ S - R \end{Bmatrix} \tag{5.7}$$

pair of   chiral   diastereomeric
enantiomers   reagent   product
(not separable)   (separable)

As a specific example, consider the separation of (*R*)- and (*S*)-lactic acids. We can allow the mixture to react with a chiral base. Many such bases, such as strychnine and quinine, occur naturally in certain plants. The acid and base react to form a salt:

$$\begin{Bmatrix} (R)\text{-acid} \\ (S)\text{-acid} \end{Bmatrix} + (S)\text{-base} \rightarrow \begin{Bmatrix} (R,S)\text{-salt} \\ (S,S)\text{-salt} \end{Bmatrix} \tag{5.8}$$

The diastereomeric salts can then be separated by fractional crystallization. Upon treatment with a strong acid such as HCl, each salt liberates one enantiomer of the original mixture:

$$(R,S)\text{-salt} + HCl \rightarrow (R)\text{-acid} + (S\text{-base})H^+Cl^- \tag{5.9}$$

$$(S,S)\text{-salt} + HCl \rightarrow (S)\text{-acid} + (S\text{-base})H^+Cl^- \tag{5.10}$$

The principle behind the resolution of racemic mixtures is the same as the principle involved in the specificity of many biological reactions. That is, a chiral reagent (in cells, usually an enzyme) can discriminate between enantiomers because the two possible products are diastereomers.

ADDITIONAL   **5.20.** Define or describe the following terms.
PROBLEMS   **a.** chiral molecule   **b.** enantiomers
                     c. polarized light   **d.** specific rotation
                     e. chiral center   **f.** plane of symmetry
                     g. racemic mixture   **h.** diastereomers
                     i. *meso* form   **j.** resolution

5.21. How can one tell whether a compound can exist in enantiomeric forms?

5.22. Which of the following substances can exist in optically active forms?
a. 2,2-dibromopropane   **b.** 1,2-dibromopropane
c. 3-ethylhexane   **d.** 2,3-dimethylhexane
e. methylcyclopentane   **f.** 1-deuterioethanol (CH_3CHDOH)

5.23. Locate with an asterisk the chiral centers in the following structures.

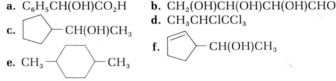

a. $C_6H_5CH(OH)CO_2H$ **b.** $CH_2(OH)CH(OH)CH(OH)CHO$
d. $CH_3CHClCCl_3$

c. ⬡—CH(OH)CH₃

f. ⬠—CH(OH)CH₃

e. CH₃—⬡—CH₃

5.24. What would happen to the *observed* rotation if, in measuring the optical activity of a solution of sugar in water, we:
a. doubled the concentration of the solution
b. doubled the length of the sample tube
What would happen to the *specific* rotation if the same changes were made?

5.25. In measuring the optical activity of a particular solution, we obtained an observed rotation of $+30°$. How could we tell whether it is really $+30°$ or $-330°$?

5.26. Draw a structural formula for an optically active compound with the molecular formula:
a. $C_4H_{10}O$ **b.** $C_5H_{11}Br$ **c.** $C_4H_8(OH)_2$ **d.** C_6H_{12}

5.27. Draw the formula of an unsaturated chloride C_5H_9Cl that can show:
a. neither *cis-trans* isomerism nor optical activity.
b. *cis-trans* isomerism but no optical activity.
c. no *cis-trans* isomerism but optical activity.
d. *cis-trans* isomerism and optical activity.

5.28. Place the members of the following groups in order of decreasing priority according to the $R-S$ convention.
a. CH_3-, $H-$, $HO-$, CH_3CH_2-
b. $H-$, CH_3-, C_6H_5-, $Cl-$
c. CH_3-, $HO-$, $-CH_2Cl$, $-CH_2OH$
d. CH_3CH_2-, $CH_3CH_2CH_2-$, $CH_2=CH-$, $-CH=O$

5.29. Assume that the four groups in each part of Problem 5.28 are attached to one carbon atom. Draw a three-dimensional formula for the R configuration of the molecule.

5.30. Tell whether the chiral centers marked with an asterisk in the following structures have the R or the S configuration.

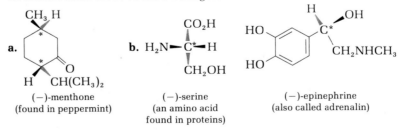

a. (−)-menthone (found in peppermint)

b. (−)-serine (an amino acid found in proteins)

(−)-epinephrine (also called adrenalin)

5.31. Name the following compounds, using the E–Z notation.

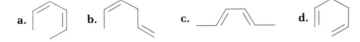

a.　　　　　b.　　　　　c.　　　　　d.

5.32. How many stereoisomers are possible for each of the following structures? Draw them and name each by the R–S and E–Z conventions.
a. 3-methyl-1,4-pentadiene　　b. 3-methyl-1,4-hexadiene
c. 4-methyl-2,5-heptadiene　　d. 2-bromo-5-chloro-3-hexene
e. 2,5-dibromo-3-hexene　　　f. 2,3,4-tribromopentane

5.33. Two possible configurations for a molecule with three different chiral centers are R–R–R and its mirror image, S–S–S. What are all the remaining possibilities? Repeat for a compound with four different chiral centers.

5.34. Addition of bromine to *trans*-2-butene gives *meso*-2,3-dibromobutane. On the other hand, addition of bromine to *cis*-2-butene gives a 50:50 mixture of (R,R)- and (S,S)-2,3-dibromobutanes. Explain. (*Hint:* Review eq. 3.26.)

5.35. When racemic 2-chlorobutane is chlorinated, we obtain some 2,3-dichlorobutane. It consists of 71% *meso*-isomer and 29% racemic isomers. Explain why the mixture need not be 50:50 *meso* and racemic 2,3-dichlorobutane. (It will help if you draw three-dimensional structures in seeking an explanation.)

5.36. The solubility of (+)-tartaric acid in water at 20°C is 139 g/100 g of water. The figure for *meso*-tartaric acid is 125 g/100 g of water. What solubility do you expect for (−)-tartaric acid?

5.37. Below are Newman projection formulas for the three tartaric acids (R,R), (S,S), and *meso*. Which is which?

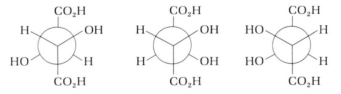

5.38. The formula for muscarine, the toxic constituent of poisonous mushrooms, is

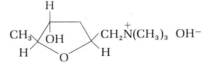

Is it chiral? How many isomers of this structure are possible? An interesting murder mystery, which you might enjoy reading, and which depends for its solution on the distinction between optically active and racemic forms of this poison, is Dorothy L. Sayers's *The Documents in the Case*, published in paperback by Avon Books [see an article by H. Hart, "Accident, Suicide, or Murder? A Question of Stereochemistry," *J. Chem. Educ.*, **52**, 444 (1975)].

5.39. The antibiotic **chloramphenicol** has the formula

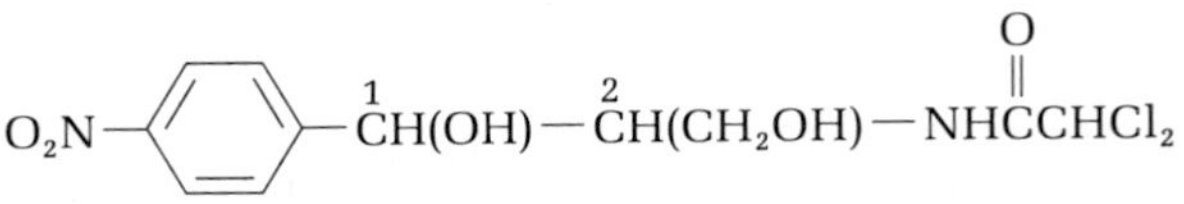

$$O_2N-\langle\ \rangle-\overset{1}{C}H(OH)-\overset{2}{C}H(CH_2OH)-NH\overset{\overset{O}{\|}}{C}CHCl_2$$

The (R,R)-isomer has the greatest antibiotic activity. Draw its formula in three dimensions (using wedges and dashed lines). Its diastereomer with the S configuration at carbon-2 shows slight biological activity. Draw its structure. The enantiomers of these forms are totally devoid of antibiotic activity. Draw their structures.

5.40. If reactions between achiral reagents always give achiral or racemic products, speculate on how the first chiral molecule might have been produced. (For a brief but fascinating discussion of several theories of the origin of optical activity, and for other discussions of the importance of stereochemistry in the biological world, see G. Natta and M. Farina, *Stereochemistry,* New York: Harper & Row, 1972.)

six

Organic Halogen Compounds; Substitution and Elimination Reactions

6.1
INTRODUCTION

In recent years, quite a few chlorine-containing and bromine-containing natural products have been isolated from various species that live in the sea—sponges, mollusks, and other ocean creatures. With these exceptions, most organic halogen compounds are the creation of the laboratory. We have already seen that they can be made by the direct halogenation of alkanes (Sec. 2.14) and aromatic compounds (Sec. 4.9) and by the addition of hydrogen halides to alkenes (Sec. 3.11) and alkynes (Sec. 3.24). In the next chapter we will learn how alkyl halides are prepared from the corresponding alcohols (Sec. 7.9, 7.10). Why so many routes to these compounds?

Halogen compounds are important for several reasons. Simple alkyl and aryl halides, especially chlorides and bromides, are the workhorses of synthetic organic chemistry. Through substitution reactions, which we will discuss in this chapter, the halogens can be replaced by many other functional groups. Organic halides can also be converted to unsaturated compounds through elimination reactions. Finally, many halogen compounds have very practical uses—as insecticides, herbicides, fire retardants, cleaning fluids, and refrigerants, for example. In this chapter, we will discuss all these aspects of halogen compounds.

6.2
NUCLEOPHILIC
SUBSTITUTION

Ethyl bromide reacts with hydroxide ion to give ethyl alcohol and bromide ion.

$$^-OH + CH_3CH_2 - Br \xrightarrow{H_2O} CH_3CH_2 - OH + Br^- \tag{6.1}$$

ethyl bromide   ethyl alcohol

This is a typical example of a **nucleophilic substitution reaction.** Hydroxide ion is the **nucleophile.** It reacts with the **substrate** (ethyl bromide) and replaces bromide ion. The bromide ion is called the **leaving group** in such reactions. In reactions of this type, one covalent bond is broken and a new covalent bond is formed. In this particular example, a carbon–bromine bond is broken and a carbon–oxygen bond is formed. The leaving group (bromide) takes with it *both* of the electrons from the C—Br bond, and the nucleophile (hydroxide ion) supplies *both* electrons for the new carbon–oxygen bond.

These ideas are summarized in the following general equations for a nucleophilic substitution reaction.

$$Nu\!: \quad + \quad R\!:\!L \quad \rightarrow \quad R\!:\!\overset{+}{Nu} \quad + \quad :\!L^- \tag{6.2}$$

nucleophile   substrate   product   leaving
(neutral)   group

or

$$Nu\!:^- \quad + \quad R\!:\!L \quad \rightarrow \quad R\!:\!Nu \quad + \quad :\!L^-$$

(anion)

A nucleophile is a reagent that can supply an electron pair to make a covalent bond.

In principle, of course, reactions like eq. 6.2 are reversible, because the leaving group is also a nucleophile. Like $Nu\!:$, it has an unshared electron pair that can be used to form a covalent bond. We can force the reaction to go in the forward direction in several ways. For example, we can have $Nu\!:$ be a *stronger* nucleophile than the leaving group $:\!L$. Or we can shift the equilibrium by using a large excess of one reagent or by removing one of the products as it is formed.

**6.3
EXAMPLES OF
NUCLEOPHILIC
SUBSTITUTIONS**

Nucleophiles can be classified according to the kind of atom that forms a new covalent bond. For example, the hydroxide ion in eq. 6.1 is an *oxygen* nucleophile. In the product, a new carbon–*oxygen* bond is formed. The most common nucleophiles are either *oxygen, nitrogen, sulfur, halogen,* or *carbon* nucleophiles. Table 6.1 shows some examples of the types of nucleophiles and the products formed when they react with an alkyl halide.

Example 6.1

Using Table 6.1 as a guide, write an equation for the reaction of sodium methoxide with ethyl bromide.

Solution

Sodium methoxide is the sodium salt of methyl alcohol (review eq. 1.7). The equation is $CH_3\overset{..}{\underset{..}{O}}\!:^- Na^+ + CH_3CH_2Br \rightarrow CH_3OCH_2CH_3 + Na^+Br^- \cdot$ The product is an ether.

Problem 6.1

Using Table 6.1 as a guide, write an equation for each of the following nucleophilic substitution reactions.
a. $NaOH + CH_3CH_2CH_2Br$ b. $NH_3 + CH_3CH_2Br$

c. $(CH_3CH_2)_3N + CH_3CH_2Br$ d. $KCN +$ [benzene ring]—CH_2Br

Problem 6.2

Write an equation for the preparation of each of the following compounds using a nucleophilic substitution reaction. In each case, label the nucleophile, the substrate, and the leaving group.
a. $CH_3CH_2CH_2CH_2SH$ b. $(CH_3)_2CHCH_2OH$
c. $(CH_3CH_2CH_2)_2NH$ d. $(CH_3CH_2)_3S^+Br^-$

TABLE 6.1 Reactions of common nucleophiles with alkyl halides;*

$$Nu: + R-X \rightarrow R-\overset{+}{Nu} + X^-$$

Nu formula	Nu name	R-Nu formula	R-Nu name	Comments
Oxygen Nucleophiles				
1. $H\ddot{O}:^-$	hydroxide	R—OH	alcohol	
2. $R\ddot{O}:^-$	alkoxide	R—OR	ether	
3. $H\ddot{O}H$	water	$R-\overset{+}{\underset{H}{\overset{H}{O}}}$	alkyloxonium ion	These ions readily lose a proton. $\xrightarrow{-H^+}$ ROH (alcohol)
4. $R\ddot{O}H$	alcohol	$R-\overset{+}{\underset{H}{\overset{R}{O}}}$	dialkyloxonium ion	$\xrightarrow{-H^+}$ ROR (ether)
5. $R\ddot{O}R$	ether	$R-\overset{+}{\underset{R}{\overset{R}{\ddot{O}}}}$	trialkyloxonium ion	
6. $R-C\overset{O}{\underset{\ddot{O}:^-}{}}$	carboxylate	$R-O\overset{O}{\overset{\|}{C}}-R$	ester	
Nitrogen Nucleophiles				
7. $\ddot{N}H_3$	ammonia	$R-\overset{+}{N}H_3$	alkylammonium ion	With a base, these ions readily lose a proton to give amines. $\xrightarrow{-H^+}$ RNH_2
8. $R\ddot{N}H_2$	primary amine	$R-\overset{+}{N}H_2R$	dialkylammonium ion	$\xrightarrow{-H^+}$ R_2NH
9. $R_2\ddot{N}H$	secondary amine	$R-\overset{+}{N}HR_2$	trialkylammonium ion	$\xrightarrow{-H^+}$ R_3N
10. $R_3\ddot{N}$	tertiary amine	$R-\overset{+}{N}R_3$	tetraalkylammonium ion	
Sulfur Nucleophiles				
11. $H\ddot{S}:^-$	hydrosulfide ion	R—SH	thiol	
12. $R\ddot{S}:^-$	mercaptide ion	R—SR	thioether (sulfide)	
13. $R_2\ddot{S}:$	thioether	$R-\overset{+}{\underset{\cdot\cdot}{S}}R_2$	trialkylsulfonium ion	
Halogen Nucleophiles				
14. $:\ddot{I}:^-$	iodide	R—I	alkyl iodide	The usual solvent for this reaction is acetone. Sodium iodide is soluble, but sodium bromide and sodium chloride are not soluble in this solvent.
Carbon Nucleophiles				
15. $^-:C\equiv N:$	cyanide	R—CN	alkyl cyanide (nitrile)	[Sometimes the isonitrile, RNC, is formed]
16. $^-:C\equiv CR$	acetylide	$R-C\equiv CR$	acetylene	

*Aryl halides and vinyl halides normally do not undergo these nucleophilic substitution reactions.

All the reactions in Table 6.1 are synthetically useful for producing the products R—Nu. Most of the reactions listed are straightforward, but some require a more detailed explanation and are amplified in the next three sections.

6.4
THE WILLIAMSON
ETHER SYNTHESIS

Reaction 2 in Table 6.1 is the second step in a general method for preparing ethers. It is known as the **Williamson ether synthesis** after the British chemist Alexander Williamson, who devised it. In the first step, an alcohol is treated with sodium metal to produce the alkoxide.

$$2\,ROH + 2\,Na \rightarrow 2\,RO^-Na^+ + H_2 \tag{6.3}$$

alcohol   a sodium
      alkoxide

The alkyl halide R′X is then added and the mixture is heated to produce the ether.

$$RO^-Na^+ + R'{-}X \rightarrow R{-}O{-}R' + Na^+X^- \tag{6.4}$$

       ether

Because R and R′ can be varied structurally within rather wide limits, the synthesis is quite versatile. For reasons that we will discuss a little later, the reaction works well if R′ is a primary or a secondary alkyl group, but not if R′ is tertiary. (Review Sec. 3.14 for definitions of primary, secondary, and tertiary groups.) There is no such restriction on the structure of R.

Example 6.2 **Write equations for the synthesis of $CH_3OCH_2CH_2CH_3$ using the Williamson method.**

Solution **There are two possibilities, depending on which alcohol and which alkyl halide are used:**

$2\,CH_3OH + 2\,Na \rightarrow 2\,CH_3O^-Na^+ + H_2$

$CH_3O^-Na^+ + CH_3CH_2CH_2X \rightarrow CH_3OCH_2CH_2CH_3 + Na^+X^-$

or

$2\,CH_3CH_2CH_2OH + 2\,Na \rightarrow 2\,CH_3CH_2CH_2O^-Na^+ + H_2$

$CH_3CH_2CH_2O^-Na^+ + CH_3X \rightarrow CH_3CH_2CH_2OCH_3 + Na^+X^-$

X is usually Cl, Br, or I.

Problem 6.3 **Write equations for the synthesis of the following ethers by the Williamson method.**
a. $(CH_3CH_2CH_2)_2O$ **b. $(CH_3)_3COCH_3$ [Reminder: The reaction fails with tertiary halides.]**

6.5
AMINES FROM
ALKYL HALIDES

Amines represent the most important type of organic bases, and their chemistry will be discussed in detail in Chapter 12. Reactions 7–9 in Table 6.1 describe one important way to synthesize amines.

Ammonia reacts with alkyl halides to give amines via a two-step process. The first step is a nucleophilic substitution reaction.

$$H_3N: + R-X \rightarrow R-\overset{+}{N}H_3 + X^- \tag{6.5}$$

alkylammonium
halide

Excess ammonia is used and, in the second step of the reaction, ammonia acts as a base to remove a proton from the alkylammonium ion, thus forming the amine.

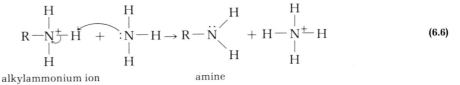

$$\tag{6.6}$$

alkylammonium ion   amine

As with the Williamson ether synthesis, the nucleophilic substitution reaction (eq. 6.5) works well only if R is a primary or secondary alkyl group. Eqs. 6.5 and 6.6 can be combined to give the overall reaction

$$2\,\overset{..}{N}H_3 + R-X \rightarrow R\overset{..}{N}H_2 + NH_4{}^+X^- \tag{6.7}$$

ammonia   alkyl   primary
             halide   amine

The primary amine formed in eq. 6.7 has an unshared electron pair on nitrogen and is therefore a nucleophile. It also can react with an alkyl halide to form a secondary amine.

$$2\,R\overset{..}{N}H_2 + R-X \rightarrow R_2\overset{..}{N}H + RNH_3{}^+X^- \tag{6.8}$$

secondary
amine

Because the secondary amine is also a nucleophile, the reaction can be continued to form tertiary amines.

$$2\,R_2\overset{..}{N}H + R-X \rightarrow R_3\overset{..}{N} + R_2NH_2{}^+X^- \tag{6.9}$$

tertiary
amine

The first step in the overall reactions expressed in eqs. 6.8 and 6.9 is a nucleophilic substitution (reactions 8 and 9, Table 6.1).

Finally, the tertiary amine can react with still another equivalent of alkyl halide to form a quaternary ammonium salt (reaction 10, Table 6.1):

$$R_3\overset{..}{N} + R-X \rightarrow R_4N^+X^- \tag{6.10}$$

quaternary
ammonium salt

All three classes of amines (primary, secondary and tertiary), as well as the quaternary ammonium salts, occur in nature.

Example 6.3   **Write an equation for the synthesis of benzylamine, $C_6H_5CH_2NH_2$.**

Solution

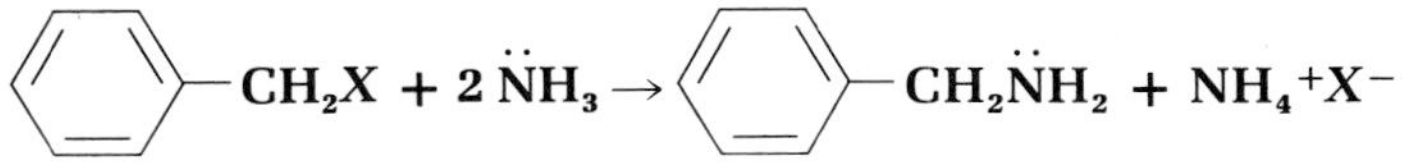

$$\text{⬡}-CH_2X + 2\ \ddot{N}H_3 \rightarrow \text{⬡}-CH_2\ddot{N}H_2 + NH_4{}^+X^-$$

(X = Cl, Br, or I)

Use of a large excess of ammonia helps prevent further substitution.

Problem 6.4 Complete the equations for the following reactions.
a. $CH_3CH_2CH_2Br + 2\ NH_3 \rightarrow$
b. $CH_3CH_2I + 2(CH_3CH_2)_2NH \rightarrow$
c. $(CH_3)_3N + CH_3I \rightarrow$

Problem 6.5 Write equations for the preparation of:
a. $(CH_3)_2CHCH_2NH_2$ from ammonia and isobutyl bromide

b. $\text{⬡}-N(CH_3)_2$ from CH_3I and $\text{⬡}-NH_2$

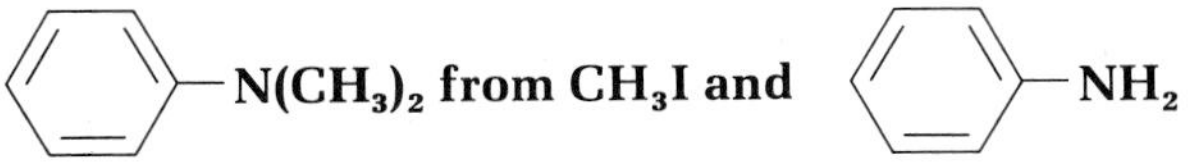

**6.6
A GENERAL
SYNTHESIS OF
ALKYNES**

Reaction 16 in Table 6.1 describes the second step in an important alkyne synthesis. The first step involves replacing the weakly acidic hydrogen on a triple bond by sodium, as previously described (Sec. 3.25).

$$R-C{\equiv}C-H + Na^+NH_2{}^- \xrightarrow{\text{liquid NH}_3} R-C{\equiv}C{:}^-Na^+ + NH_3 \qquad \text{(6.11)}$$

The resulting organosodium compound can react with alkyl halides via nucleophilic substitution.

$$R-C{\equiv}C{:}^-Na^+ + R'-X \rightarrow R-C{\equiv}C-R' + Na^+X^- \qquad \text{(6.12)}$$

The net result of eqs. 6.11 and 6.12 is to replace the acetylenic hydrogen by an alkyl group. Once again, the reaction works well only if R′ is a primary or secondary alkyl group.

Example 6.4 Show how acetylene can be converted to 2-pentyne.

Solution $H-C{\equiv}C-H \xrightarrow{?} CH_3-C{\equiv}C-CH_2CH_3$

We must replace one acetylenic hydrogen by a methyl group and the other by an ethyl group. This can be accomplished by carrying out the sequence of eq. 6.11 and 6.12 twice, as follows:

$$H-C{\equiv}C-H + Na^+NH_2{}^- \xrightarrow{NH_3} Na^+{}^-{:}C{\equiv}C-H + NH_3$$

$$Na^+{}^-{:}C{\equiv}C-H + CH_3I \rightarrow CH_3-C{\equiv}C-H + Na^+I^-$$

$$CH_3-C{\equiv}C-H + Na^+NH_2{}^- \xrightarrow{NH_3} CH_3-C{\equiv}C{:}^-Na^+ + NH_3$$

$$CH_3-C{\equiv}C{:}^-Na^+ + CH_3CH_2Br \rightarrow CH_3-C{\equiv}C-CH_2CH_3 + Na^+Br^-$$

The reactions could, of course, be carried out in the reverse order. That is, we could first replace an acetylenic hydrogen with an ethyl group and later replace the other hydrogen with a methyl group.

Problem 6.6 **Show equations that accomplish the following syntheses.**

a.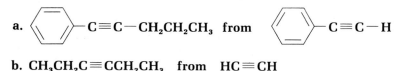

b. $CH_3CH_2C\equiv CCH_2CH_3$ from $HC\equiv CH$

Let us take a look at the mechanisms by which the substitution reactions listed in Table 6.1 take place. As a result of experiments that began over fifty years ago, we now understand the mechanisms of nucleophilic substitution reactions rather well. We use the plural because such reactions may occur by more than one mechanism. The specific mechanism observed depends on the structures of the nucleophile and the alkyl halide, the solvent, the reaction temperature, and other factors.

In the main, there are two extremes in nucleophilic substitution mechanisms. These are described by the symbols S_N2 and S_N1, respectively. The S_N part of each symbol indicates "substitution, nucleophilic." The meaning of the numbers 2 and 1 will become apparent as we discuss each mechanism.

The S_N2 mechanism is a *one-step* process that can be represented by the following equation:

$$Nu: + \overset{\diagdown}{\underset{\diagup}{C}}-L \rightarrow \left[\overset{\delta+}{Nu} \cdots \overset{|}{\underset{\diagdown}{C}} \cdots \overset{\delta-}{L} \right] \rightarrow \overset{+}{Nu}-\overset{\diagup}{\underset{\diagdown}{C}} + :L^- \qquad (6.13)$$

nucleophile   substrate   transition state

The nucleophile attacks the *rear* of the C—L bond. At some stage (the transition state) the nucleophile *and* the leaving group are *both* associated with the carbon at which substitution occurs. As the leaving group departs with its electron pair, the nucleophile supplies another electron pair to the carbon atom.

The symbol 2 is used to describe this mechanism because the reaction is *bimolecular*. That is, two molecules—the nucleophile and the substrate—are involved in the key step (the only step!) in the reaction mechanism.

How can we recognize when a particular nucleophile and substrate are reacting by the S_N2 mechanism? There are several tell-tale signs.

1. *Because both the nucleophile and the substrate are involved, the rate of the reaction depends on the concentration of each.* The reaction of hydroxide ion with ethyl bromide (eq. 6.1) is an example of an S_N2 reaction. If we double the base concentration (OH^-), we find that the reaction goes twice as fast. We obtain the same result if we double the ethyl bromide concentration. We will see shortly that this rate behavior is not observed in the S_N1 process.

2. *The reaction occurs with* **inversion** *of configuration.* For example, if we treat (R)-2-bromobutane with sodium hydroxide, we obtain (S)-2-butanol.

$$HO^- + \begin{array}{c} CH_3 \\ \diagdown \\ H^{\cdots} C-Br \\ \diagup \\ CH_3CH_2 \end{array} \rightarrow HO-C\begin{array}{c} CH_3 \\ \diagup \\ \cdots H \\ \diagdown \\ CH_2CH_3 \end{array} + Br^- \qquad \textbf{(6.14)}$$

(R)-2-bromobutane   (S)-2-butanol

Hydroxide ion must therefore attack the rear side of the C—Br bond. As substitution occurs, the three groups attached to the sp^3 carbon *invert*, something like an umbrella caught in a strong wind. If, somehow, the OH had taken the exact position occupied by the Br, (R)-2-butanol would have resulted. At first this stereochemical result came as quite a surprise, because it seemed easier to assume that the nucleophile and the leaving group exchanged places. We know now, however, that *every S_N2 displacement occurs with inversion of configuration.*

3. When a substrate R—L reacts by the S_N2 mechanism, *the reaction is fastest when R is methyl or primary and slowest when R is tertiary.* Secondary R groups react at an intermediate rate. The reason for this reactivity order is fairly obvious if we think about the S_N2 mechanism (eq. 6.13). The rear side of the carbon, where displacement occurs, becomes more crowded the more alkyl groups are attached to the carbon bearing the leaving group, thus slowing down the reaction rate.

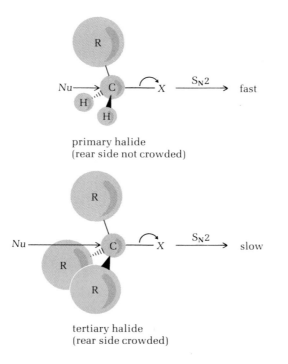

primary halide
(rear side not crowded)

tertiary halide
(rear side crowded)

To summarize, the S_N2 mechanism is a one-step process favored when the alkyl halide is methyl > primary > secondary ≫ tertiary. It occurs with inversion of configuration, and its rate depends on the concentration of *both* the nucleophile and the substrate (the alkyl halide).

Now let us see how these features are different for the S_N1 mechanism.

**6.9
THE S_N1
MECHANISM**

The S_N1 mechanism is a two-step process. In the first step, the bond between the carbon and the leaving group breaks as the substrate dissociates.

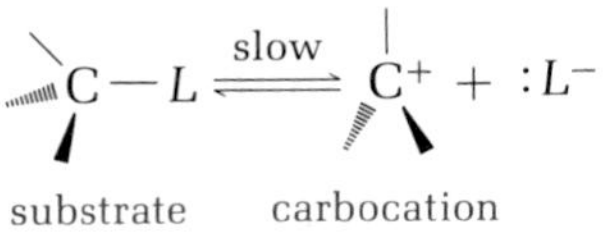

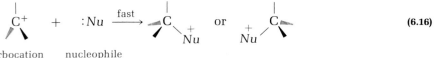

$$\text{substrate} \qquad \text{carbocation}$$

(6.15)

The electrons of the bond go with the leaving group, and a carbocation is formed. In the second, fast step, the carbocation combines with the nucleophile to give the product.

(6.16)

carbocation   nucleophile

In the S_N1 mechanism, substitution occurs in two steps. The symbol 1 is used because the slow step (eq. 6.15) involves *only one* of the two reactants, the substrate. It does not involve the nucleophile at all. That is, the first step is *unimolecular*.

How can we recognize when a particular nucleophile and substrate are reacting by the S_N1 mechanism? Here are the tell-tale signs:

1. *The rate of the reaction does not depend on the concentration of the nucleophile.* Only the first step is rate-determining and the nucleophile is not involved in the first step. As soon as it is formed, the carbocation reacts with the nucleophile.

2. If the carbon bearing the leaving group is chiral, *the reaction occurs mainly with loss of optical activity* (that is, with racemization). In carbocations only three groups are attached to the positive carbon. Therefore, the positive carbon is sp^2-hybridized and planar. As shown in eq. 6.16, the nucleophile can react at either "face" of the positive carbon to give a 50:50 mixture of two enantiomers, a racemic mixture. For example, the reaction of (S)-3-bromo-3-methylhexane with water gives the racemic alcohol.

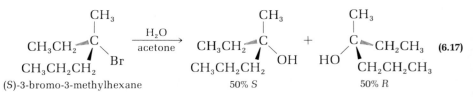

(6.17)

(S)-3-bromo-3-methylhexane   50% S   50% R

The intermediate is the planar carbocation:

$$H_2O \searrow \underset{\underset{CH_3CH_2 \quad CH_2CH_2CH_3}{\diagdown}}{\overset{\overset{CH_3}{|}}{C^+}} \swarrow H_2O$$

Reaction with water from either face is equally probable, giving racemic product.

3. When a substrate $R{-}L$ reacts by the S_N1 mechanism, *the reaction is fastest when R is tertiary and slowest when R is primary.* S_N1 reactions proceed via carbocations, so the reactivity order is the same as the order of carbocation stability, $3° > 2° \gg 1°$. That is, the reactions proceed faster the easier it is to form the carbocation.

To summarize, the S_N1 mechanism is a two-step process favored when the alkyl halide is tertiary $>$ secondary $\gg$ primary. It occurs with racemization and its rate is independent of nucleophile concentration.

**6.10
THE S_N1 AND S_N2
MECHANISMS
COMPARED**

Table 6.2 summarizes what we have said so far about substitution mechanisms and compares them with respect to a few other variables, such as solvent and nucleophile structure.

Note first that primary halides almost always react by the S_N2 mechanism, whereas tertiary halides react by the S_N1 mechanism. Only with secondary halides are we likely to encounter both possibilities.

The first step of the S_N1 mechanism involves the formation of ions, so that mechanism is highly favored by polar solvents. Thus, for a secondary halide, which might react by either mechanism, we may change the mechanism simply by adjusting the solvent polarity. For example, we might change the mechanism by which a secondary halide reacts with water

TABLE 6.2		S_N2	S_N1
Comparison of S_N2 and S_N1 substitutions			
	Halide structure		
	Primary or CH$_3$	common	never
	Secondary	sometimes	sometimes
	Tertiary	never	common
	Stereochemistry	inversion	racemization
	Nucleophile	rate depends on nucleophile concentration; mechanism is favored when nucleophile is an anion	rate is independent of nucleophile concentration; mechanism is more likely with neutral nucleophiles
	Solvent	rate is only mildly affected by solvent polarity	rate is highly favored with polar solvents

(to form an alcohol) from S_N2 to S_N1 by changing the solvent from 95% acetone–5% water (relatively nonpolar) to 50% acetone–50% water (more polar, and a better ionizing solvent).

Example 6.5   The reaction of (R)-2-iodobutane with 95% acetone–5% water gives mainly (S)-2-butanol. With 30% acetone–70% water, the product 2-butanol has a much lower optical activity, being only about 60% (S)-isomer and 40% (R)-isomer. Explain.

Solution   In 95% acetone–5% water, a relatively nonpolar solvent, substitution occurs by an S_N2 mechanism, with inversion.

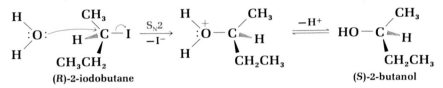

(R)-2-iodobutane   (S)-2-butanol

When the percentage of water in the solvent is increased, the solvent is more polar and some reaction occurs by the S_N1 process.

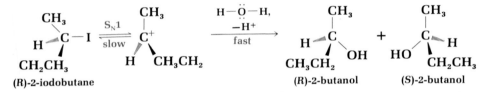

(R)-2-iodobutane   (R)-2-butanol   (S)-2-butanol

There is a slight preference for the (S)-isomer for one of two reasons. Either some of the reaction is still occurring by the S_N2 mechanism, or the departing iodide ion in the S_N1 mechanism is still close enough to the carbocation to partially block that "face" from nucleophilic attack.

As we have seen, the rate of an S_N2 reaction depends on the nucleophile. If a reagent is a *strong* nucleophile, the S_N2 mechanism will be favored. How can we tell whether a nucleophile is strong or weak, or whether one nucleophile is stronger than another? Here are a few generalizations that are useful.

1. Negative ions are more nucleophilic, better electron suppliers, than the corresponding neutral molecules. Thus

$$HO^- > HOH \qquad RS^- > RSH$$
$$RO^- > ROH \qquad R-\overset{\|}{\underset{O}{C}}-O^- > R-\overset{\|}{\underset{O}{C}}-OH$$

2. Elements lower in the periodic table tend to be more nucleophilic than elements above them in the same column. Thus

$$HS^- > HO^- \qquad I^- > Br^- > Cl^- > F^-$$
$$R\ddot{S}H > R\ddot{O}H \qquad (CH_3)_3\ddot{P}: > (CH_3)_3N:$$

3. Elements in the same row in the periodic table tend to be less nucleophilic the more electronegative the element (that is, the more tightly it holds electrons to itself). Thus

$$\begin{matrix} R \\ | \\ R-C^- \\ | \\ R \end{matrix} > \begin{matrix} R \\ \diagdown \\ N^- \\ \diagup \\ R \end{matrix} > R-O^- > F^- \qquad \text{and} \qquad H_3N: > H_2\ddot{O}: > H\ddot{F}:$$

Because C and N are in the same row, it is perhaps not surprising that the cyanide ion, $:C{\equiv}N:^-$, reacts mainly at carbon when it acts as a nucleophile (see Table 6.1, reaction 15).

Problem 6.7   **Which mechanism, S_N1 or S_N2, would you predict for each of the following reactions?**
a. $(CH_3)_3CBr + CH_3OH \rightarrow (CH_3)_3COCH_3 + HBr$
b. $CH_3CH_2I + NaCN \rightarrow CH_3CH_2CN + NaI$
c. $\underset{\underset{Br}{|}}{CH_3CHCH_2CH_2CH_3} + Na^+SH^- \rightarrow \underset{\underset{SH}{|}}{CH_3CHCH_2CH_2CH_3} + NaBr$

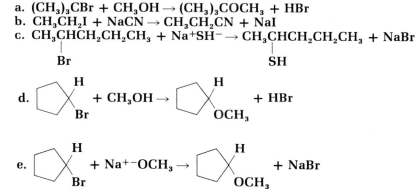

d.   (cyclopentane with H and Br) $+ CH_3OH \rightarrow$ (cyclopentane with H and OCH$_3$) $+ HBr$

e.   (cyclopentane with H and Br) $+ Na^{+-}OCH_3 \rightarrow$ (cyclopentane with H and OCH$_3$) $+ NaBr$

Problem 6.8   **Explain why the Williamson ether synthesis (Sec. 6.4), the formation of amines from alkyl halides (Sec. 6.5), and the general alkyne synthesis (Sec. 6.6) work well only for primary or secondary alkyl halides.**

**6.11
ELIMINATION
REACTIONS; THE
E2 AND E1
MECHANISMS**

When an alkyl halide with a hydrogen atom attached to the carbon *adjacent* to the halogen-bearing carbon reacts with a nucleophile, two competing reaction paths are possible, **substitution** or **elimination**.

$$\overset{H}{\underset{\diagup}{\underset{\textstyle C-C}{\overset{\textstyle 2\ \ 1}{}}}} \begin{matrix} \\ \diagdown \\ X \end{matrix} + Nu:$$

$\xrightarrow{\text{substitution}(S)} \quad \underset{|}{-}\overset{H}{\underset{|}{C}}-\overset{|}{\underset{|}{C}}-Nu + X^-$ (6.18)

$\xrightarrow{\text{elimination}(E)} \quad \begin{matrix} \diagdown & & \diagup \\ & C{=}C & \\ \diagup & & \diagdown \end{matrix} + Nu\,H + X^-$ (6.19)

In the substitution reaction, the nucleophile substitutes for or replaces the halogen X (eq. 6.18). In the elimination reaction (eq. 6.19), the halogen X and a hydrogen from an adjacent carbon atom are *eliminated* and a new bond (a π bond) is formed between the carbons that originally bore

X and H. The symbol E is used to designate this elimination process. Elimination reactions are standard ways of preparing compounds with double or triple bonds.

Often S and E reactions occur simultaneously with the same set of reactants—the nucleophile and the substrate. One reaction type or the other may predominate, depending on the structure of the nucleophile, the structure of the substrate, and other reaction conditions. As with substitution reactions, there are two main mechanisms for elimination reactions. They are designated E2 and E1. To learn how to control these reactions, we must understand their mechanisms.

The E2 mechanism, like the S_N2 mechanism, is a one-step process. *The nucleophile acts as a base* and removes the proton (hydrogen) on the carbon atom *adjacent* to the one that bears the leaving group. At the same time the leaving group departs and a double bond is formed. The flow of electron pairs is shown by the curved arrows:

$$\tag{6.20}$$

The preferred conformation for an E2 reaction is shown in eq. 6.20. The H—C—C—L atoms lie in a single plane, with H and L in a *transoid or anti* arrangement. The reason for this preference is that the C—H and C—L orbitals are properly aligned in this conformation to overlap and form the new π bond.

The E1 mechanism has the same first step as the S_N1 mechanism, the slow and rate-determining ionization of the substrate to give a carbocation (compare with eq. 6.15).

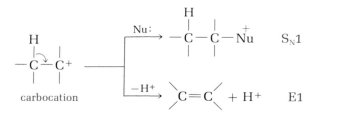

$$\tag{6.21}$$

Two reactions are then possible for the carbocation. It may either combine with a nucleophile (the S_N1 process) or lose a proton from a carbon atom adjacent to the positive carbon, as shown by the curved arrow, to give an alkene (the E1 process).

$$\tag{6.22}$$

Now let us see how substitution and elimination reactions compete with one another in some specific examples.

6.12
SUBSTITUTION
AND ELIMINATION
IN COMPETITION

Consider first the reactions of an alkyl halide with potassium hydroxide dissolved in methyl alcohol. The nucleophile is hydroxide ion, OH⁻, a strong nucleophile and a strong base. The solvent is moderately polar but less polar, for example, than water. These conditions favor S_N2 and E2 processes over S_N1 and E1.

Suppose the alkyl group of the alkyl halide is primary, as in 1-bromobutane. Both processes may occur:

$$HO^- + CH_3CH_2CH_2CH_2 \overset{\frown}{-} Br \xrightarrow{S_N2} CH_3CH_2CH_2CH_2OH + Br^- \quad \text{(6.23)}$$

•1-butanol

$$HO^- + CH_3CH_2CH \overset{\rightarrow H}{-} CH_2 \overset{\frown}{-} Br \xrightarrow{E2} CH_3CH_2CH = CH_2 + H_2O + Br^- \quad \text{(6.24)}$$

1-butene

The product will be a mixture of 1-butanol and 1-butene. The S_N2 reaction can be favored by using a more polar solvent (water), a modest concentration of base, and a moderate reaction temperature. The E2 reaction can be favored by using a less polar solvent, a higher base concentration, and a higher reaction temperature.

Consider now the effect of changing from a primary to a tertiary alkyl halide. The substitution reaction will be retarded (remember, the reactivity order for S_N2 reactions is $1° > 2° \gg 3°$). However, the elimination reaction will actually be favored because the product is a more substituted alkene. In fact, with *t*-butyl bromide, *only* the E2 process occurs.

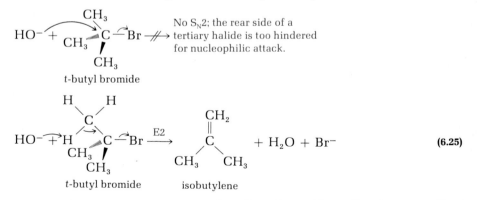

How, then, can we convert tertiary butyl bromide to the corresponding alcohol? We use water as the nucleophile instead of hydroxide ion. Water is a weaker base than hydroxide ion, so the E2 reaction (eq. 6.25) is suppressed. Water is also a polar solvent, favoring the two-step ionization mechanism. Can we avoid the elimination reaction altogether? No, because some E1 in competition with S_N1 is inescapable.

$$(CH_3)_3CBr \xrightarrow{H_2O} (CH_3)_3C^+ + Br^-$$

$$\xrightarrow{H_2O,\ S_N1} (CH_3)_3COH \quad \text{(about 80\%)}$$

$$\xrightarrow{E1} (CH_3)_2C{=}CH_2 + H^+ \quad \text{(about 20\%)}$$

(6.26)

t-butyl bromide

Most of the product (80%) is the substitution product, but some elimination (about 20%) still occurs.

To summarize, tertiary halides react with strong bases in nonpolar solvents to give only elimination (E2), not substitution. With weaker bases and weaker nucleophiles, and in more polar solvents, tertiary halides give mainly substitution (S_N1), but some elimination (E1) also occurs. Primary halides react only by the S_N2 and E2 mechanisms, because they do not ionize to carbocations. Secondary halides occupy an intermediate position, and the mechanism by which they react is particularly sensitive to the choice of reaction conditions. Secondary halides may react by S_N1 and S_N2 mechanisms simultaneously.

Example 6.6   **What product do you expect from the reaction of 1-chloro-1-methylcyclopentane with:**
a. sodium ethoxide in ethanol
b. refluxing ethanol

Solution   **The alkyl chloride is tertiary:**

a. **The first set of conditions favors the E2 process, because sodium ethoxide is a strong base. Two elimination products are possible:**

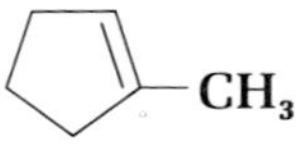

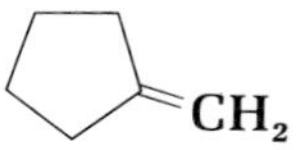

The former actually predominates.
b. **This set of conditions favors ionization, and the S_N1 process predominates. The main product is**

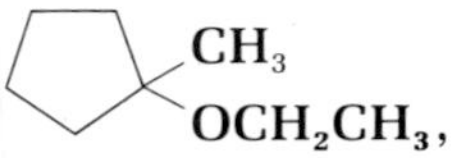

although some alkene will also be formed, by the E1 mechanism.

Problem 6.9 Reaction of *t*-butyl bromide with ammonia does not give any *t*-butylamine, $(CH_3)_3CNH_2$. What is the product and how is it formed?

Problem 6.10 Give *all* of the possible E2 products when 2-bromobutane reacts with concentrated sodium hydroxide.

Problem 6.11 When $CH_3CH_2CH_2CH_2Br$ reacts with sodium methoxide ($Na^+{}^-OCH_3$) in methanol, the product is mainly $CH_3CH_2CH_2CH_2OCH_3$, but when it reacts with potassium *t*-butoxide $[K^+{}^-OC(CH_3)_3]$ in tertiary butyl alcohol, the product is mainly $CH_3CH_2CH=CH_2$ and not $CH_3CH_2CH_2CH_2OC(CH_3)_3$. Explain.

a word about ──────────────────────────────

8. S$_N$2 Reactions in the Cell: Biochemical Methylations

Substitution and elimination reactions are so useful that it is not surprising to find them employed in living matter. However, alkyl halides are not compatible with cytoplasm, being hydrocarbon-like and therefore water-insoluble. In the cell, <u>alkyl phosphates play the same role as alkyl halides in the laboratory.</u> Adenosine triphosphate (ATP) is an example of a biological equivalent of an alkyl halide. We will abbreviate its structure here as Ad-O-Ⓟ-Ⓟ-Ⓟ (the full structure is given in Sec. 16.13). Ad— can be considered a primary alkyl group, and -O-Ⓟ-Ⓟ-Ⓟ is the triphosphate group. It acts as a leaving group, just like the halogens.

There are many compounds in nature with a methyl group attached to an oxygen or nitrogen atom. Examples include *mescaline* (a hallucinogen from the peyote cactus), which has three —OCH$_3$ groups; *morphine* (the pain-relieving drug from opium), which has an $>$N—CH$_3$ group; and codeine (a close relative of morphine, used as an anticough agent), which has an $>$N—CH$_3$ group and an —OCH$_3$ group.

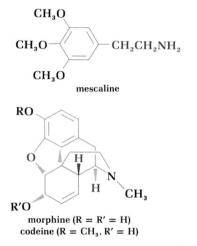

mescaline

morphine (R = R' = H)
codeine (R = CH$_3$, R' = H)

How do the methyl groups get there? Two steps are involved, and both are nucleophilic substitutions.

The methyl carrier is a sulfur-containing amino acid called *methionine*. In the first step of biochemical methylations, methionine is alkylated by ATP to form *S*-adenosylmethionine (shown in Figure 6.1). This reaction is just a biological example of reaction 13 in Table 6.1. The methionine acts as a sulfur nucleophile in an S$_N$2 reaction and displaces the triphosphate ion.

FIGURE 6.1 The formation of *S*-adenosylmethionine

In the second step, the oxygen or nitrogen atom to be methylated acts as a nucleophile and becomes methylated.

$$-\overset{..}{\underset{..}{O}}H \text{ or } -\overset{..}{N}H_2 + CH_3 \overset{Ad}{-}S^+ - CH_2CH_2\underset{NH_2}{CH}CO_2H \xrightarrow{S_N2}$$

$$-OCH_3 \text{ or } -NHCH_3 + AdSCH_2CH_2\underset{NH_2}{CH}CO_2H$$

S-adenosylhomocysteine

The S-adenosylmethionine is acting just like a methyl halide. Eventually the S-adenosylhomocysteine formed in the second step is converted, by enzymes, back to ATP and methionine for re-use.

Biochemical methylations are just one of many examples of nucleophilic substitutions that occur in various metabolic processes.

6.13 POLY-HALOGENATED ALIPHATIC COMPOUNDS

Because of their special properties, several polyhalogen compounds are produced commercially. A few of the more important compounds of this type are described here.

Chlorinated methanes are made by several routes, the most direct being the chlorination of methane (eq. 2.13). **Carbon tetrachloride** (CCl_4, bp 77°C) is a colorless liquid with a mild, somewhat pleasant odor. It is insoluble in water, but it is a good solvent for oils and greases and has been used in dry cleaning of clothes. Because of its high density and nonflammability, carbon tetrachloride has been used as a fire extinguisher, although in recent years it has largely been replaced in this use by the more effective brominated compounds $CBrClF_2$ and $CBrF_3$. **Chloroform** ($CHCl_3$, bp 62°C) and **methylene chloride** (CH_2Cl_2, bp 40°C) are both widely used as solvents for organic substances. Both carbon tetrachloride and chloroform are suspected carcinogens, and adequate ventilation is essential when they are used as solvents. Chloroform was once used in surgery as a general anesthetic, but it is too toxic and can cause severe liver damage. Since 1956, the most general anesthetic of this type used world-wide is **halothane**, $CF_3CHClBr$.

Tri- and **tetrachloroethylenes** are important solvents for dry cleaning and as degreasing agents in metal and textile processing. They are made from 1,2-dichloroethane via a combination of chlorination, dehydrochlorination, and oxidation.

$$ClCH_2CH_2Cl + Cl_2 + O_2 \xrightarrow[CuCl_2]{420-450°C} Cl_2C{=}CHCl + Cl_2C{=}CCl_2 + H_2O \quad \textbf{(6.27)}$$

trichloroethylene   tetrachloroethylene
bp 87°C   bp 121°C

Perhaps the most important polyhalogenated methanes and ethanes commercially are the **chlorofluorocarbons.** The term **Freon,** which is the DuPont company's registered trademark for its chlorofluorocarbon products, is now widely used to describe this class of compounds. Freons are colorless gases or low-boiling liquids that are nontoxic, nonflammable, and relatively noncorrosive. They are used as industrial and domestic refrigerants in air-conditioning and deepfreeze units. Freons are produced commercially from chlorocarbons by fluorination.

$$CCl_4 \xrightarrow[SbF_5]{HF} CCl_3F \xrightarrow[SbF_5]{HF} CCl_2F_2 \xrightarrow[SbF_5]{HF} CClF_3 \qquad (6.28)$$

CCl_4	CCl_3F	CCl_2F_2	$CClF_3$
bp 76.5°C	trichlorofluoro-methane (Freon-11) bp 23.7°C	dichlorodifluoro-methane (Freon-12) bp −29.8°C	chlorotrifluoro-methane (Freon-13) bp −81.1°C

$$CHCl_3 \xrightarrow[SbF_5]{HF} CHCl_2F \xrightarrow[SbF_5]{HF} CHClF_2 \qquad (6.29)$$

$CHCl_3$	$CHCl_2F$	$CHClF_2$
bp 61.7°C	(Freon-21) bp 9°C	(Freon-22) bp −40.8°C

The production of chlorofluorocarbons soared during the early 1970s due to their use as aerosol propellants. The use was first developed in the early 1940s, when U.S. troops fighting in the Pacific war zone were suffering more casualties from tropical insect bites (leading to malaria) than from enemy action. It was found that the most effective and convenient way of dispensing insecticides such as DDT was to dissolve them in a little oil and mix this solution with CCl_2F_2 in a self-pressurized container. When the solution was discharged into the atmosphere, the CCl_2F_2 immediately evaporated, breaking up the insecticide droplets into an extremely fine mist (or aerosol) that was much longer lasting and more effective at insect control than previous conventional sprays.

Eventually Freons came to be used as aerosol propellants not only for insecticides but also in hair sprays, deodorants, polishes, paints, shaving foams, oven cleaners, and so on. Excessive use led to concern that these inert materials might accumulate and eventually pollute the atmosphere. In particular, they can diffuse upward and have a damaging effect on the earth's ozone layer, a section of the stratosphere that screens out much of the dangerous ultraviolet radiation coming from the sun. This could result in climate modification, crop damage, and (possibly) additional cases of skin cancer. In the late 1970s, the manufacture and use of Freons for non-essential propellant uses was banned in the United States and they have been replaced with other materials.

The case of chlorofluorocarbons is one of many examples of the tradeoffs that have been made between the beneficial and the harmful effects of new research products. Chlorofluorocarbons and DDT can be credited with having saved many lives during World War II, but their indiscriminate and careless use can have serious undesirable effects on our environment.*

Chlorodifluoromethane (Freon-22, $CHClF_2$) is converted at high temperatures to **tetrafluoroethylene**, the raw material for the fluorinated polymer **Teflon**.

---

*Water can save your life if you are dying of thirst on a desert; it can take your life if you fall overboard at sea and cannot swim. Chemicals themselves are neither good nor evil, but we must be careful how we use them.

$$2\ CHClF_2 \xrightarrow{600-800°C} CF_2{=}CF_2 + 2\ HCl \qquad (6.30)$$
$$\text{tetrafluoroethylene}$$

$$nCF_2{=}CF_2 \xrightarrow[\text{catalyst}]{\text{peroxide}} {+}CF_2CF_2{\,\mathord{+}}_n \qquad (6.31)$$
$$\text{Teflon}$$

This polymer is resistant to almost all chemicals and is now widely used as a nonstick coating for pots, pans, and other cooking utensils. These coatings stand up to heat and prevent food from sticking to the utensil surface, making it easier to clean. Teflon film lenses for high-intensity lamps provide more illumination and easier maintenance than glass lenses and are used in modern industrial plants and sports arenas.

a word about

9. Artificial Blood

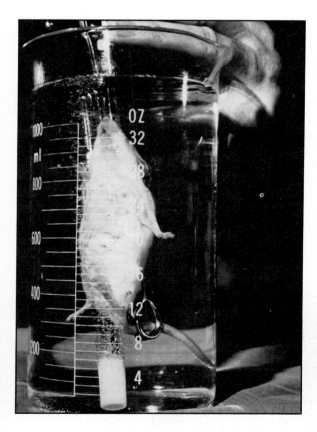

Fantastic as it may seem, it has now been proved that animals can survive after having most or all of

their natural blood replaced by a suitable emulsion of perfluorochemicals—hydrocarbons, ethers, or amines in which all of the hydrogens are replaced by fluorine atoms! This discovery, which dates back to the mid-1960s, holds great promise for medical use and has already saved human lives. Here is a brief account of this fascinating discovery.

Perfluorochemicals can dissolve as much as 60% of oxygen by volume. By contrast, whole blood dissolves only about 20% and blood plasma about 3%. Rats totally submerged in fluorochemicals saturated with oxygen continue to "breath" for long periods. Although perfluorochemicals are immiscible with blood, it was found that they could be emulsified (broken up into microdroplets that remain suspended in solution) by adding various surfactants (compounds related to soaps) and other materials. Here is a typical example of an artificial blood recipe: perfluorotributylamine, $(CF_3CF_2CF_2CF_2)_3N$, 11–13 mL; Pluronic F-68 (a polymeric emulsifier), 2.3–2.7g; hydroxyethyl starch, 2.5–3.2g; NaCl, 54 mg; KCl, 32 mg; $MgCl_2$, 7 mg; $CaCl_2$, 10 mg; Na_2HPO_4, 9.6 mg; enough Na_2CO_3 to adjust the pH to 7.44; enough water to dilute to 100 mL of artificial blood. Other perfluorochemicals, such as F-decalin $(C_{10}F_{18})$, can be used. Several different techniques are employed to convert these mixtures into emulsions whose particle size must be carefully regulated if the blood substitute is to be effective.

Unlike ordinary blood, artificial blood is unaffected by certain poisons. "Bloodless" rats have breathed oxygen containing 10% to 50% carbon monoxide for hours and continued to live normally on return to an ordinary nitrogen–oxygen atmosphere. Of course, the artificial blood is gradually replaced by normal blood again as the animal continues to live and produce blood in the usual way. This experiment proved, however, that the artificial blood is fully capable of carrying out the oxygen-transport function of blood, because any red blood cells that might have been present would have been prevented by the carbon monoxide from transporting oxygen.

The perfluorochemicals that function temporarily as artificial blood are eventually excreted over time (days to weeks). Because of their chemical inertness, they apparently do no permanent harm.

What uses are foreseen for artificial blood? It can be used in emergencies and disasters when blood banks are depleted and natural blood is in short supply. No typing is required, so it can be used even with patients who have a rare blood type. Artificial blood can also be used to preserve organs for transplant surgery (kidneys, hearts), to treat certain diseases (sickle-cell anemia, cancer-associated anemia), and to administer drugs that would otherwise react with ordinary blood. Complete body washout of ordinary blood could be used to rid the body of viruses, toxins, or drug overdoses. In Japan doctors have used artificial blood on patients threatened with death due to excessive bleeding, and artificial blood has been used to save the life of a patient who refused a conventional transfusion for religious reasons.

When fluorochemicals were first prepared they were chemical curiosities. No one could have foreseen their properties or uses. They provide an outstanding example of why research in basic science should be financed without regard to any immediate practical benefit.

---

a word about

10. Insecticides and Herbicides

Many polyhalogen compounds have been used as insecticides and herbicides. Perhaps the best-known insecticide is DDT (dichlorodiphenyltrichloroethane). It is manufactured by the acid-catalyzed reaction of chlorobenzene with trichloroacetaldehyde.

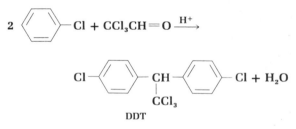

DDT was used during World War II to control malaria by killing mosquitoes. It is also effective against flies and many agricultural pests. Unfortunately, DDT is not easily degraded biochemically, and excessive use has resulted in its accumulation in the environment. It tends to accumulate in fat tissues and can cause harm, particularly to fish and birds. Currently the use of DDT is permitted but restricted, and annual world production is about 80,000 tons.

The problems that weeds create for agriculture are formidable. They consume nutrients and moisture needed by crops. They also rob crops of sunlight and space, thus reducing crop yields. U.S. agricultural production is diminished by about 10% because of weeds. The annual financial loss is about $12 billion and an additional $6 billion is spent annually on weed control.

One way to control weeds is through the use of herbicides. About 85–90% of U.S. acreage for corn, soybeans, cotton, peanuts, and rice is sprayed with herbicides to control weeds. Some herbicides are sprayed before the crop is planted, others are applied after planting but before emergence of the crop, and still others are applied to the weed foliage itself. In 1981 over 600 million pounds of active ingredients in herbicides were used in the United States, and the figure grows annually. Our rising world population constantly makes increasing demands on food production and, ultimately, on herbicide use to meet that demand for food. Although most herbicides are used for this purpose some are also employed for industrial purposes—along railway and powerline rights-of-way, on roadsides, rangeland, vacant lots, and so on—as well as for home lawns and gardens.

Farmers have used weed killers for many years; common salt was used for this purpose even in ancient times. Before World War II most chemical herbicides were not very selective (they killed weeds but also damaged crops) and they had to be used in large quantities per acre of land. The breakthrough came with the discovery that 2,4-dichlorophenoxyacetic acid (2,4-D) killed broadleaf weeds but allowed narrowleaf plants to grow unharmed and in greater yield. In addition, only 0.25–2.0 lb were needed per acre (compared to over 200 lb per acre for inorganic herbicides such as sodium chlorate). 2,4-D is still the most popular herbicide used on wheat fields.

A few years later 2,4,5-T came on the market. It is superior to 2,4-D for combatting brush and weeds in forests. It was used on pastures and grazing lands, as well as in rice, wheat, and sugarcane fields. However, 2,4,5-T incurred severe criticism when the U.S. military began using agent orange (so named after the color of its storage drums), a 50 : 50 mixture of esters of 2,4,5-T and 2,4-D, to defoliate the jungles of Vietnam. Although the matter is under study, it seems that 2,4,5-T can contain trace amounts of dioxin, a by-product in its manufacture. Dioxin is one of the most powerful poisons known. When the manufacture of 2,4,5-T is carefully controlled, the dioxin content is less than 1 part per million, but still there is some risk. At present, use of 2,4,5-T is severely restricted.

At least 40 herbicides are in large-scale use. Some of the most widely used, all of which contain halogens, are

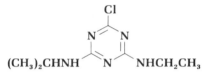

atrazine
(corn, sugarcane, pineapples)

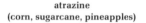

fluometuron
(cotton, sugarcane)

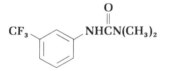

trifluralin
(cotton, lima beans, cantaloupes,
tomatoes, sugar beets)

A new herbicide recently developed at the DuPont Company seems to represent another breakthrough. The compound is a chlorine-containing urea derivative:

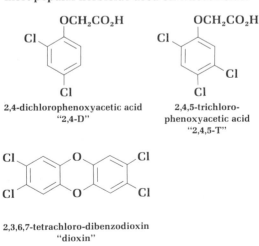

2,4-dichlorophenoxyacetic acid
"2,4-D"

2,4,5-trichloro-
phenoxyacetic acid
"2,4,5-T"

2,3,6,7-tetrachloro-dibenzodioxin
"dioxin"

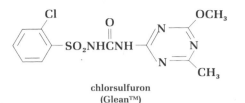

chlorsulfuron
(Glean™)

It is effective against a broad spectrum of weeds in cereal grains (wheat, barley, oats) at the remarkably low application level of less than one ounce per acre.

The achievements of modern agriculture and the feeding of our ever-growing world population would not be possible without the herbicides that scientists have developed.

ADDITIONAL PROBLEMS

6.12. Using Table 6.1 as a guide, write an equation for each of the following substitution reactions.

a. 1-bromobutane + sodium iodide (in acetone)
b. 2-chlorobutane + sodium ethoxide
c. *t*-butyl bromide + water
d. *p*-chlorobenzyl chloride + sodium cyanide
e. *n*-propyl iodide + sodium acetylide
f. 2-chloropropane + sodium hydrosulfide
g. allyl chloride + ammonia (2 equivalents)
h. 1,4-dibromobutane + sodium cyanide
i. 1-methyl-1-bromocyclohexane + methanol

6.13. Select an alkyl halide and a nucleophile that would give each of the following products.

a. $CH_3CH_2CH_2NH_2$ **b.** $CH_3CH_2SCH_2CH_3$
c. $HC{\equiv}CCH_2CH_2CH_3$ **d.** $(CH_3)_2CHOCH(CH_3)_2$

e. NCCH$_2$—⟨benzene ring⟩—CH$_2$CN   **f.** ⟨benzene ring⟩—OCH$_2$CH$_3$

6.14. Give equations for two different combinations of reagents that might be used to synthesize methyl *sec*-butyl ether by the Williamson method. Which combination would be preferred?

6.15. Using the general equations in Sec. 6.5 as a guide, write out each step in the reaction of ethyl bromide with ammonia to form successively ethylamine, diethylamine, triethylamine, and (eventually) tetraethylammonium bromide.

6.16. Draw out each of the following equations in a way that shows clearly the stereochemistry of the reactants and products.

a. (S)-2-bromobutane + sodium methoxide (in methanol) $\xrightarrow{S_N2}$ 2-methoxybutane

b. (R)-3-bromo-3-methylhexane + methanol $\xrightarrow{S_N1}$ 3-methoxy-3-methylhexane
c. *cis*-2-bromo-1-methylcyclopentane + NaSH → 2-methylcyclopentanethiol

6.17. Determine the order of reactivity when $(CH_3)_2CHCH_2Br$, $(CH_3)_3CBr$, and $CH_3\underset{\underset{Br}{|}}{CH}CH_2CH_3$ react with:

a. sodium cyanide   **b.** 50% aqueous acetone

6.18. When treated with sodium iodide, a solution of (R)-2-iodooctane in acetone gradually loses all its optical activity. Explain.

6.19. Eq. 6.26 shows that hydrolysis of t-butyl bromide gives about 80% $(CH_3)_3COH$ and 20% $(CH_3)_2C{=}CH_2$. The same ratio of alcohol to alkene is obtained when the starting halide is t-butyl chloride or t-butyl iodide. Explain.

6.20. Tell what product you expect, and by what mechanism it is formed, for each of the following reactions.
a. 1-chloro-1-methylcyclopentane + ethyl alcohol
b. 1-chloro-1-methylcyclopentane + sodium ethoxide (in ethanol)

6.21. Give the structures of all possible products when 2-chloro-2-methylbutane reacts by the E1 mechanism.

6.22. Explain the difference between the products in the following two equations by considering the mechanism by which each reaction proceeds.

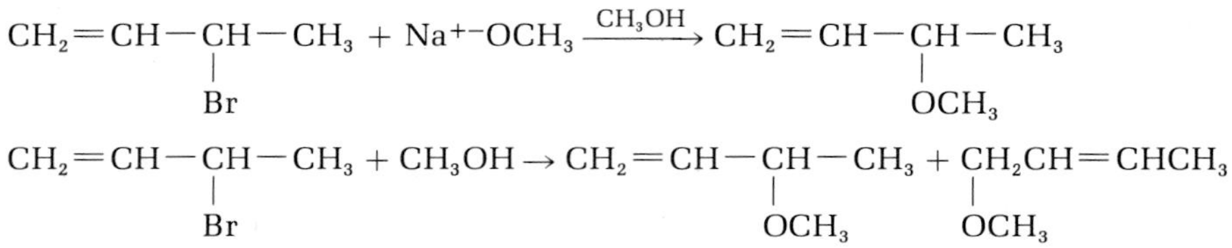

$$CH_2{=}CH{-}\underset{\underset{Br}{|}}{CH}{-}CH_3 + Na^+{}^-OCH_3 \xrightarrow{CH_3OH} CH_2{=}CH{-}\underset{\underset{OCH_3}{|}}{CH}{-}CH_3$$

$$CH_2{=}CH{-}\underset{\underset{Br}{|}}{CH}{-}CH_3 + CH_3OH \rightarrow CH_2{=}CH{-}\underset{\underset{OCH_3}{|}}{CH}{-}CH_3 + \underset{\underset{OCH_3}{|}}{CH_2}CH{=}CHCH_3$$

6.23. Menthyl chloride (derived from the natural product menthol) and neomenthyl chloride differ only in the stereochemistry of the C$-$Cl bond.

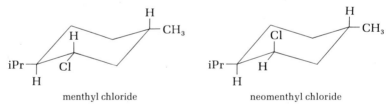

menthyl chloride   neomenthyl chloride

When each is treated with a strong base ($Na^+{}^-OC_2H_5$ in C_2H_5OH), menthyl chloride gives 100% 2-menthene, whereas neomenthyl chloride gives 75% 3-menthene and only 25% 2-menthene.

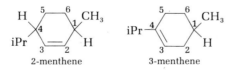

2-menthene   3-menthene

Show how these results are consistent with the transoid coplanar geometry for the transition state in E2 eliminations.

6.24. Write out the steps by which Teflon is formed from tetrafluoroethylene (eq. 6.31).

seven

Alcohols, Phenols, and Thiols

7.1
INTRODUCTION

Alcohols have the general formula **R—OH.** They are structurally similar to water, but they have one of the hydrogens replaced by an alkyl group. The functional group of alcohols is the **hydroxyl group,** —OH. **Phenols** have the same functional group as alcohols, but the hydroxyl group is directly attached to an aromatic ring. **Thiols** have structures similar to alcohols and phenols, but the oxygen is replaced by sulfur.

$$H—\overset{..}{\underset{..}{O}}—H \qquad R—\overset{..}{\underset{..}{O}}—H \qquad Ar—\overset{..}{\underset{..}{O}}—H \qquad R—\overset{..}{\underset{..}{S}}—H \qquad Ar—\overset{..}{\underset{..}{S}}—H$$

water   an alcohol   a phenol   a thiol   a thiophenol

Alcohols, phenols, and thiols all occur commonly in nature. In this chapter we will discuss the physical properties and main chemical reactions of each of these classes of compounds. We will also describe their industrial and laboratory syntheses and give examples of their importance in biology.

7.2
NOMENCLATURE OF ALCOHOLS

Common names for alcohols are derived by naming the alkyl group attached to the —OH and then adding the word *alcohol.* In the IUPAC system, the *-ol* ending indicates the presence of a hydroxyl group. The following examples illustrate the use of IUPAC rules. (Common names are shown in parentheses.)

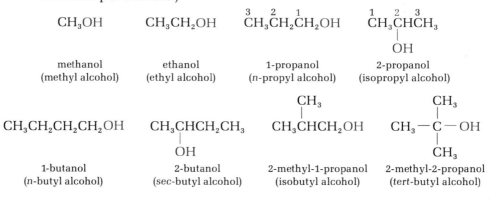

$$CH_3OH \qquad CH_3CH_2OH \qquad \overset{3}{C}H_3\overset{2}{C}H_2\overset{1}{C}H_2OH \qquad \overset{1}{C}H_3\overset{2}{C}HCH_3$$

methanol   ethanol   1-propanol   2-propanol

(methyl alcohol)   (ethyl alcohol)   (n-propyl alcohol)   (isopropyl alcohol)

$$CH_3CH_2CH_2CH_2OH \qquad CH_3CHCH_2CH_3 \qquad CH_3CHCH_2OH \qquad CH_3—C—OH$$

1-butanol   2-butanol   2-methyl-1-propanol   2-methyl-2-propanol

(n-butyl alcohol)   (sec-butyl alcohol)   (isobutyl alcohol)   (tert-butyl alcohol)

Example 7.1   Name the following alcohols by the IUPAC system.
a. $ClCH_2CH_2OH$ b.   OH   c. $CH_2{=}CH{-}CH_2OH$

Solution   a. 2-chloroethanol (Number from the carbon that bears the hydroxyl group.)
b. cyclobutanol
c. 2-propen-1-ol (Number from the alcohol carbon; two suffixes are used (-en for the double bond and -ol for the hydroxyl group) and a number must be used to locate each). This compound also has a common name, allyl alcohol.

Problem 7.1   Name these alcohols by the IUPAC system.
a. $BrCH_2CH_2CH_2OH$ b. H OH   c. $CH_2{=}CHCH_2CH_2OH$

Problem 7.2   Write structural formulas for:
a. 2-pentanol   b. 1-phenylethanol
c. cyclopentylmethanol   d. 3-penten-2-ol

7.3
CLASSIFICATION
OF ALCOHOLS

Alcohols are classified as primary (1°), secondary (2°), or tertiary (3°) depending on whether one, two, or three organic groups are connected to the hydroxyl-bearing carbon atom.

$$R{-}CH_2OH \qquad R{-}\overset{\displaystyle R}{\underset{}{C}}HOH \qquad R{-}\overset{\displaystyle R}{\underset{\displaystyle R}{C}}{-}OH$$

primary (1°)   secondary (2°)   tertiary (3°)

Methyl alcohol, which does not strictly fit into this classification, is usually considered a primary alcohol. Note that this classification is similar to that of carbocations (Sec. 3.14). We will see that the chemistry of an alcohol sometimes depends on its class.

Problem 7.3   Classify the first eight alcohols (shown in Sec. 7.2) as 1°, 2°, or 3°.

7.4
NOMENCLATURE
OF PHENOLS

Phenols are usually named as derivatives of the parent compounds (review Sec. 4.7).

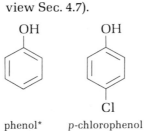

phenol*   p-chlorophenol   2,4,6-tribromophenol

*The name *benzenol* has recently been introduced for phenol and its derivatives. Although this name is used by Chemical Abstracts Service, it is not yet in common use among organic chemists.

Because they were discovered rather early in the history of organic chemistry, many phenols have common names. The methylphenols, for example, are known as **cresols** (the name comes from *creosote*, a coal or wood tar that contains these compounds).

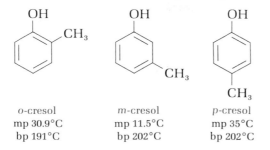

o-cresol	m-cresol	p-cresol
mp 30.9°C	mp 11.5°C	mp 35°C
bp 191°C	bp 202°C	bp 202°C

Problem 7.4 **Write the structures for:**
a. *p*-ethylphenol **b. pentachlorophenol (an insecticide for termite control, also used as a fungicide)**

7.5 HYDROGEN BONDING IN ALCOHOLS AND PHENOLS

The boiling points of alcohols are abnormally high relative to those of ethers or hydrocarbons with similar molecular weights.

	CH_3CH_2OH	CH_3OCH_3	$CH_3CH_2CH_3$
mol wt	46	46	44
bp	+78.5°C	−24°C	−42°C

The reason for the relatively high boiling points of alcohols is that the molecules form **hydrogen bonds** with one another. The $O-H$ bond is highly polarized due to the high electronegativity of the oxygen atom (review Sec. 1.5). Because of the high partial positive charge on the hydrogen atom and its small size, it can link together two electronegative oxygen atoms, as shown in the following equation:

$$\underset{\text{two separate alcohol molecules}}{\overset{R}{\underset{}{\searrow}}\ \underset{\delta-}{O}\!-\!\underset{\delta+}{H} \ + \ \overset{R}{\underset{}{\searrow}}\ \underset{\delta-}{O}\!-\!\underset{\delta+}{H}} \ \rightleftharpoons \ \underset{\text{a hydrogen bond}}{\overset{R}{\underset{}{\searrow}}\ \underset{\delta-}{O}\!-\!\underset{\delta+}{H}\cdots\overset{R}{\underset{}{\searrow}}\underset{\delta-}{O}\!-\!\underset{\delta+}{H}} \qquad (7.1)$$

Two or more alcohol molecules thus become associated with one another through hydrogen bonds.

The strength of a hydrogen bond is much less than that of a covalent bond—perhaps only 10%. It is about 5–10 kcal/mol (20–40 kJ/mol). Nevertheless, its strength is significant. Consequently, the reason why alcohols and phenols have relatively high boiling points is that we must not only supply enough heat to vaporize each molecule, but we must also supply enough heat to dissociate the hydrogen bonds.

Water, of course, is also a hydrogen-bonded liquid. The lower-molecular-weight alcohols can readily replace water molecules in the hydrogen-bonded network.

TABLE 7.1	Name	Formula	bp, °C	Solubility in H_2O, g/100 g at 20°C
Boiling points and water solubility of some alcohols	methanol	CH_3OH	65	completely miscible
	ethanol	CH_3CH_2OH	78.5	completely miscible
	1-propanol	$CH_3CH_2CH_2OH$	97	completely miscible
	1-butanol	$CH_3CH_2CH_2CH_2OH$	117.7	7.9
	1-pentanol	$CH_3CH_2CH_2CH_2CH_2OH$	137.9	2.7
	1-hexanol	$CH_3CH_2CH_2CH_2CH_2CH_2OH$	155.8	0.59

This accounts for the complete miscibility of the lower alcohols with water. However, as the organic chain lengthens and the alcohol becomes relatively more hydrocarbon-like, the solubility in water decreases. Table 7.1 illustrates these properties.

**7.6
THE ACIDITY OF
ALCOHOLS AND
PHENOLS**

Like water, alcohols and phenols are weak acids. Alcohols are about 10–100 times *weaker* acids than water. (The ionization constant of water, for the equilibrium $H_2O \rightleftharpoons H^+ + OH^-$, is 10^{-14}. For most alcohols, the corresponding ionization constant for $ROH \rightleftharpoons H^+ + {}^-OR$ is about 10^{-15} to 10^{-16}.)

Alcohols react with metals such as sodium or potassium to liberate hydrogen and form **alkoxides.**

$$2\,RO{-}H + 2\,Na \rightarrow 2\,RO^-Na^+ + H_2 \tag{7.2}$$

alcohol   a sodium alkoxide

Example 7.2 Write the equation for the reaction of methanol with sodium metal, and name the product.

Solution $2\,CH_3OH + 2\,Na \rightarrow 2\,CH_3O^-Na^+ + H_2$

The product is sodium methoxide.

Problem 7.5 Write the equation for the reaction of *t*-butyl alcohol with potassium metal, and name the product.

Metal alkoxides dissolved in the corresponding alcohol are strong bases, just as sodium hydroxide dissolved in water is a strong base. Indeed, alkoxides are somewhat stronger bases than the corresponding hydroxides (because alcohols are weaker acids than water). For this reason, when an alkoxide is added to water, it hydrolyzes to the alcohol.

$$RO^-Na^+ + H{-}OH \rightarrow ROH + Na^+OH^-$$ (7.3)

Consequently, one cannot ordinarily convert an alcohol to its alkoxide by treating it with sodium hydroxide.

In contrast to alcohols, phenols are *stronger* acids than water. Phenol itself is 10,000 times more acidic than water. (Indeed, an old name for phenol is **carbolic acid** and, in dilute solution, it is an effective antiseptic and disinfectant).

The principal reason why phenols are stronger acids than water or alcohols is because phenoxide ions are stabilized by resonance. Whereas the negative charge in a hydroxide or alkoxide is fixed on the oxygen atom, the negative charge on a phenoxide ion can be delocalized to the *ortho* and *para* positions of the benzene ring, through resonance.

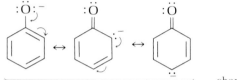

charge delocalized in phenoxide ion

charge localized
on the oxygen atom
in alkoxide ions

Because they are resonance-stabilized, phenoxide ions are more easily formed from phenols than are alkoxide ions from alcohols. Hence the greater acidity of phenols than alcohols. The counterpart of this argument is, of course, that phenoxide ions are weaker bases than hydroxide or alkoxide ions.

Because of their acidity, phenols can be easily converted to phenoxides by treatment with aqueous base.

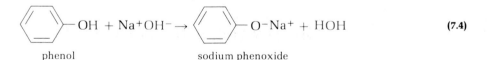

(7.4)

phenol   sodium phenoxide

Problem 7.6   **Write an equation for the reaction, if any, between:**
a. *p*-cresol and aqueous potassium hydroxide
b. cyclohexanol and aqueous sodium hydroxide

We can make good use of these acidity differences to separate alcohols from phenols. For example, suppose we had a mixture of 1-octanol (bp 194°C) and o-cresol (bp 191°C). Because their boiling points are so similar, separation by distillation would be difficult. But we could treat the mixture with aqueous sodium hydroxide.

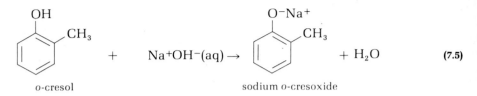

$$CH_3(CH_2)_6CH_2OH + Na^+OH^-(aq) \rightarrow \text{no reaction}$$
1-octanol

(7.6)

The o-cresol would react to form sodium o-cresoxide, which, being ionic, would dissolve in the water layer (eq. 7.5). The 1-octanol is too weak an acid and would not react with the aqueous base (eq. 7.6). Once the organic and aqueous phases were separated, the o-cresol could be recovered from the aqueous phase by acidification with a strong acid such as hydrochloric acid.

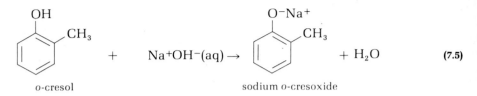

(7.7)

sodium o-cresoxide   o-cresol

Thus, by using the difference in their acidities, we can easily separate alcohols from phenols.

7.7
THE BASICITY OF ALCOHOLS AND PHENOLS

Water is not only a weak acid but also a weak base. The unshared electron pairs on the oxygen atom can accept a proton and form an oxonium ion.

$$H-\overset{..}{\underset{..}{O}}-H + H^+ \rightleftharpoons \left[H-\overset{\overset{H}{|}}{\underset{..}{O}}-H \right]^+$$
oxonium ion

(7.8)

In a similar manner, alcohols (and phenols) can react with strong acids to form substituted oxonium ions.

$$R-\overset{..}{\underset{..}{O}}-H + H^+ \rightleftharpoons \left[R-\overset{\overset{H}{|}}{\underset{..}{O}}-H \right]^+$$
alcohol acting   alkyloxonium ion
as a base

(7.9)

This reaction is the first step in two important reactions of alcohols that we shall discuss in the following two sections: their dehydration to alkenes and their conversion to alkyl halides.

7.8
DEHYDRATION OF ALCOHOLS TO ALKENES

Alcohols can be dehydrated by heating them with a strong acid. For example, when ethanol is heated at 180°C with a small amount of concentrated sulfuric acid, a good yield of ethylene is obtained.

$$H-CH_2CH_2-OH \xrightarrow{\text{H}^+, 180°C} CH_2{=}CH_2 + H-OH$$
ethanol   ethylene

(7.10)

The reaction, which can be used to synthesize alkenes, is the reverse of the hydration of alkenes (Sec. 3.10). It is an elimination reaction and can occur by either an E1 or an E2 mechanism, depending on the class of alcohol ($3°, 2°$ or $1°$; review Sec. 6.11).

Tertiary alcohols dehydrate by the E1 mechanism, via a relatively easily formed tertiary carbocation. Let us consider t-butyl alcohol as a typical example. The first step involves reversible protonation of the hydroxyl group.

$$(CH_3)_3C-\overset{..}{\underset{..}{O}}H + H^+ \rightleftharpoons (CH_3)_3C-\overset{+}{\underset{|}{\overset{..}{O}}}-H \qquad (7.11)$$
$$\underset{H}{}$$

Ionization, with water as the leaving group, gives the t-butyl cation.

$$(CH_3)_3C\overset{+}{\underset{|}{\overset{..}{O}}}-H \rightleftharpoons (CH_3)_3C^+ + H_2O \qquad (7.12)$$
$$\underset{H}{}$$

$$t\text{-butyl cation}$$

Proton loss from a carbon atom adjacent to the positive carbon completes the reaction.

$$\underset{\underset{CH_3}{|}}{\overset{\overset{H}{|}\quad\overset{CH_3}{|}}{CH_2-C^+}} \rightarrow CH_2=C\overset{CH_3}{\underset{CH_3}{}} + H^+ \qquad (7.13)$$

The overall dehydration reaction is the sum of all three steps.

$$\underset{\underset{CH_3}{|}}{\overset{\overset{H}{|}\quad\overset{CH_3}{|}}{CH_2-C-OH}} \xrightarrow[\text{heat}]{H^+} CH_2=C\overset{CH_3}{\underset{CH_3}{}} + H-OH \qquad (7.14)$$

$$t\text{-butyl alcohol} \qquad\qquad \text{isobutylene}$$
$$\text{(2-methylpropene)}$$

In the case of a primary alcohol, a relatively unstable, primary carbocation intermediate is avoided by combining the last two steps of the mechanism. The loss of water and the adjacent proton occur simultaneously. Thus the two steps in the E2 mechanism for dehydration of ethanol (eq. 7.10) are

$$CH_3CH_2\overset{..}{\underset{..}{O}}H + H^+ \rightleftharpoons CH_3CH_2-\overset{+}{\underset{|}{\overset{..}{O}}}-H \qquad (7.15)$$
$$\underset{H}{}$$

$$\overset{\overset{H}{|}}{CH_2}-CH_2\overset{+}{\underset{|}{\overset{..}{O}}}-H \rightarrow CH_2=CH_2 + H^+ + H_2O \qquad (7.16)$$
$$\underset{H}{}$$

The important things to remember about alcohol dehydrations are (1) they all begin by protonation of the hydroxyl group (that is, the alcohol acts as a base, as in eq. 7.9) and (2) the ease of alcohol dehydration is $3° > 2° > 1°$ (that is, the rates follow the order of carbocation stability).

Sometimes a single alcohol may give two or more alkenes, because the proton lost during dehydration can come from any carbon atom adjacent to the hydroxyl-bearing carbon. For example, 2-methyl-2-butanol can give two alkenes:

$$
\underset{\text{2-methyl-2-butanol}}{\overset{\overset{\displaystyle H \quad OH \ H}{\displaystyle | \qquad | \quad |}}{CH_2-C-CH-CH_3}} \underset{CH_3}{\overset{}{|}} \xrightarrow[\substack{\text{heat} \\ -H_2O}]{H^+} \underset{\text{2-methyl-1-butene}}{\overset{}{CH_2=C-CH_2CH_3}}\underset{CH_3}{\overset{}{|}} \ \text{and/or} \ \underset{\text{2-methyl-2-butene}}{\overset{}{CH_3-C=CHCH_3}}\underset{CH_3}{\overset{}{|}} \qquad (7.17)
$$

In these cases, the alkene with the most substituted double bond usually predominates. In eq. 7.17, the major product is 2-methyl-2-butene.

Problem 7.7 **Write the structures for all the possible dehydration products of:**

a. 3-methyl-3-hexanol   **b.** [structure: cyclopentane ring with OH and CH_3 substituents]

In each case, which product do you expect to predominate?

Problem 7.8 **Write out the steps in the mechanism for the acid-catalyzed dehydration of 2-propanol to propene.**

**7.9
THE REACTION
OF ALCOHOLS
WITH HYDROGEN
HALIDES**

Alcohols react with hydrogen halides to give alkyl halides.

$$
\underset{\text{alcohol}}{R-OH} + H-X \rightarrow \underset{\text{alkyl halide}}{R-X} + H-OH \qquad (7.18)
$$

This reaction provides a general synthesis for alkyl halides. The reaction rate and mechanism depend on the alcohol structure (3°, 2°, or 1°), the mechanism being either S_N1 or S_N2 (review Sec. 6.7–6.9).

Tertiary alcohols react fastest. For example, we can convert t-butyl alcohol to t-butyl chloride simply by shaking it for a few minutes at room temperature with concentrated hydrochloric acid.

$$
\underset{\text{t-butyl alcohol}}{(CH_3)_3COH} + H-Cl \xrightarrow[\text{15 min}]{\text{rt,}} \underset{\text{t-butyl chloride}}{(CH_3)_3C-Cl} + H-OH \qquad (7.19)
$$

On the other hand, 1-butanol, a primary alcohol, must be heated for several hours with a mixture of concentrated hydrochloric acid and zinc chloride to accomplish the same type of reaction.

$$
\underset{\text{1-butanol}}{CH_3CH_2CH_2CH_2OH} + H-Cl \xrightarrow[\text{several hours}]{\text{heat, ZnCl}_2} CH_3CH_2CH_2CH_2-Cl + H-OH \qquad (7.20)
$$

Of course, the difference in reaction conditions is related to the change in mechanism as one goes from a tertiary to a primary alcohol.

Example 7.3   Write out the steps in the mechanism for eq. 7.19.

Solution   In the first step, the alcohol is protonated by the acid (see eq. 7.11). Ionization (eq. 7.12) is the slow and rate-determining step, typical of the S_N1 mechanism. In the final step, the *t*-butyl cation is captured by chloride ion.

$$(CH_3)_3C^+ + Cl^- \xrightarrow{\text{fast}} (CH_3)_3CCl$$

Example 7.4   Write out the steps in the mechanism for eq. 7.20, and explain the role played by the zinc chloride catalyst.

Solution   In the first step, the alcohol is protonated by the acid, just as in the previous example.

$$CH_3CH_2CH_2CH_2 - \overset{..}{\underset{..}{O}}H + H^+ \rightleftharpoons CH_3CH_2CH_2CH_2 - \overset{+}{\underset{\underset{H}{|}}{O}} - H$$

In the second step, chloride ion displaces water in a typical S_N2 process.

$$Cl^- \quad \overset{CH_3CH_2CH_2}{\underset{H}{\diagdown}} C \overset{+}{\underset{\underset{H}{|}}{\overset{..}{O}}} - H \rightarrow ClCH_2CH_2CH_2CH_3 + H_2O$$

The zinc chloride increases the chloride ion concentration, thus speeding up the S_N2 displacement.

Problem 7.9   Explain why *t*-butyl alcohol reacts at about equal rates with HCl, HBr, and HI (to form, in each case, the corresponding *t*-butyl halide).

Problem 7.10   Explain why *n*-butyl alcohol reacts with hydrogen halides in the rate order HI > HBr > HCl (to form, in each case, the corresponding *n*-butyl halide).

The **Lucas test,** used to distinguish between primary, secondary, and tertiary alcohols, is based on the different rates at which these classes of alcohols are converted to the corresponding chlorides. The Lucas reagent is a solution of zinc chloride in concentrated hydrochloric acid. Tertiary alcohols react immediately and the tertiary alkyl chloride forms as a cloudy dispersion or separate layer. Secondary alcohols dissolve because of oxonium ion formation (eq. 7.9), and within about five minutes the alkyl chloride forms and separates from solution. Primary alcohols are not converted to the chlorides by the reagent at room temperature. They simply dissolve.

7.10
OTHER WAYS TO
PREPARE ALKYL
HALIDES FROM
ALCOHOLS

As we saw in Chapter 6, alkyl halides are extremely useful in synthesis. It is not surprising, then, that many ways to prepare alkyl halides from alcohols have been devised. Besides HX (Sec. 7.9), several inorganic acid halides can be used. For example, **thionyl chloride** reacts with alcohols to give alkyl chlorides.

$$ROH + Cl-\overset{\overset{\displaystyle O}{\|}}{S}-Cl \xrightarrow{\text{heat}} RCl + HCl\uparrow + SO_2\uparrow \qquad (7.21)$$

thionyl chloride

The special advantage this method offers is that two other reaction products, hydrogen chloride and sulfur dioxide, are gases and evolve from the reaction mixture, leaving behind only the desired alkyl chloride. The method is not good, however, for preparing low-boiling alkyl chlorides (when R has only a few carbon atoms), because they are easily lost with the gaseous products. Thionyl bromide can be used with alcohols to make alkyl bromides.

Phosphorus halides also convert alcohols to alkyl halides.

$$3\ ROH + PX_3 \rightarrow 3\ RX + H_3PO_3 \quad (X = Cl\ or\ Br) \qquad (7.22)$$

In this case the other reaction product, phosphorous acid, has a rather high boiling point. Thus the alkyl halide is usually the lowest-boiling component of the reaction mixture and can be isolated from it by distillation.

Problem 7.11   **Tell how you would prepare each of the following alkyl halides from the corresponding alcohols without using HX.**
a. $CH_3(CH_2)_6CH_2Cl$ b. $(CH_3)_2CHBr$

**7.11
INORGANIC ESTERS**

Esters are compounds in which the proton of an acid is replaced by an organic group. The most common inorganic esters are nitrates, sulfates, and phosphates (from nitric, sulfuric, and phosphoric acid, respectively).

Alkyl nitrates may be prepared by the reaction of alcohols with nitric acid, in the cold.

$$ROH + HONO_2 \xrightarrow{\text{cold}} RONO_2 + H_2O \qquad (7.23)$$

alcohol   nitric   alkyl
            acid   nitrate

Alkyl nitrates are explosive and must be handled with great care. Nitroglycerine (Sec. 7.14) and nitrocellulose (guncotton, Sec. 14.12b) are prepared by this method.

Alkyl nitrites can be similarly prepared, from alcohols and nitrous acid.

$$ROH + HONO \rightarrow RONO + H_2O \qquad (7.24)$$

nitrous   alkyl
acid   nitrite

Alkyl nitrites and nitrates are used in medicine to dilate coronary blood vessels in the treatment of certain heart diseases (angina pectoris).

Problem 7.12   **Write an equation for the preparation of isopentyl nitrite, [$(CH_3)_2CHCH_2CH_2ONO$], the first nitrite to be used to treat heart disease.**

When sulfuric acid reacts *in the cold* with alcohols (especially primary alcohols), the product is an **alkyl hydrogen sulfate.**

$$\text{ROH} + \text{HOSO}_3\text{H} \xrightarrow{\text{cold}} \text{ROSO}_3\text{H} + \text{H}_2\text{O} \qquad (7.25)$$

alcohol   sulfuric acid   alkyl hydrogen
sulfate

Alkyl hydrogen sulfates are used to prepare synthetic detergents (Sec. 11.13).
Alcohols also form a variety of **phosphates.**

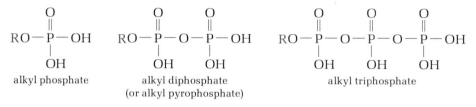

alkyl phosphate   alkyl diphosphate   alkyl triphosphate
(or alkyl pyrophosphate)

Alkyl phosphates, diphosphates, and triphosphates are the workhorses of the biological laboratory (the cell), just as alkyl halides are the workhorses of the organic laboratory. (See A Word About S$_N$2 Reactions in the Cell, page 161.)

Esters of organic acids are discussed later (Sec. 10.10 and beyond).

**7.12
OXIDATION OF
ALCOHOLS TO
ALDEHYDES AND
KETONES**

Alcohols with at least one hydrogen attached to the hydroxyl-bearing carbon can be oxidized to carbonyl compounds. Primary alcohols give aldehydes that may be further oxidized to acids. Secondary alcohols give ketones.

$$
\underset{\text{primary alcohol}}{
\begin{array}{c}
\text{H} \\
| \\
\text{R}-\text{C}-\text{OH} \\
| \\
\text{H}
\end{array}}
\xrightarrow[\text{agent}]{\text{oxidizing}}
\underset{\text{aldehyde}}{
\begin{array}{c}
\text{H} \\
| \\
\text{R}-\text{C}=\text{O}
\end{array}}
\xrightarrow[\text{agent}]{\text{oxidizing}}
\underset{\text{acid}}{
\begin{array}{c}
\text{OH} \\
| \\
\text{R}-\text{C}=\text{O}
\end{array}}
\qquad (7.26)
$$

$$
\underset{\text{secondary alcohol}}{
\begin{array}{c}
\text{R}' \\
| \\
\text{R}-\text{C}-\text{OH} \\
| \\
\text{H}
\end{array}}
\xrightarrow[\text{agent}]{\text{oxidizing}}
\underset{\text{ketone}}{
\begin{array}{c}
\text{R}' \\
| \\
\text{R}-\text{C}=\text{O}
\end{array}}
\qquad (7.27)
$$

Tertiary alcohols cannot undergo this type of oxidation.

Common laboratory oxidizing agents for this purpose are chromic acid, H_2CrO_4 (derived from potassium dichromate, $K_2Cr_2O_7$, and strong acid), and chromic anhydride, CrO_3, both of which contain Cr^{6+}. A typical example of this reaction is the oxidation of cyclohexanol to cyclohexanone.

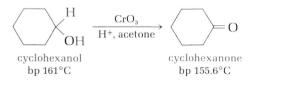

cyclohexanol   cyclohexanone
bp 161°C   bp 155.6°C

(7.28)

With primary alcohols, oxidation can be stopped at the aldehyde stage by special reagents. The 2:1 complex of pyridine with chromic anhydride in a nonpolar solvent gives a good yield of aldehyde.

$$CH_3(CH_2)_6CH_2OH \xrightarrow[CH_2Cl_2,\ 25°C]{(pyridine)_2CrO_3} CH_3(CH_2)_6CH=O \qquad (7.29)$$

1-octanol   octanal
bp 194–195°C   bp 163–164°C

Problem 7.13 **Write an equation for the oxidation of:**
a. 3-pentanol   b. 1-pentanol

Although the oxidation mechanism is complex, it is known that the reactions proceed via chromate esters of the alcohols. These esters then undergo an elimination reaction with loss of a proton, to give the products.

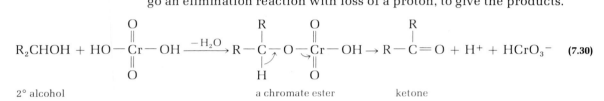

2° alcohol   a chromate ester   ketone

a word about

11. Biological Oxidation of Alcohols

The oxidation of alcohols to aldehydes or ketones and the reverse reduction are involved in many metabolic processes. A key oxidizing agent in many such reactions is nicotinamide adenine dinucleotide, abbreviated NAD⁺. (For the complete structure of NAD, see Sec. 16.13.) In the presence of an appropriate enzyme, NAD⁺ oxidizes alcohols to carbonyl compounds, and in the process it is reduced to NADH. For example, ethanol can be oxidized to acetaldehyde.

This reaction, which occurs in the liver, is a key step in the body's attempt to rid itself of imbibed alcohol. Acetaldehyde, like ethanol, is toxic. It is further oxidized to acetate ion and ultimately to carbon dioxide and water.

NAD⁺ is also involved in the oxidation of lactic acid to pyruvic acid, an important step in carbohydrate metabolism.

$$\underset{\text{lactic acid}}{CH_3\overset{\overset{\displaystyle OH}{|}}{C}HCO_2H} \underset{\text{enzyme}}{\overset{NAD^+}{\rightleftharpoons}} \underset{\text{pyruvic acid}}{CH_3\overset{\overset{\displaystyle O}{\|}}{C}CO_2H} + NADH$$

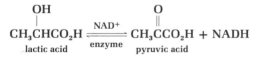

During vigorous muscular activity, glucose is consumed to produce energy. The process is called *glycolysis*, and in this process glucose is converted to pyruvate, which is further reduced (by NADH) to lactate. Lactate is the dead end of this metabolism and, as its concentration in the muscles increases, they become fatigued. To accomplish recovery, the lactate diffuses from the muscle into the bloodstream and is transported to the liver. There it is reoxidized by NAD$^+$ to pyruvate, which is further converted to glucose by a process called *gluconeogenesis*. This glucose produced in the liver diffuses into the bloodstream and is transported to the muscles where it can be reused, thus completing the cycle.

**7.13
ALCOHOLS WITH
MORE THAN ONE
HYDROXYL GROUP**

Compounds with two adjacent alcohol groups are called **glycols.** The most important example is ethylene glycol. Compounds with more than two hydroxyl groups are also known, and several, such as **glycerol** and **sorbitol,** are important commercial chemicals.

$$CH_2 - CH_2$$
$$| \quad\quad |$$
$$OH \quad OH$$

ethylene glycol
(1,2-ethanediol)
bp 198°C

$$CH_2 - CH - CH_2$$
$$| \quad\quad | \quad\quad |$$
$$OH \quad OH \quad OH$$

glycerol (glycerine)
(1,2,3-propanetriol)
bp 290°C (decomposes)

$$CH_2 - CH - CH - CH - CH - CH_2$$
$$| \quad\quad | \quad\quad | \quad\quad | \quad\quad | \quad\quad |$$
$$OH \quad OH \quad OH \quad OH \quad OH \quad OH$$

sorbitol
(1,2,3,4,5,6-hexanehexaol)
mp 110–112°C

We will describe the commercial syntheses of these compounds later in the text.

Ethylene glycol is used directly as the "permanent" antifreeze in automobile radiators. It is also a raw material used in the manufacture of Dacron. Ethylene glycol is completely miscible with water and, because of its increased hydrogen-bonding opportunities, it has an exceptionally high boiling point for its molecular weight—much higher than ethanol.

Glycerol is a syrupy, colorless, water-soluble, high-boiling liquid with a distinctly sweet taste. It is used for its soothing qualities in shaving and toilet soaps, and in cough drops and syrups. It is also used as a moistening agent in tobacco.

Nitration of glycerol gives **glyceryl trinitrate** (nitroglycerine), a powerful and shock-sensitive explosive.

$$\begin{array}{l} CH_2OH \\ | \\ CHOH + 3\ HONO_2 \xrightarrow{H_2SO_4} \\ | \\ CH_2OH \end{array} \begin{array}{l} CH_2ONO_2 \\ | \\ CHONO_2 + 3\ H_2O \\ | \\ CH_2ONO_2 \end{array} \qquad (7.31)$$

glycerol

glyceryl trinitrate
(nitroglycerine)

Alfred Nobel, the inventor of dynamite (1866), found that glyceryl trinitrate could be controlled by absorbing it on an inert, porous material. At present, dynamite contains only about 15% of glyceryl (and glycol) nitrate. The main explosive is ammonium nitrate (55%), and the other components are sodium nitrate and wood pulp (about 15% each). Dynamite is used mainly in mining operations and for construction purposes.

Nitroglycerine is also used in medicine as a vasodilator, to prevent heart attacks in patients who suffer from angina.

Triesters of glycerol are fats and oils, whose chemistry is discussed in Chapter 11.

Sorbitol, with its many hydroxyl groups, is also water-soluble. It is almost as sweet as ordinary cane sugar and is used in candy manufacture and as a sugar substitute for diabetics.

a word about _____

12. Some Alcohols Used Every Day

The lower alcohols, those with up to four carbon atoms, are manufactured on a very large scale. They have important uses in their own right and are also used as raw materials for other valuable chemicals.

Methanol was at one time produced from wood by distillation and is still sometimes referred to as wood alcohol. The word *methyl* originates from the Greek (*methy*, wine, and *yle*, wood). At present, however, methanol is manufactured from carbon monoxide and hydrogen.

$$CO + 2 H_2 \xrightarrow[400°C,\ 150\ atm]{ZnO - Cr_2O_3} CH_3OH$$

The world capacity for methanol production is approximately 10 million tons per year. Most methanol serves as a raw material for the production of formaldehyde (Sec. 9.3) and other chemicals, but it is also used as a solvent and as an antifreeze. As petroleum sources dwindle, methanol may find a use as a fuel in internal combustion engines, where it offers the advantage that the exhaust gases are low in air pollutants. Recently methanol has been used as the carbon source in the commercial production of single-cell proteins. Some yeasts and bacteria (single cells) can synthesize proteins from methanol and other carbon sources in the presence of aqueous nutrient salt solutions containing certain essential sulfur, phosphorus, and nitrogen compounds. This protein is used as an animal food supplement and may eventually also play a part in human nutrition. Methanol itself, however, is highly toxic and can cause permanent blindness or death if taken internally.

Ethanol is prepared by the fermentation of blackstrap molasses, the residue that results from the purification of cane sugar.

$$C_{12}H_{22}O_{11} + H_2O \xrightarrow{yeast} 4\ CH_3CH_2OH + 4\ CO_2$$
cane sugar   ethyl alcohol

The starch in grain, potatoes, and rice can similarly be fermented to ethanol, which is sometimes referred to as grain alcohol.

Besides fermentation, ethanol is also manufactured by the acid-catalyzed hydration of ethylene (eq. 3.11). This method, using sulfuric acid or other acid catalysts, results in an annual world production of over 1 million tons.

Commercial alcohol is a constant-boiling mixture, containing 95% ethanol and 5% water, that cannot be further purified by distillation. To remove the remaining water and obtain absolute alcohol, one adds quicklime (CaO), which reacts with water to form calcium hydroxide, but which does not react with the ethanol.

Ethanol has been known since early times as an ingredient of fermented beverages (beer, wine, whiskey, and so on). The term *proof*, as used in the United States in reference to alcoholic beverages, is approximately twice the volume percentage of alcohol present. For example 100-proof whiskey contains 50% ethanol.

Ethanol is used as a solvent, as a topical antiseptic, and as a starting material for the manufacture of ether and of ethyl esters. Like methanol, it can be used as a fuel (gasohol) and as a carbon source for single-cell proteins.

2-Propanol (isopropyl alcohol) is manufactured commercially by the acid-catalyzed hydration of propene (eq. 3.17). It is the main component of rubbing alcohol. More than half the isopropyl alcohol produced (over 1 million tons annually) is used to manufacture acetone (Sec. 7.12).

7.14
A COMPARISON
OF ALCOHOLS
AND PHENOLS

Because they both have the same functional group, the hydroxyl group, alcohols and phenols have many similar properties. We have already mentioned that both classes of compounds form hydrogen bonds (Sec. 7.5). They are both weakly acidic, although phenols are much more acidic than alcohols (Sec. 7.6). Both types of compounds are also weakly basic (Sec. 7.7). How, then, do alcohols and phenols differ?

The main difference is in reactions that involve breaking the C—OH bond. Whereas it is relatively easy, with acid catalysis, to break the C—OH bond of alcohols, this bond is difficult to break in phenols. Protonation of the phenolic hydroxyl group and loss of a water molecule would give a phenyl cation.

$$\langle\!\langle\;\rangle\!\rangle\!-\overset{+}{\underset{H}{\ddot{O}}}-H \xrightarrow{\;/\!\!/\;} \langle\!\langle\;\rangle\!\rangle^{+} + H_2O \qquad\qquad (7.32)$$

a phenyl
cation

With only two attached groups, the positive carbon in a phenyl cation would want to be sp-hybridized and linear. But this geometry is prevented by the benzene ring. Hence phenyl cations are exceedingly difficult to form. Consequently, phenols cannot undergo replacement of the hydroxyl group by an S_N1 mechanism. Neither can phenols undergo displacement by the S_N2 mechanism (the geometry of the ring makes the usual inversion mechanism impossible). Therefore, whereas hydrogen halides, phosphorus halides, and thionyl halides can bring about replacement of the hydroxyl group by halogens in alcohols (Sec. 7.9, 7.10), analogous reactions do not occur with phenols.

Problem 7.14   **Compare the reactions of cyclohexanol and phenol with:**
a. HBr   b. H_2SO_4, heat   c. PCl_3

On the other hand, phenols can undergo some interesting reactions that involve the aromatic ring.

7.15
AROMATIC
SUBSTITUTION
IN PHENOLS

The hydroxyl group is strongly ring-activating (Table 4.2), so phenols undergo electrophilic aromatic substitution under very mild conditions. For example, phenol can be nitrated with dilute aqueous nitric acid to give predominantly *para*-nitrophenol.

$$\langle\!\langle\;\rangle\!\rangle\!-OH + HONO_2 \rightarrow O_2N\!-\!\langle\!\langle\;\rangle\!\rangle\!-OH + H_2O \qquad\qquad (7.33)$$

phenol   p-nitrophenol

Phenol can also be brominated rapidly with *bromine* water, to produce 2,4,6-tribromophenol.

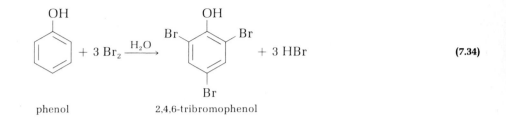

phenol   2,4,6-tribromophenol

Example 7.5 Draw the intermediate in electrophilic aromatic substitution *para* to a hydroxyl group, and show how the intermediate benzenonium ion is stabilized by the hydroxyl group.

Solution

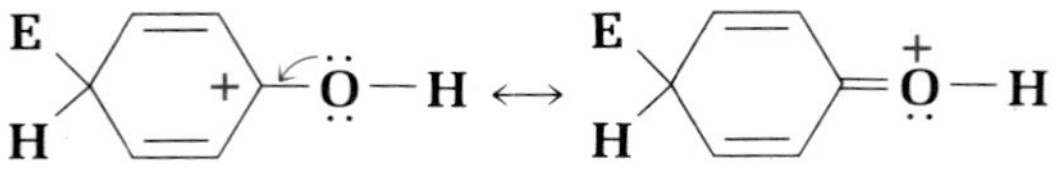

intermediate in electrophilic substitution
para to a phenolic hydroxyl group

An unshared electron pair on the oxygen atom can help delocalize the positive charge.

Problem 7.15 Explain why phenoxide ion would undergo electrophilic aromatic substitution even more easily than phenol.

Problem 7.16 Write an equation for:
a. *p*-cresol + $HONO_2$ (1 mol)
b. *o*-chlorophenol + Br_2 (1 mol)

7.16 OXIDATION OF PHENOLS

Phenols are easily oxidized. Samples of phenols that stand in the air for some time often become highly colored due to the formation of oxidation products. With **hydroquinone** (1,4-dihydroxybenzene), the reaction is easily controlled to give **1,4-benzoquinone** (commonly called *quinone*).

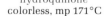

(7.35)

hydroquinone   benzoquinone
colorless, mp 171°C   yellow, mp 116°C

Hydroquinone and related compounds are used in photographic developers, where they reduce unexposed silver ion to metallic silver (and, in turn, are themselves oxidized to quinones). The oxidation of hydroquinones to quinones is reversible, and their interconversion plays an important role in several biological oxidation–reduction reactions. (See A Word About Quinones on page 231.)

Problem 7.17   **Draw the structures of the quinones expected from the oxidation of:**

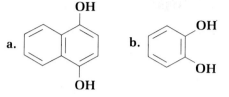

Substances that might otherwise be damaged through air oxidation can be protected from such oxidation by phenolic additives. The added phenols are oxidized instead and thus function as **antioxidants.** Two commercial phenolic antioxidants are **BHA** (butylated hydroxy anisole) and **BHT** (butylated hydroxy toluene):

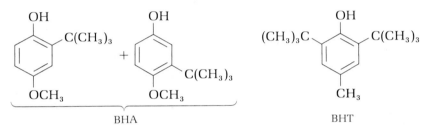

These antioxidants succeed because they react with and "trap" peroxy radicals, ROO·, the main culprits in air oxidations. They trap these radicals in the following way:

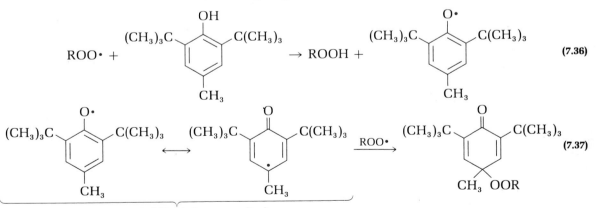

resonance-stabilized phenoxy radical

In the first step (eq. 7.36), the phenolic hydrogen is abstracted. The resulting phenoxy radical is stabilized by resonance, but it can react at the 4-position to capture a peroxy radical. In this way, peroxy radicals are "destroyed" and prevented from oxidizing the substance being protected.

BHA is used as an antioxidant in foods, especially meats. BHT is used not only in foods, animal feeds, and animal or vegetable oils, but also in lubricating oils, synthetic rubber, and various plastics.

Vitamin E (α-tocopherol) is a widespread naturally occurring phenol. One of its biological functions appears to be acting as a radical scavenger and natural antioxidant.

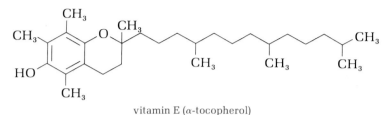

vitamin E (α-tocopherol)

Problem 7.18   **BHT is manufactured commercially from p-cresol, isobutylene, and an acid catalyst. Write equations that show the steps in this Friedel–Crafts synthesis.**

a word about ─────────────────────────────────

13. Biologically Important Alcohols and Phenols

The hydroxyl group appears in many biologically important molecules, both as an alcohol and as a phenol. We describe a few such compounds here.

Four metabolically important unsaturated primary alcohols are

3-methyl-2-buten-1-ol

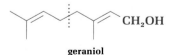

3-methyl-3-buten-1-ol

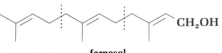

geraniol

farnesol

The two 5-carbon alcohols contain the basic 5-carbon isoprene unit (Sec. 3.18) present in many natural products. These alcohols, which occur in cells mainly as diphosphate esters, can combine to give geraniol, which then can add yet another 5-carbon unit to give farnesol. Note the isoprene units, marked off by faint dashed lines, in the structures of geraniol and farnesol.

Compounds of this type are called terpenes. Terpenes occur in the *essential oils* of many plants and flowers. They have 10, 15, 20 and so on carbon atoms, formed by linking isoprene units in various ways.

Geraniol, as its name implies, occurs in oil of geranium, but it also constitutes about 50% of rose oil, the extract of rose petals. Geraniol is also the biological precursor of α-pinene (Figure 1.12), a terpene that is the main component of turpentine.

Farnesol, which occurs in the essential oils of rose and cyclamen, has a pleasing lily-of-the-valley odor. Both geraniol and farnesol are used in making perfume.

Combination of two farnesol units (15 carbons each) leads to squalene, a 30-carbon hydrocarbon present in small amounts in the liver of most higher animals (Figure 7.1). Squalene is the biological precursor of the steroids (Sec. 11.14).

Cholesterol, a typical steroidal alcohol, has the structure

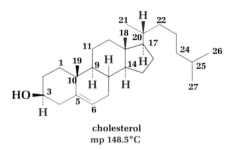

cholesterol
mp 148.5°C

Although it has 27 carbon atoms (instead of 30), and is therefore not strictly a terpene, cholesterol is synthesized in the body from the terpene squalene via a complex process that, in its final stages, involves the loss of 3 carbon atoms.

Cholesterol was first isolated from and is the chief constituent of gallstones, but it is also present in rather large amounts in the brain and spinal cord and in smaller amounts in all cells of animal organisms. The total amount of cholesterol extractable from all body tissues and blood of an average person amounts to about half a pound. If excess cholesterol is present in the body, it may precipitate in the gall bladder as gallstones and in blood vessels, where it constricts their diameter and increases blood pressure.

Vitamin A, sometimes called retinol, is an unsaturated primary alcohol that is enzymatically oxidized to retinal (page 69), a substance essential for vision. Its biological precursor is β-carotene, found in many vegetables, including carrots (Figure 7.2).

Phenols are somewhat less involved than alcohols in fundamental metabolic processes. Three phenolic alcohols do, however, form the basic building blocks of lignins, complex polymeric substances that, together with cellulose, form the "woody" parts of trees and shrubs. They have very similar structures:

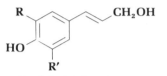

coniferyl alcohol (R = OCH₃, R' = H)
sinapyl alcohol (R = R' = OCH₃)
p-coumaryl alcohol (R = R' = H)

Phenolic natural products to be avoided are the urushiols, the active allergenic ingredients in poison ivy and poison oak.

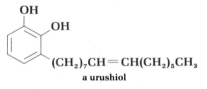

a urushiol

The long side chain may be saturated or may have additional double bonds. It may also have two more carbon atoms.

FIGURE 7.1   Structural formula of squalene.

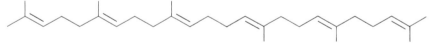

vitamin A

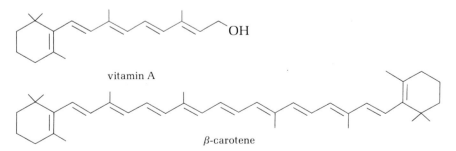

β-carotene

FIGURE 7.2   Structural formulas of vitamin A and β-carotene.

**7.17
THIOLS, THE
SULFUR ANALOGS
OF ALCOHOLS
AND PHENOLS**

Sulfur is immediately below oxygen in the periodic table (Table 1.3) and can often take the place of oxygen in organic structures. The —SH group, called the **sulfhydryl group**, is the functional group of **thiols**. Thiols are named as follows:

$$CH_3SH \qquad\qquad CH_3CH_2CH_2CH_2SH$$

methanethiol   1-butanethiol   thiophenol
(methyl mercaptan)   (n-butyl mercaptan)   (phenyl mercaptan)

Thiols are sometimes called **mercaptans** because of their reaction with mercuric ion to form **mercaptides**.

$$2\,RSH + HgCl_2 \rightarrow (RS)_2Hg + 2\,HCl \tag{7.38}$$

a mercaptide

Problem 7.19   **Draw the structures for:**
a. 2-butanethiol   b. isopropyl mercaptan

Alkyl thiols can be made from alkyl halides by nucleophilic displacement with sulfhydryl ion.

$$R-X + {}^-SH \rightarrow R-SH + X^- \tag{7.39}$$

Perhaps the most distinctive feature of thiols is their intense and disagreeable odor. The thiols, $CH_3CH{=}CHCH_2SH$ and $(CH_3)_2CHCH_2CH_2SH$, for example, are responsible for the odor of skunk.

Thiols have lower boiling points and are less soluble in water than the corresponding alcohols. The reason is that the sulfur atom is larger and less electronegative than oxygen, so that —SH compounds form weaker hydrogen bonds than do —OH compounds.

Thiols are more acidic than alcohols, and they readily form **thiolates** on treatment with aqueous base.

$$RSH + Na^+OH^- \rightarrow RS^-Na^+ + HOH \tag{7.40}$$

a sodium thiolate

Problem 7.20   **Write an equation for the reaction of ethanethiol with:**
a. KOH   b. HgCl₂

Thiols are easily oxidized to **disulfides** by mild oxidizing agents such as hydrogen peroxide or iodine.

$$2\,RS-H \underset{\text{reduction}}{\overset{\text{oxidation}}{\rightleftharpoons}} RS-SR \tag{7.41}$$

thiol   disulfide

The reaction can be reversed with a variety of reducing agents. Proteins contain disulfide links, so this reversible oxidation–reduction can be used to manipulate their structures.

a word about ————————————————————————

14. Hair, Curly or Straight

Hair consists of a fibrous protein called <u>keratin</u>, which, as proteins go, contains an unusually large percentage of the sulfur-containing amino acid <u>cystine</u>. Horse hair, for example, contains about 8% cystine:

$$HO_2CCHCH_2S-SCH_2CHCO_2H$$

$$\quad\;\; | \qquad\qquad\qquad\quad |$$

$$\quad NH_2 \qquad\qquad\qquad NH_2$$

cystine (CyS—SCy)

The disulfide link in cystine serves to cross-link the chains of amino acids that make up the protein (Figure 7.3).

The chemistry used in waving or straightening hair is the simple oxidation–reduction chemistry of the disulfide bond (eq. 7.41). First the hair is treated with a reducing agent, which breaks the —S—S— bonds, converting each to two —SH groups. This breaks the cross-links between the long protein chains. The reduced hair can now be shaped as desired, either waved or straightened. Finally, the reduced and rearranged hair is treated with an oxidizing agent to re-form the disulfide cross-links. Of course, the new disulfide bonds are no longer in their original positions and they hold the hair in its new shape.

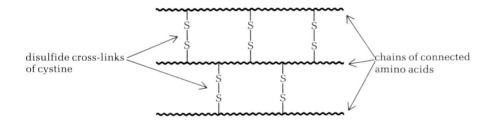

disulfide cross-links of cystine — chains of connected amino acids

FIGURE 7.3 Schematic structure of hair.

ADDITIONAL PROBLEMS

7.21. Write a structural formula for each of the following compounds.
a. 2,2-dimethyl-1-butanol
b. o-bromophenol
c. 2,3-pentanediol
d. 2-phenylethanol
e. ethyl hydrogen sulfate
f. tricyclopropylmethanol
g. sodium ethoxide
h. 1-methylcyclopentanol
i. *trans*-2-methylcyclopentanol
j. (R)-2-butanol

7.22. Classify the alcohols in parts a, d, f, h and i of Problem 7.21 as primary, secondary or tertiary.

7.23. Name each of the following compounds.

a. $CH_3C(CH_3)_2CH(OH)CH_3$

b. $CH_3CHBrC(CH_3)_2OH$

c.
OH
Cl
Cl

d.

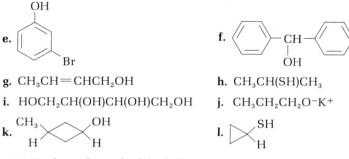

e.

f.

g. $CH_3CH=CHCH_2OH$ **h.** $CH_3CH(SH)CH_3$

i. $HOCH_2CH(OH)CH(OH)CH_2OH$ **j.** $CH_3CH_2CH_2O^-K^+$

k.   **l.**

7.24. Explain why each of the following names is unsatisfactory, and give a correct name.
a. 2,2-dimethyl-3-butanol   **b.** 2-ethyl-1-propanol
c. 1-propene-3-ol   **d.** 5-chlorocyclohexanol
e. 6-bromo-p-cresol   **f.** 2,3-propanediol

7.25. Arrange the compounds in each of the following groups in order of increasing solubility in water, and briefly explain your answer.
a. ethanol; ethyl chloride; 1-hexanol
b. 1-pentanol; 1,5-pentanediol; $CH_2OH(CHOH)_3CH_2OH$

7.26. Arrange the following compounds in order of increasing acidity, and explain the reasons for your choices.
a. phenol   **b.** p-chlorophenol   **c.** p-cresol   **d.** cyclohexanol

7.27. p-Nitrophenol is 1000 times stronger as an acid than phenol itself. Compare the resonance contributors of p-nitrophenoxide ion with those of phenoxide ion, and suggest one reason for the greater acidity of p-nitrophenol.

7.28. Explain, with the aid of equations, what would happen if a solution of cyclohexanol and p-cresol in an inert solvent were successively (a) shaken with 10% aqueous sodium hydroxide, (b) subjected to separation of the organic and aqueous layers, and (c) subjected to acidification of the aqueous layer.

7.29. Tell how the following mixtures could be separated without the use of distillation.
a. benzene and phenol
b. phenol and 1-hexanol
c. 1-propanol and 1-heptanol

7.30. Show the structures of all possible acid-catalyzed dehydration products of:
a. cyclohexanol   **b.** 2-butanol
c. 1-methylcyclopentanol   **d.** 2-phenylethanol
If more than one alkene is possible, predict which one will be formed in the largest amount.

7.31. Explain why the reaction shown in eq. 7.12 occurs much more easily than the reaction $(CH_3)_3C-OH \rightleftharpoons (CH_3)_3C^+ + OH^-$. (That is, why is it necessary to protonate the alcohol before ionization can occur?)

7.32. Write out all of the steps in the mechanism of eq. 7.17, showing how each product is formed.

7.33. Although the reaction shown in eq. 7.19 occurs faster than that shown in eq. 7.20, the yield of product is lower. The yield of t-butyl chloride is only 80%, whereas the yield of n-butyl chloride is nearly 100%. What by-product is formed in eq. 7.19, and by what mechanism is it formed? Why is a similar by-product *not* formed in eq. 7.20?

7.34. Write an equation for each of the following reactions.
a. 2-methyl-2-butanol + HCl   **b.** 1-pentanol + Na
c. cyclopentanol + PBr$_3$ **d.** 1-phenylethanol + SOCl$_2$
e. 1-butanol + cold, conc. H$_2$SO$_4$ **f.** ethylene glycol + HONO$_2$
g. 1-pentanol + aqueous NaOH   **h.** 1-octanol + HBr + ZnBr$_2$
i. 2-pentanol + CrO$_3$, H$^+$ **j.** benzyl alcohol + acetic acid

7.35. Treatment of 3-buten-2-ol with concentrated hydrochloric acid gives a mixture of two products, 3-chloro-1-butene and 1-chloro-2-butene. Write a reaction mechanism that explains how both products are formed.

7.36. Which four-carbon alcohols can be manufactured commercially by acid-catalyzed hydration of alkenes? (Remember Markovnikov's rule!)

7.37. Write equations for each of the following two-step syntheses.
a. cyclohexene to cyclohexanone
b. 1-bromobutane to butanal
c. 1-butanol to 1-butanethiol

7.38. What product would you expect from the oxidation of cholesterol with CrO$_3$ and H$^+$?

7.39. Mark off the isoprene units in squalene, vitamin A, and β-carotene. For the structures, see A Word About Biologically Important Alcohols and Phenols, page 186.

7.40. Give a correct IUPAC name for the two thiols that are responsible for skunk odor (Sec. 7.17).

7.41. Dimethyl disulfide, CH$_3$S—SCH$_3$, found in the vaginal secretion of female hamsters, acts as a sexual attractant for the male hamster. Write an equation for its synthesis from methanethiol.

7.42. The disulfide [(CH$_3$)$_2$CHCH$_2$CH$_2$S]$_2$ is a component of the odorous secretion of mink. Describe a synthesis of this disulfide, starting with 3-methyl-1-butanol.

eight
Ethers, Epoxides, and Sulfides

8.1
INTRODUCTION

To most people the word *ether* is synonymous with the well-known anesthetic. That particular ether, however, is only one member of the general class of organic compounds known as **ethers,** compounds that have two organic groups connected to a single oxygen atom. The general formula for ethers is R—O—R′, where R and R′ may be identical or different, and where they may be alkyl or aryl groups. In the common anesthetic, both R's are ethyl groups, CH_3CH_2—O—CH_2CH_3.

In this chapter we will describe the physical and chemical properties of ethers. Their excellent solvent properties are applied in the preparation of Grignard reagents, organometallic compounds with a carbon–magnesium bond. We will give special attention to **epoxides,** cyclic three-membered ring ethers that have special industrial utility. Finally, we will briefly describe the chemistry of organic **sulfides,** which have structures analogous to ethers but with sulfur in place of oxygen.

8.2
NOMENCLATURE OF ETHERS

Ethers are usually named by giving the name of each alkyl or aryl group, in alphabetical order, followed by the word *ether.* For example,

CH_3—O—CH_2CH_3

ethyl methyl ether

CH_3CH_2—O—CH_2CH_3

diethyl ether (the prefix *di-* is sometimes omitted)

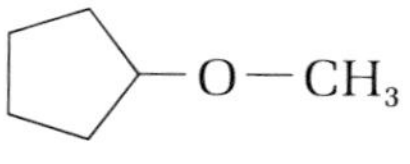

cyclopentyl methyl ether

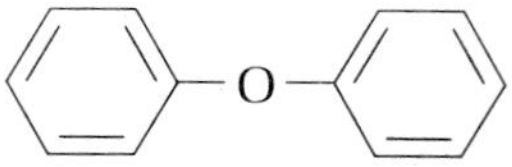

diphenyl ether

In ethers with more complex structures, it may be necessary to name the —**OR** group as an **alkoxy group.** For example, in the IUPAC system, ethers may be named as alkoxy-substituted hydrocarbons:

$CH_3CHCH_2CH_2CH_3$
|
OCH_3

2-methoxypentane

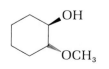

trans-2-methoxycyclohexanol

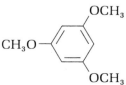

1,3,5-trimethoxybenzene

Problem 8.1   **Give a correct name for:**

a. $(CH_3)_2CHOCH_3$ b. 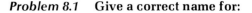$O-CH_2CH_2CH_3$ c. $CH_3CHCH(CH_3)_2$
$$\overset{\displaystyle |}{OCH_3}$$

Problem 8.2   **Write the structural formula for:**
a. dicyclopropyl ether   b. 2-ethoxyoctane

8.3
PHYSICAL
PROPERTIES
OF ETHERS

Ethers are colorless compounds with characteristic, relatively pleasant odors. They have lower boiling points than alcohols with an equal number of carbon atoms and, in fact, have nearly the same boiling point as the corresponding hydrocarbon in which a $-CH_2-$ group replaces the ether oxygen. The data in the following table illustrate these facts:

		bp	mol wt
1-butanol	$CH_3CH_2CH_2CH_2OH$	118°C	74
diethyl ether	$CH_3CH_2-O-CH_2CH_3$	35°C	74
pentane	$CH_3CH_2-CH_2-CH_2CH_3$	36°C	72

Ether molecules cannot form hydrogen bonds with one another. This is why they boil so much lower than their isomeric alcohols.

   Although ethers cannot form hydrogen bonds with one another, they can and do form hydrogen bonds with $-OH$ compounds:

$$R-O\cdots\cdots H-O$$
$$\overset{\displaystyle |}{R}\qquad\quad\overset{\displaystyle |}{R}$$

For this reason, alcohols and ethers are usually mutually soluble. Indeed, the lower-molecular-weight ethers, such as dimethyl ether, are quite soluble in water. The water solubility of diethyl ether is 7 g per 100 mL of water. The greater the number of carbon atoms in an ether, the lower its solubility in water. Ethers are less dense than water.

Problem 8.3   **Write structures for each of the following *isomers*, and arrange them in order of decreasing boiling points: 3-methoxy-1-propanol, 1,2-dimethoxyethane, 1,4-butanediol.**

8.4
ETHERS AS
SOLVENTS

Ethers do not usually react with dilute acids, with dilute bases, or with the common oxidizing and reducing agents. They do not react with metallic sodium—a property that distinguishes them from alcohols. They do not, in general, react with other classes of organic compounds.

   This general inertness of ethers, coupled with the fact that most organic compounds are ether-soluble, makes ethers excellent solvents in which to carry out organic reactions.

   Ethers are also used frequently to extract organic compounds from their natural sources. Ordinary diethyl ether is particularly good for this purpose. Its low boiling point makes it easy to remove from an extract and easy to recover. Diethyl ether is highly flammable, however, and must not be used when there are any flames in the same laboratory. Also, ethers that have been in the laboratory for a long time, exposed to air, contain organic peroxides as a result of oxidation. These peroxides are extremely explosive and must be removed before the ether can be used safely. Shaking with aqueous ferrous sulfate destroys these peroxides by reduction.

**8.5
THE GRIGNARD
REAGENT; AN
ORGANOMETALLIC
COMPOUND**

One of the most striking examples of the solvating power of ethers is in the preparation of **Grignard reagents.** These reagents, which are exceedingly useful in organic synthesis, were discovered by the French organic chemist Victor Grignard [pronounced "greenyar(d)"]. In 1912 he received the Nobel Prize for this contribution to organic synthesis.

Grignard found that, when a dry ether solution of an alkyl or aryl halide was stirred with magnesium turnings, the metal gradually dissolved. The resulting solutions contain Grignard reagents:

$$R-X + Mg \xrightarrow{\text{dry} \atop \text{ether}} R-MgX \tag{8.1}$$

a Grignard reagent

Note that the magnesium inserts itself into the carbon–halogen bond.

Although the ether used as a solvent for this reaction is normally not shown as part of the Grignard reagent structure, it does play an important role. The unshared electron pairs on the ether oxygen help to stabilize the magnesium through coordination.

R$\diagdown$ $\diagup$R
$\ddot{\underset{\cdot\cdot}{O}}$

R$-$Mg$-$X

$\ddot{\underset{\cdot\cdot}{O}}$
R$\diagup$ $\diagdown$R

The two ethers most commonly used in Grignard preparations are diethyl ether and the cyclic ether tetrahydrofuran (Sec. 8.10). The ether must be scrupulously dry—free of traces of water or alcohols. Otherwise the Grignard reagent will not form.

Grignard reagents are usually named as shown in the following equations:

$$CH_3-I + Mg \xrightarrow{\text{ether}} CH_3MgI \tag{8.2}$$

methyl iodide   methylmagnesium
                                       iodide

$$\text{\Large$\langle$}\underset{=}{\bigcirc}\text{\Large$\rangle$}-Br + Mg \xrightarrow{\text{ether}} \text{\Large$\langle$}\underset{=}{\bigcirc}\text{\Large$\rangle$}-MgBr \tag{8.3}$$

phenylmagnesium bromide

Although the exact nature of Grignard solutions is still a subject of research, Grignard reagents usually react as though the alkyl or aryl group were negatively charged (a carbanion, or nucleophile), and the magnesium were positive:

$$\overset{\delta-}{R}\!-\!\overset{\delta+}{MgX}$$

Grignard reagents react vigorously with water or with any other compound with an O—H, S—H, or N—H bond:

$$\overset{\delta-}{R}\!-\!MgX + \overset{\delta+}{H}\!-\!OH \rightarrow RH + Mg^{2+}(OH)^-X^- \tag{8.4}$$

This is the reason why the ether used to prepare the Grignard reagent must be entirely free of water.

The reaction of a Grignard reagent with water can be used to convert an organic halide to a hydrocarbon. If heavy water (D_2O) is used, deuterium can be introduced in place of the halogen.

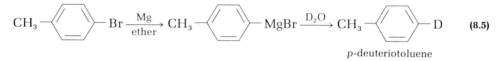

$$\tag{8.5}$$

p-deuteriotoluene

Example 8.1   Show how to prepare CH_3CHDCH_3 from $CH_2\!=\!CHCH_3$.

Solution $CH_2\!=\!CHCH_3 \xrightarrow{HBr} CH_3CHCH_3$
              |
              Br
              | Mg
              ↓ ether
$CH_3CHDCH_3 \xleftarrow{D_2O} CH_3CHCH_3$
                             |
                           MgBr

Problem 8.4   Show how to prepare CH_3CHDCH_3 from $(CH_3)_2CHOH$.

Grignard reagents are **organometallic compounds;** they contain a carbon–metal bond. Many other types of organometallic compounds are known. Among the more useful in synthesis are **organolithium compounds,** which can be prepared in a similar manner to Grignard reagents.

$$R\!-\!X + 2 Li \xrightarrow{ether} R\!-\!Li + Li^+X^- \tag{8.6}$$

an alkyllithium

Problem 8.5   Write an equation for the preparation of *n*-propyllithium and for its reaction with D_2O.

Later in this chapter and elsewhere in the text, we shall see examples of the synthetic utility of organometallic reagents.

8.6
PREPARATION OF
ETHERS

The most important commercial ether is diethyl ether. It is prepared from ethanol and sulfuric acid.

$$2\ CH_3CH_2OH \xrightarrow[140°C]{H_2SO_4} CH_3CH_2OCH_2CH_3 + H_2O \qquad (8.7)$$
$$\text{ethanol} \qquad\qquad\qquad \text{diethyl ether}$$

Note that ethanol can be dehydrated by sulfuric acid to give either ethylene (eq. 7.10) or diethyl ether (eq. 8.7). Of course, the reaction conditions are different in each case. These reactions provide a good example of how important it is to control reaction conditions and to specify them in equations.

Although it can be adapted to other ether syntheses, the alcohol–sulfuric acid method is most commonly used to make symmetric ethers from primary alcohols.

Another method for ether synthesis is the Williamson procedure (review Sec. 6.4).

a word about ────────────────────────────────

15. Ether and Anesthesia

Prior to the 1840s, relief of pain during surgery was attempted by various methods (asphyxiation, pressure on nerves, administration of narcotics or alcohol), but on the whole it was almost worse torture to undergo an operation than to endure the disease. Modern use of anesthesia during surgery has changed all that. It stems from the work of several physicians in the mid-nineteenth century. The earliest experiments used nitrous oxide, ether, or chloroform. Perhaps the best known of these was the removal of a tumor from the jaw of a patient anesthetized by ether, performed by the Boston dentist William T. G. Morton in 1846.

Anesthetics fall into two major categories, general and local. *General anesthetics* are usually administered to accomplish three ends: insensitivity to pain (analgesia), loss of consciousness, and muscle relaxation. Gases such as nitrous oxide and cyclopropane and volatile liquids such as ether are administered by inhalation, but other general anesthetics such as the barbiturates (Sec. 12.13c) may be injected intravenously.

The exact mechanism by which anesthetics affect the central nervous system is not completely known. Unconsciousness may result from several factors: changes in the properties of nerve cell membranes, suppression of certain enzymatic reactions, and solubility of the anesthetic in lipid membranes.

A good inhalation anesthetic should vaporize readily and have appropriate solubility in the blood and tissues. It should also be stable, inert, nonflammable, potent, and minimally toxic. It should have an acceptable odor and cause minimal side effects such as nausea or vomiting. No anesthetic that meets all of these specifications has yet been developed. *Diethyl ether,* although it is perhaps the best-known general anesthetic to the layperson, fails on several counts (flammability, side effects of nausea or vomiting, and relatively slow action). It is quite potent, however, and produces good analgesia and muscle relaxation. The use of ether at present is rather limited, mainly because of its undesirable side effects. Halothane (Sec. 6.13) comes closest to an ideal inhalation anesthetic at the moment.

Local anesthetics are either applied to body surfaces or injected near nerves to desensitize a particular region of the body to pain. The best known of these anesthetics is procaine (Novocain), an aromatic amino-ester (p. 313).

The discovery of anesthetics enabled physicians to perform surgery with deliberation and care, leading to many of the advances of modern medicine.

8.7
CLEAVAGE OF
ETHERS

Ethers are weak bases because of the unshared electron pairs on the oxygen atom. They react with strong proton acids and with Lewis acids such as the boron halides.

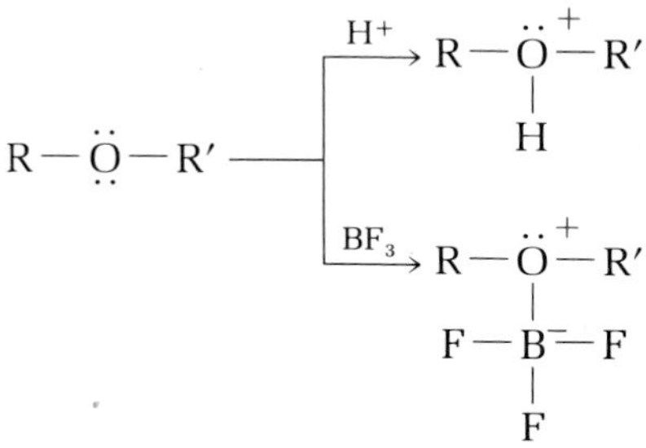

$$
R-\overset{..}{\underset{..}{O}}-R' \quad
\begin{cases}
\xrightarrow{H^+} & R-\overset{..}{\overset{+}{O}}-R' \\
& \quad | \\
& \quad H \\
\\
\xrightarrow{BF_3} & R-\overset{..}{\overset{+}{O}}-R' \\
& \quad | \\
& F-\overset{-}{\underset{|}{B}}-F \\
& \quad | \\
& \quad F
\end{cases}
$$

(8.8)

If the negative ion associated with these acids is a good nucleophile (such as bromide ion), the ethers can be cleaved. For example,

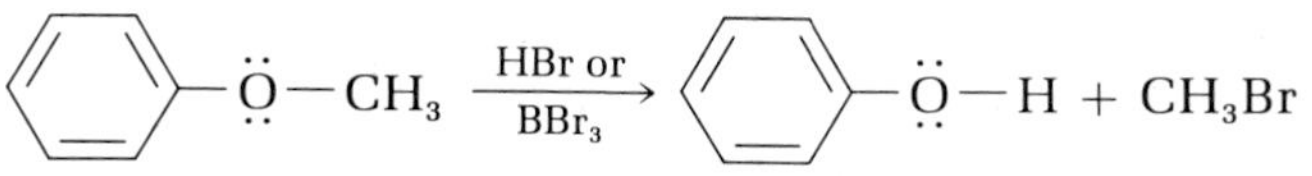

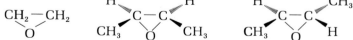

methyl phenyl ether   phenol   methyl bromide
(anisole)

(8.9)

The cleavage of ethers is a useful reaction when one is trying to determine the structure of a complex naturally occurring ether, because it allows one to break the large molecule into more easily handled smaller fragments. The reaction mechanisms are similar to the comparable reactions of alcohols (Sec. 7.9).

Problem 8.6   Write out the steps in the mechanism of eq. 8.9.

8.8
EPOXIDES
(OXIRANES)

Epoxides (or oxiranes) are cyclic ethers with a three-membered ring with one oxygen atom.

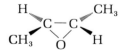

ethylene oxide
(oxirane)
bp 13.5°C

cis-2-butene oxide
(cis-2,3-dimethyloxirane)
bp 60°C

trans-2-butene oxide
(trans-2,3-dimethyloxirane)
bp 54°C

The most important epoxide commercially is ethylene oxide, which is produced by the silver-catalyzed air oxidation of ethylene.

$$CH_2{=}CH_2 + O_2 \xrightarrow[250°, \text{ pressure}]{\text{silver catalyst}} CH_2{-}CH_2 \qquad \qquad \text{(8.10)}$$
$$\underset{O}{\diagdown \diagup}$$

ethylene oxide
bp 13.5°C

Annual U.S. production of ethylene oxide exceeds 4 billion pounds. Only rather small amounts are used as such (for example, as a fumigant in grain storage). Mostly, ethylene oxide is used as a versatile raw material for the manufacture of other products, the main one being ethylene glycol (see Sec. 8.9).

In the laboratory, epoxides are usually prepared by the reaction of an alkene with an organic peracid. For example,

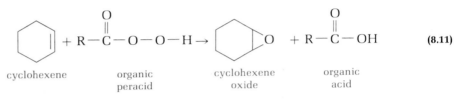

| cyclohexene | organic peracid | cyclohexene oxide | organic acid |

$$\qquad \qquad \qquad \qquad \qquad \qquad \qquad \qquad \qquad \qquad \qquad \qquad \text{(8.11)}$$

Problem 8.7 Write an equation for the preparation of the corresponding epoxides, using peracetic acid (eq. 8.11, R = CH$_3$) and:
a. *cis*-2-butene   b. *trans*-2-butene

a word about

16. The Gypsy Moth's Epoxide

It is rapidly becoming clear that the main mode of communication between insects is via the emission and detection of specific chemical substances. These substances are called <u>pheromones</u>. The word is from the Greek (*pherein*, to carry, and *horman*, to excite). Although they are emitted and detected in exceedingly small amounts, pheromones have profound biological effects. One of their main effects is sexual attraction and stimulation, but pheromones are also used as alarm substances to alert members of the same species to danger, as aggregation substances to call together both sexes of the species, and as trail substances to lead members of the species to a food supply.

Often pheromones are chemically relatively simple compounds—alcohols, esters, aldehydes, ketones, ethers, epoxides, and even hydrocarbons. They have molecular weights that are low enough so that the substances are volatile, yet not so low that they disperse too rapidly. They must also have such a distinctive molecular structure that they are species-specific: Survival of the species would not be served by attracting another insect species. Often this specificity is attained through stereoisomerism (at double bonds or chiral centers), but it can also be achieved by using specific ratios of two or more chemical substances for a particular communication purpose.

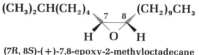

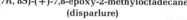

$(7R, 8S)-(+)-7,8$-epoxy-2-methyloctadecane
(disparlure)

Let us consider a specific pheromone, <u>disparlure</u>, the sex attractant of the gypsy moth (*Lymantria dispar*). The gypsy moth is a serious despoiler of forest and shade trees as well as fruit orchards. Gypsy moth larvae, which hatch each spring, are voracious eaters and can strip a tree bare of leaves in just a few weeks.

The abdominal tips (last two segments) of the virgin female moth contain the sex attractant. Extraction of 78,000 tips led to isolation of the main sex attractant, which was the following *cis*-epoxide:

The active isomer has the R configuration at C-7 and the S configuration at C-8. Although this isomer can be detected by the male gypsy moth at a concentration as low as 10^{-10} g/mL, its enantiomer is inactive in solutions 10^6 times as concentrated.

Disparlure has been synthesized in the laboratory. The synthetic material can be used to lure the male to traps and, in that way, control the insect population. This form of insect control sometimes has distinct advantages over chemical spraying.

8.9
REACTIONS OF
EPOXIDES

Because of the strain in the three-membered ring, epoxides are more reactive than ordinary ethers, and they give products in which the ring has opened. For example, they undergo acid-catalyzed ring opening with water to give glycols.

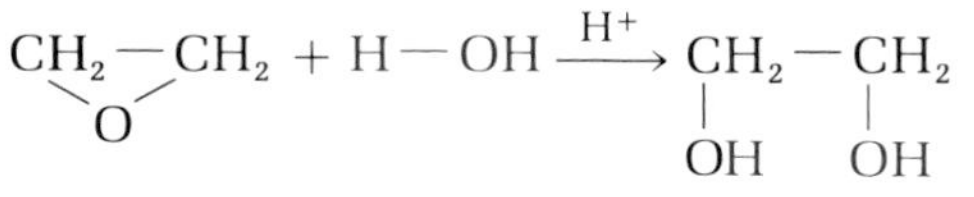

$$CH_2\!-\!CH_2 + H\!-\!OH \xrightarrow{\;H^+\;} CH_2\!-\!CH_2 \qquad\qquad (8.12)$$

ethylene oxide   ethylene glycol

with OH and OH groups.

In this way, about 3 billion pounds of ethylene glycol are produced annually in the United States alone. Approximately half of the glycol is used in automobile cooling systems as an antifreeze. Most of the rest is used to prepare polyesters such as Dacron (Sec. 11.4).

Problem 8.8   **Write an equation for the acid-catalyzed reaction of cyclohexene oxide with water. Predict the stereochemistry of the product.**

Other nucleophiles besides water add to epoxides in a similar way. For example,

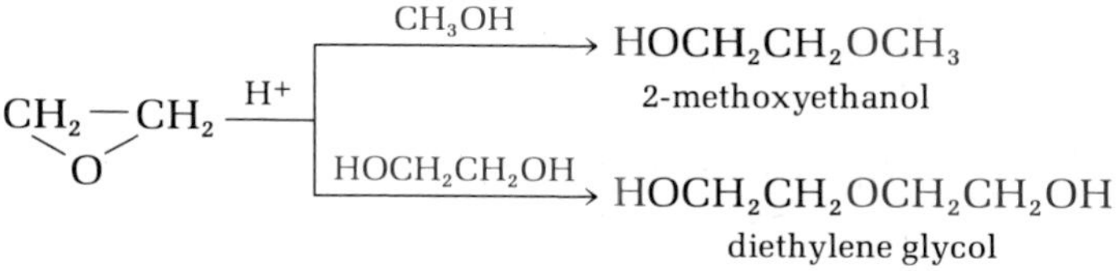

$$CH_2\!-\!CH_2 \xrightarrow{\;H^+\;}
\begin{cases}
\xrightarrow{\;CH_3OH\;} HOCH_2CH_2OCH_3 & \text{2-methoxyethanol} \\[2mm]
\xrightarrow{\;HOCH_2CH_2OH\;} HOCH_2CH_2OCH_2CH_2OH & \text{diethylene glycol}
\end{cases} \qquad (8.13)$$

2-Methoxyethanol is manufactured as an additive to jet fuels, to prevent water from freezing out in fuel lines. Being both an alcohol and an ether, it is soluble in both water and organic solvents. *Diethylene glycol* is useful as a plasticizer (softener) in cork gaskets and tiles.

   Unlike ordinary ethers, epoxides can also be cleaved in neutral or alkaline solution by nucleophilic attack on the epoxide itself. The reactions occur by the S_N2 mechanism. For example, ethylene oxide reacts with ammonia as follows:

$$H_3N: + CH_2-CH_2 \rightarrow \underset{NH_2}{CH_2}-\underset{OH}{CH_2} \tag{8.14}$$
$$\underset{\text{ethanolamine}}{}$$

The product, **ethanolamine,** is a water-soluble organic base that is used for absorbing and concentrating carbon dioxide in the manufacture of dry ice.

Problem 8.9   **Write an equation for the reaction of sodium ethoxide in ethanol with ethylene oxide.**

   Grignard reagents react with ethylene oxide in a similar way. The product, after hydrolysis, is a primary alcohol with two more carbon atoms than the Grignard reagent:

$$\overset{\delta-}{R}-\overset{\delta+}{MgX} + CH_2-CH_2 \rightarrow R-CH_2CH_2-OMgX \xrightarrow{H-OH} RCH_2CH_2OH + Mg^{2+}OH^-X^- \tag{8.15}$$

Problem 8.10   **Write an equation for the preparation of 1-pentanol from a Grignard reagent and ethylene oxide.**

a word about _____

17. Epoxy Resins

Most people hear the word *epoxy* in connection with *epoxy resins,* materials used as adhesives for bonding to metal, glass, and ceramics. Epoxy resins are also used in surface coatings (for example, paints) because of their exceptional inertness, hardness, and flexibility.

   Two raw materials for the manufacture of epoxy resins are <u>epichlorhydrin</u> and <u>bisphenol-A</u>.

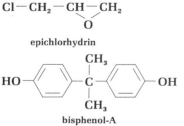

$$Cl-CH_2-CH-CH_2$$
$$\underset{\text{epichlorhydrin}}{O}$$

bisphenol-A

Reaction of a mixture of these two raw materials with base gives an epoxy resin. Although the structures and equations appear complex, we have in fact discussed all the reaction types that are involved.

The base converts bisphenol-A to its dianion (eq. 7.4) which then reacts at each end with the epoxide ring of an epichlorhydrin molecule (Sec. 8.9). The resulting alkoxide displaces chloride ion intramolecularly (an S_N2 reaction) to re-form an epoxide ring at each end of the molecule. Continued sequential reaction with bisphenol-A and epichlorhydrin produces a "linear" epoxy resin, a long-chain polymer that may still have epoxide rings at each end of the molecule (see Figure 8.1). Commercial resins of this type range from liquids (where n is small)

to viscous adhesives to solids used in surface coatings (where n may be as large as 25).

It is possible to take advantage of the remaining epoxide rings and hydroxyl groups in the "linear" polymer to form cross-links between polymer chains, thus substantially increasing the molecular weight of the polymer. This is especially important when the end use is as a surface coating.

Epoxy resins can be made to exhibit a variety of structures by varying n and by forming various types of cross-links. All or part of the bisphenol-A can be replaced by other di- or polyhydroxy compounds, and epoxides other than epichlorhydrin can also be used. Annual world production of these very versatile epoxy resins runs to about a billion pounds.

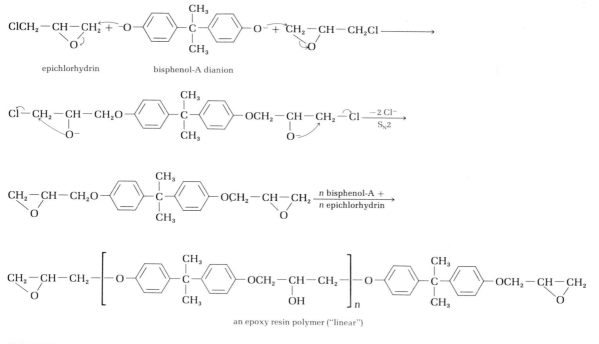

FIGURE 8.1 Production of a "linear" epoxy resin.

<table>
<tr><td>8.10
CYCLIC ETHERS</td><td>Many cyclic ethers with larger rings than epoxides are known. The most common have five- and six-membered rings. They usually have common names.</td></tr>
</table>

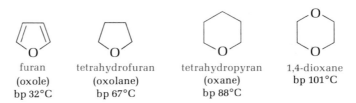

furan	tetrahydrofuran	tetrahydropyran	1,4-dioxane
(oxole)	(oxolane)	(oxane)	bp 101°C
bp 32°C	bp 67°C	bp 88°C	

Furan is an aromatic compound (Sec. 4.16). Hydrogenation of furan gives **tetrahydrofuran** (THF), a particularly useful solvent because it not only dissolves many organic compounds but is miscible with water. Tetrahydrofuran is an excellent solvent—often superior to diethyl ether—in which to prepare Grignard reagents. Although it has the same number of carbon atoms as diethyl ether, they are "pinned back" in a ring. The oxygen in THF is therefore more basic and better at coordinating the magnesium in a Grignard reagent. **Tetrahydropyran** and **1,4-dioxane** are also soluble in both water and organic solvents.

The most common cyclic ethers are the carbohydrates. They are usually either pyranoses (six-membered ring) or furanoses (five-membered ring). Because of their special importance, we will devote a full chapter to their chemistry (Chapter 14).

In recent years there has been much interest in macrocyclic (large-ring) polyethers. Some examples are

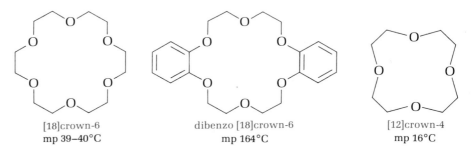

[18]crown-6	dibenzo [18]crown-6	[12]crown-4
mp 39–40°C	mp 164°C	mp 16°C

These compounds are called **crown ethers** because of their crownlike shape.

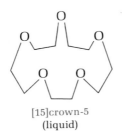

[15]crown-5
(liquid)

The first number in their common names gives the ring size and the second number gives the number of oxygens.

Crown ethers have the unique property of forming complexes with positive ions (Na^+, K^+, and so on). The positive ions fit "inside" the macrocyclic rings selectively, depending on the size of the ring and the size of the ion. For example, [18]crown-6 binds K$^+$ more tightly than it does the smaller Na$^+$ (too loose a fit) or the larger Cs$^+$ (too large to fit in the hole).

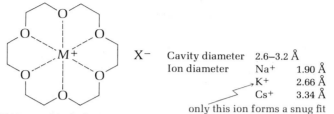

X$^-$ Cavity diameter 2.6–3.2 Å
Ion diameter Na$^+$ 1.90 Å
K$^+$ 2.66 Å
Cs$^+$ 3.34 Å
only this ion forms a snug fit

M$^+$ trapped in [18]crown-6

On the other hand, [15]crown-5 binds Na$^+$, and [12]crown-4 binds Li$^+$.

This complexing ability is so strong that ionic compounds can be dissolved in organic solvents that contain some crown ether. For example, potassium permanganate (KMnO$_4$) is soluble in water but insoluble in benzene. However, if some dicyclohexyl[18]crown-6 is dissolved in the benzene, it is possible to extract the potassium permanganate from the water into the benzene! The resulting "purple benzene" contains free, essentially unsolvated permanganate ions and is a powerful oxidizing agent.

Recently chiral crowns have been synthesized. They may coordinate, for example, with one enantiomer of an amino acid but not the other. This chiral recognition can be used to resolve racemic mixtures. Enzymes also make this distinction between enantiomers, and crown ethers have been studied as enzyme models.

The selective binding of metallic ions by macrocyclic compounds is thought to be important in nature. Several antibiotics, such as **nonactin,** have large rings that contain regularly spaced oxygen atoms. Nonactin (which contains four tetrahydrofuran rings joined through four ester links) selectively binds K$^+$ (compared to Na$^+$) in aqueous media, and it can allow the selective transport of K$^+$ (but not Na$^+$) through cell membranes.

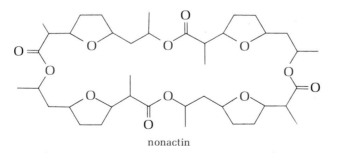

nonactin

8.11 SULFIDES Organic sulfides are **thioethers.** Their structures are analogous to ethers, but they have sulfur in place of oxygen. They are commonly named in a manner similar to ethers.

$$CH_3-S-CH_3 \qquad CH_3CH_2-S-CH_3 \qquad \begin{array}{c} CH_3CHCH_2CH_3 \\ | \\ S-CH_3 \end{array}$$

dimethyl sulfide   ethyl methyl sulfide   2-(methylthio)butane

Sulfides can be prepared from thiols and alkyl halides, with base (see Table 6.1, reaction 12). For example,

$$CH_3SH + CH_3Br \xrightarrow[S_N2]{\text{dil. NaOH}} CH_3SCH_3 + Na^+Br^- + H_2O \qquad (8.16)$$

Problem 8.11   **The reaction in eq. 8.16 really occurs in two steps. What are they? (*Hint:* Review eq. 7.40.)**

Certain sulfides occur in nature. Others may have marked physiological activity. Examples are **diallyl sulfide,** which is a constituent of garlic and onions, and **mustard gas,** which produces severe blisters and has been used in chemical warfare.

$$CH_2=CHCH_2-S-CH_2CH=CH_2 \qquad\qquad ClCH_2CH_2-S-CH_2CH_2Cl$$

diallyl sulfide   di-2-chloroethyl sulfide
bp 139°C   (mustard gas)
                        mp 13°C, bp 217°C

Sulfides can be easily oxidized to sulfoxides or sulfones.

$$R-S-R \xrightarrow[25°C]{H_2O_2} \overset{\overset{O}{\|}}{R-S-R} \xrightarrow[90-100°C]{H_2O_2} \overset{\overset{O}{\|}}{\underset{\underset{O}{\|}}{R-S-R}} \qquad (8.17)$$

sulfide   sulfoxide   sulfone

Dimethyl sulfoxide is a colorless liquid, bp 189°C, which is miscible with water and a good solvent for organic substances as well. It diffuses rapidly through the skin and has been tested medically as an anti-inflammatory agent to treat arthritis.

Sulfides react with alkyl halides to form sulfonium salts.

$$R-\overset{..}{\underset{..}{S}}-R + R'-X \xrightarrow{S_N2} \overset{R'}{\underset{|}{R-S-R}}{}^+ \quad X^- \qquad (8.18)$$

a trialkylsulfonium salt

Sulfonium salts often play a role in biological processes (see A Word About S_N2 Reactions in the Cell, page 161).

ADDITIONAL PROBLEMS

8.12. Write a structural formula for each of the following compounds.
a. di-*n*-propyl ether   **b.** *t*-butyl methyl ether
c. 3-methoxyhexane   **d.** diallyl ether
e. *p*-bromophenyl ethyl ether   **f.** *cis*-2-ethoxycyclopentanol
g. ethylene glycol dimethyl ether   **h.** diethyl sulfide
i. propylene oxide   **j.** ethyloxirane

8.13. Name each of the following compounds.

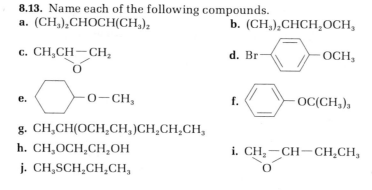

a. $(CH_3)_2CHOCH(CH_3)_2$

b. $(CH_3)_2CHCH_2OCH_3$

c. $CH_3CH\!-\!CH_2$ with O bridge

d. Br—⟨ ⟩—OCH_3

e. ⟨ ⟩—$O\!-\!CH_3$

f. ⟨ ⟩—$OC(CH_3)_3$

g. $CH_3CH(OCH_2CH_3)CH_2CH_2CH_3$

h. $CH_3OCH_2CH_2OH$

i. $CH_2\!-\!CH\!-\!CH_2CH_3$ with O bridge

j. $CH_3SCH_2CH_2CH_3$

8.14. Ethers and alcohols may be isomeric. Write the structures and give the names for all possible isomers with the molecular formula $C_4H_{10}O$.

8.15. Consider four compounds with nearly the same molecular weights: 1,2-dimethoxyethane, ethyl n-propyl ether, hexane, and 1-pentanol. Which would you expect to have the highest boiling point? Which would be most soluble in water? Explain the reasons for your choices.

8.16. 2-Phenylethanol has the aroma of oil of roses and is used in perfumes. Write equations to show how 2-phenylethanol can be synthesized from bromobenzene and ethylene oxide, using a Grignard reagent.

8.17. Ethers are soluble in cold, concentrated sulfuric acid, whereas alkanes are not. This difference can be used as a simple chemical test to distinguish between these two classes of compounds. What chemistry (show an equation) is the basis for this difference?

8.18. Write an equation for each of the following reactions. If no reaction occurs, say so.
a. n-butyl ether + boiling aqueous $NaOH \rightarrow$
b. methyl n-propyl ether + excess HBr (hot) $\rightarrow$
c. n-propyl ether + Na $\rightarrow$
d. ethyl ether + cold concentrated $H_2SO_4 \rightarrow$
e. ethyl phenyl ether + $BBr_3 \rightarrow$

8.19. When heated with excess HBr, a cyclic ether gave 1,4-dibromobutane as the only organic product. Write the structure of the ether and an equation for the reaction.

8.20. Write equations for the reaction of each of the following with (1) Mg in ether and (2) addition of D_2O to the resulting solution.
a. $CH_3CH_2CH_2CH_2Br$ b. $CH_3OCH_2CH_2CH_2Br$

8.21. Write an equation for the reaction of ethylene oxide with:
a. 1 mol of HBr   b. excess HBr   c. phenol + H^+

8.22. $CH_3CH_2OCH_2CH_2OH$ (ethyl cellosolve) and $CH_3CH_2OCH_2CH_2OCH_2CH_2OH$ (ethyl carbitol) are solvents used in the formulation of lacquers. They are produced commercially from ethylene oxide and certain other reagents. Show, via equations, how this might be done.

8.23. The first commercial method for producing ethylene oxide involved treatment of ethylene with hypochlorous acid (HO—Cl), followed by reaction of the product with dilute base. Write equations for these reactions, and describe the mechanism of each step.

8.24. Write out the steps in the reaction mechanisms for the reactions given in eq. 8.13.

8.25. What chemical test would distinguish the compounds in each of the following pairs? Indicate what is visually observed in each test.
a. n-propyl ether and hexane
b. ethyl phenyl ether and allyl phenyl ether
c. 2-butanol and methyl n-propyl ether
d. phenol and methyl phenyl ether

8.26. An organic compound with the molecular formula $C_4H_{10}O_3$ shows properties of both an alcohol and an ether. When treated with an excess of hydrogen bromide, it yields only one organic compound, 1,2-dibromoethane. Draw a structural formula for the original compound.

8.27. 1,4-Dioxane (Sec. 8.10) can be synthesized by slowly distilling a mixture of ethylene glycol ($HOCH_2CH_2OH$) and dilute sulfuric acid. Write a series of equations that describe the mechanism of this reaction.

8.28. Furan (Sec. 8.10) is insoluble in water, whereas the closely related tetrahydrofuran is completely miscible with water. Suggest a possible explanation.

8.29. Draw electron-dot formulas for dimethyl sulfide and dimethyl sulfoxide. Show any formal charges that may be present. Explain the sharp rise in boiling point in going from dimethyl sulfide (bp 38°C) to dimethyl sulfoxide (bp 189°C).

nine

Aldehydes and Ketones

We now come to the structure and reactions of possibly the most important functional group in organic chemistry, the **carbonyl group,** $\diagdown C=O$.

This group is present in aldehydes, ketones, carboxylic acids, esters, and several other classes of compounds. These compounds are important in many biological processes and often have commercial significance as well. In this chapter we will discuss aldehydes and ketones and in the next chapter carboxylic acids and related compounds.

    Aldehydes have at least one hydrogen atom attached to the carbonyl group. The remaining group may be another hydrogen atom or an alkyl or aryl group.

$$\underset{\text{aldehyde group}}{-\overset{\overset{\text{O}}{\|}}{C}-H \text{ or } -CHO} \qquad \underset{\text{formaldehyde}}{H-\overset{\overset{\text{O}}{\|}}{C}-H} \qquad \underset{\text{aliphatic aldehyde}}{R-\overset{\overset{\text{O}}{\|}}{C}-H} \qquad \underset{\text{aromatic aldehyde}}{Ar-\overset{\overset{\text{O}}{\|}}{C}-H}$$

    In **ketones,** the carbonyl carbon atom is connected to two other carbon atoms.

$$\underset{\text{aliphatic ketone}}{R-\overset{\overset{\text{O}}{\|}}{C}-R} \qquad \underset{\text{alkyl aryl ketone}}{R-\overset{\overset{\text{O}}{\|}}{C}-Ar} \qquad \underset{\text{aromatic ketone}}{Ar-\overset{\overset{\text{O}}{\|}}{C}-Ar} \qquad \underset{\text{a cyclic ketone}}{C=O}$$

In the IUPAC system, the characteristic ending for aldehydes is -al (from the first syllable of aldehyde). The following examples illustrate the system:

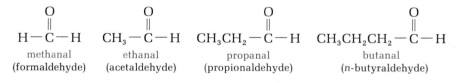

$$\underset{\substack{\text{methanal}\\\text{(formaldehyde)}}}{H-\overset{\overset{\text{O}}{\|}}{C}-H} \qquad \underset{\substack{\text{ethanal}\\\text{(acetaldehyde)}}}{CH_3-\overset{\overset{\text{O}}{\|}}{C}-H} \qquad \underset{\substack{\text{propanal}\\\text{(propionaldehyde)}}}{CH_3CH_2-\overset{\overset{\text{O}}{\|}}{C}-H} \qquad \underset{\substack{\text{butanal}\\\text{(n-butyraldehyde)}}}{CH_3CH_2CH_2-\overset{\overset{\text{O}}{\|}}{C}-H}$$

Of course, aldehydes have been known for such a long time that many have common names. These are shown below the IUPAC names. They are still in common use, so you should learn them.

For substituted aldehydes, we number the chain starting with the aldehyde carbon, as the following examples illustrate:

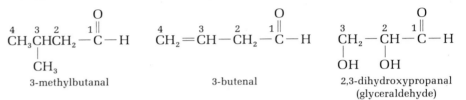

3-methylbutanal   3-butenal   2,3-dihydroxypropanal
(glyceraldehyde)

For cyclic aldehydes, the suffix -*carbaldehyde* is used. Aromatic aldehydes often have common names:

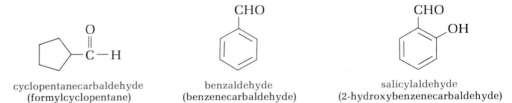

cyclopentanecarbaldehyde   benzaldehyde   salicylaldehyde
(formylcyclopentane)   (benzenecarbaldehyde)   (2-hydroxybenzenecarbaldehyde)

In the IUPAC system, the ending for ketones is -*one* (from the last syllable of ketone). The chain is numbered to give the carbonyl carbon the lowest possible number. Commonly, ketones are named by adding the word *ketone* to the names of the alkyl or aryl groups attached to the carbonyl carbon. In still other cases, common names are used. The following examples illustrate these methods:

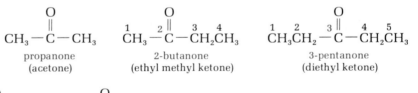

propanone   2-butanone   3-pentanone
(acetone)   (ethyl methyl ketone)   (diethyl ketone)

    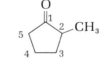   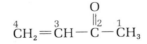

cyclohexanone   2-methylcyclopentanone   3-buten-2-one
(methyl vinyl ketone)

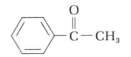   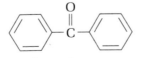   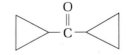

acetophenone   benzophenone   dicyclopropyl ketone
(methyl phenyl ketone)   (diphenyl ketone)

Problem 9.1   **Write the structures for the following compounds.**
 a. pentanal   b. *p*-bromobenzaldehyde
 c. 2-pentanone   d. *t*-butyl methyl ketone

Problem 9.2   **Write a correct name for:**
 a. $(CH_3)_2CHCH_2CH{=}O$ b. $CH_3CH{=}CHCH{=}O$

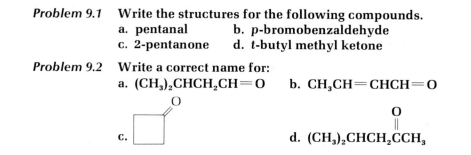

 c.

 d. $(CH_3)_2CHCH_2\overset{\overset{\displaystyle O}{\|}}{C}CH_3$

9.3
SOME COMMON
ALDEHYDES AND
KETONES

Formaldehyde, the simplest aldehyde, is manufactured on a very large scale via the oxidation of methanol (Sec. 7.12).

$$CH_3OH \xrightarrow[600–700°C]{\text{Ag catalyst}} CH_2{=}O + H_2 \tag{9.1}$$

Annual world production is nearly 2 million tons. Although formaldehyde is a gas (bp $-21°C$), it cannot be stored in the free state because it readily polymerizes. Normally it is supplied as a 37% aqueous solution called **formalin.** In this form it is used as a disinfectant and preservative, but most formaldehyde is used in the manufacture of plastics. Formaldehyde is a suspected carcinogen. Special safety precautions should always be taken when handling it.

Acetaldehyde boils close to room temperature (bp 20°C). It was at one time manufactured via the hydration of acetylene (eq. 3.49). Now it is manufactured mainly by the Wacker Process, which involves direct selective oxidation of ethylene over a palladium–copper catalyst.

$$2\,CH_2{=}CH_2 + O_2 \xrightarrow[100–130°C]{\text{Pd–Cu}} 2\,CH_3CH{=}O \tag{9.2}$$

Acetaldehyde is also manufactured by oxidation of ethanol. About half of the 2 million tons of acetaldehyde produced annually is oxidized to acetic acid. The rest is used for the production of 1-butanol (Sec. 9.21) and other commercial chemicals.

Acetone, the simplest ketone, is also produced on a large scale—about 2 million tons annually. The most common methods for its commercial synthesis are the Wacker oxidation of propene (analogous to eq. 9.2), the oxidation of isopropyl alcohol (eq. 7.27, R = R′ = CH₃), and the oxidation of isopropylbenzene (page 115). In the United States some acetone is also produced through starch fermentation. About 30% of the acetone is used directly as such, for it is not only completely miscible with water but also an excellent solvent for many organic substances (such as resins, paints, dyes, and nail polish). The rest is used to manufacture other commercial chemicals, including bisphenol-A for epoxy resins (see A Word About Epoxy Resins, page 200).

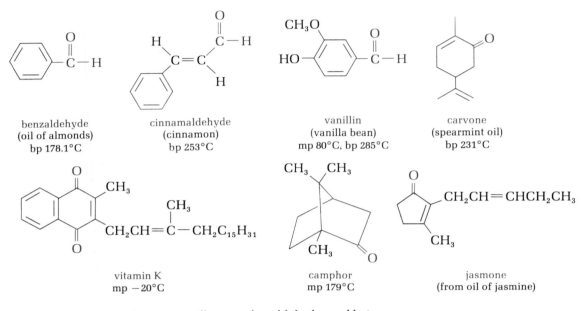

FIGURE 9.1 Some naturally occurring aldehydes and ketones.

9.4
ALDEHYDES AND
KETONES IN
NATURE

Aldehydes and ketones occur very widely in nature (Figures 1.11 and 1.12). Figure 9.1 gives a few additional examples. Many have very pleasant odors and flavors.

9.5
THE CARBONYL
GROUP

To best understand the reactions of aldehydes, ketones, and other carbonyl compounds, we must first appreciate the structure and properties of the carbonyl group.

The carbon–oxygen double bond consists of a σ bond and a π bond (Figure 9.2). The carbon atom is sp²-hybridized. The σ bond results from overlap of a carbon sp² orbital with an oxygen p orbital. As expected for sp² hybridization, *the three atoms attached to the carbonyl carbon lie in a plane with bond angles of 120°*. The π bond is formed by overlap of the remaining p orbital on carbon with an oxygen p orbital. There are also two unshared electron pairs on the oxygen atom, in orbitals that lie in the same plane as the carbonyl carbon and the three atoms attached to it. The $C=O$ bond distance is 1.24 Å, compared with the $C-O$ distance in alcohols and ethers of 1.43 Å.

Oxygen is much more electronegative than carbon. Consequently the electrons in the $C=O$ bond are attracted to the oxygen, and the bond is highly polarized. This effect is especially pronounced for the less firmly held π electrons. It can be expressed in the following ways:

FIGURE 9.2

Bonding in the carbonyl group.

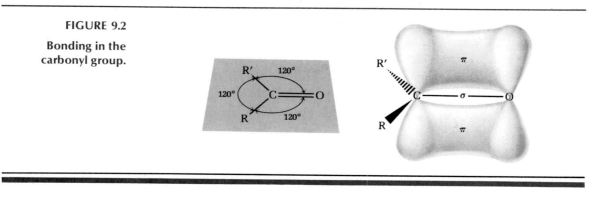

resonance contributors to the carbonyl group

polarization of the carbonyl group

As a consequence of this polarization, many reactions of carbonyl compounds involve attack of a nucleophile (a supplier of electrons) at the carbonyl carbon atom.

Example 9.1 Explain the fact that carbonyl compounds boil at higher temperatures than hydrocarbons but at lower temperatures than alcohols of comparable molecular weight (for example, $CH_3(CH_2)_3CH_3$, bp 36°C; $CH_3(CH_2)_2CH=O$, bp 75°C; $CH_3(CH_2)_2CH_2OH$, bp 118°C).

Solution Being polar, carbonyl compounds tend to associate: the positive part of one molecule is attracted to the negative part of another molecule. This attractive force, not significant in hydrocarbons, requires energy (heat) to overcome when the substance is converted from liquid to vapor. Having no O—H bonds, however, carbonyl compounds cannot form hydrogen bonds with one another, as alcohols can.

Example 9.2 Explain why carbonyl compounds of lower molecular weight are water-soluble.

Solution Although they cannot form hydrogen bonds with themselves, carbonyl compounds readily form hydrogen bonds with other O—H or N—H compounds:

$$\overset{\delta+}{C}=\overset{\delta-}{\underset{\cdot\cdot}{O}}:\cdots H-O\overset{H}{\diagup}$$

Problem 9.3 Arrange benzaldehyde (mol. wt. 106), benzyl alcohol (mol. wt. 108), hydroquinone (mol. wt. 110), and p-xylene (mol. wt. 106) in order of:
a. increasing boiling point
b. increasing water solubility

**9.6
NUCLEOPHILIC
ADDITION TO
CARBONYL
GROUPS;
MECHANISTIC
CONSIDERATIONS**

Nucleophiles attack the carbon atom of a carbon–oxygen double bond because that carbon has a partial positive charge. The π electrons of the C=O bond move to the oxygen atom, which (because of its electronegativity) can easily accommodate the negative charge that it acquires. When these reactions are carried out in a hydroxylic solvent (such as alcohol or water), the reaction is usually completed by addition of a proton to the negative oxygen.

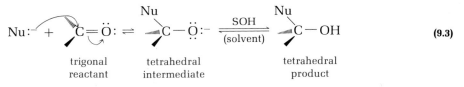

$$\text{(9.3)}$$

The carbonyl carbon, which is trigonal and sp^2-hybridized in the starting aldehyde or ketone, becomes tetrahedral and sp^3-hybridized in the reaction product.

Because of the unshared electron pairs on the oxygen atom, carbonyl compounds are weakly basic and can be protonated. Acids can catalyze the addition of weak nucleophiles to carbonyl compounds by protonating the carbonyl oxygen atom.

$$\ce{>C=O: + H+ ->} \left[\ce{>C\overset{+}{=}OH <-> >\overset{+}{C}-OH} \right] \xrightarrow{\ce{Nu:-}} \ce{Nu\atop >C-OH} \qquad \text{(9.4)}$$

a resonance-stabilized
carbocation

In this way the carbonyl carbon is made more positive, thus enhancing its susceptibility to attack by nucleophiles.

In general, *ketones are somewhat less reactive than aldehydes toward nucleophiles*. There are two main reasons for this reactivity difference. The first is steric. The carbonyl carbon atom is more crowded in ketones (two R groups) than in aldehydes (one R group and one H). Since in nucleophilic addition we bring these attached groups closer together (the hybridization changes from sp^2 to sp^3 and the bond angles decrease from 120° to 109.5°), less strain is involved in additions to aldehydes than in additions to ketones. The second reason is electronic. Simple R groups (alkyl) are usually electron-donating relative to hydrogen (Sec. 3.14). Consequently, they tend to neutralize the partial positive charge on the carbonyl carbon, decreasing its reactivity toward nucleophiles. If the R groups are strongly electron-withdrawing (halogens, for example), they can have the opposite effect and increase the carbonyl reactivity toward nucleophiles.

In the following discussion, we will classify nucleophilic additions to aldehydes and ketones according to the type of new bond formed to the carbonyl carbon. We will consider oxygen, carbon, and nitrogen nucleophiles.

9.7
ADDITION OF ALCOHOLS; FORMATION OF HEMIACETALS AND ACETALS

Alcohols are oxygen nucleophiles. They can attack the carbonyl carbon of aldehydes or ketones, resulting in addition to the C=O bond.

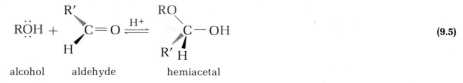

alcohol   aldehyde   hemiacetal

(9.5)

Because alcohols are weak nucleophiles, an acid catalyst is generally used. The product is a **hemiacetal,** which contains both an alcohol and an ether function at the same carbon atom. The addition is reversible, however, and most attempts to isolate hemiacetals give the starting alcohol and carbonyl compound.

Example 9.3   Show the steps in the mechanism for eq. 9.5.

Solution   First the carbonyl carbon is protonated by the acid catalyst, as in eq. 9.4. The alcohol oxygen then attacks the carbonyl carbon, and a proton is lost from the resulting positive oxygen. *Each step is reversible.*

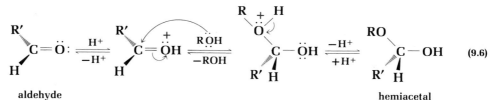

aldehyde   hemiacetal

(9.6)

Problem 9.4   Write an equation for the formation of a hemiacetal from acetaldehyde, ethanol, and H+. Show each step in the reaction mechanism.

In the presence of *excess alcohol,* hemiacetals react further to form **acetals.**

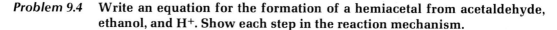

hemiacetal   acetal

(9.7)

The hydroxyl group of the hemiacetal is replaced by an alkoxyl group. The reaction is analogous to the acid-catalyzed formation of ethers from alcohols (eq. 8.7). Acetals have two ether functions at the same carbon atom.

Example 9.4   Show the steps in the mechanism for eq. 9.7.

Solution

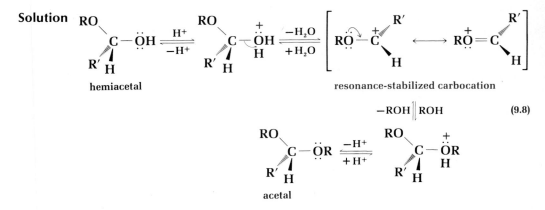

hemiacetal   resonance-stabilized carbocation

$$-ROH \parallel ROH \qquad (9.8)$$

acetal

Either oxygen of the hemiacetal can be protonated. If the hydroxyl group is protonated, loss of water leads to a resonance-stabilized carbocation. Reaction of this carbocation with the alcohol, which is usually the solvent and is present in large excess, gives (after proton loss) the acetal. The mechanism is like an S_N1 reaction. *Each step is reversible.*

Example 9.5   What would happen if the "ether" oxygen of the hemiacetal were protonated instead of the "alcohol" oxygen?

Solution   This can happen because both oxygens have unshared electron pairs and are basic. Protonation of the ether oxygen would be the first step in the reversal of eq. 9.6. Loss of alcohol from this intermediate would continue the reversal of eq. 9.6 back to aldehyde and alcohol. *Excess alcohol is present, however, so all the equilibria in eq. 9.6 would be driven forward again to the hemiacetal.* Protonation of the "ether" oxygen is therefore nonproductive.

Problem 9.5   Write an equation for the reaction of the hemiacetal $CH_3\overset{\overset{\displaystyle OH}{\displaystyle |}}{C}HOCH_2CH_3$ with excess ethanol and H^+. Show each step in the mechanism.

Ketones react with alcohols in an analogous manner to aldehydes, to form **hemiketals** and **ketals**. If a glycol is used as the alcohol component, the product has a cyclic structure.

$$\underset{\substack{\text{acetone}}}{\overset{\substack{CH_3}}{\underset{\substack{CH_3}}{>}}C=O} + \underset{\substack{\text{ethylene glycol}}}{\overset{\substack{HO-CH_2}}{\underset{\substack{HO-CH_2}}{|}}} \overset{H^+}{\rightleftharpoons} \underset{\substack{\text{acetone–ethylene}\\\text{glycol ketal}}}{\overset{\substack{CH_3}}{\underset{\substack{CH_3}}{>}}C\overset{\substack{O-CH_2}}{\underset{\substack{O-CH_2}}{|}}} + H_2O \qquad (9.9)$$

To summarize, aldehydes and ketones react with alcohols to form first hemiacetals (or hemiketals) and then, if excess alcohol is present, acetals (or ketals).

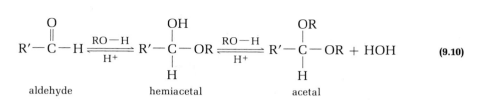

$$(9.10)$$

aldehyde　　　　　　　hemiacetal　　　　　　　acetal

The equilibria are driven in the forward direction by excess alcohol. On the other hand, an acetal can be hydrolyzed to its component aldehyde and alcohol by treatment with *excess water* in the presence of acid (the reverse of eq. 9.10). The hemiacetal intermediate in both the forward and the reverse processes normally cannot be isolated when R and R' are simple alkyl or aryl groups.

Example 9.6　**Write an equation for the reaction of benzaldehyde dimethylacetal with H_2O, H^+.**

Solution

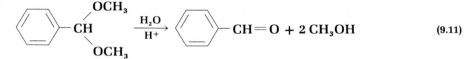

$$(9.11)$$

Problem 9.6　**Show the steps in the mechanism for eq. 9.11.**

The acid-catalyzed cleavage of acetals (and ketals) occurs much more readily than the acid-catalyzed cleavage of simple ethers (Sec. 8.7) because the intermediate carbocation (eq. 9.8) is resonance-stabilized. Acetals and ketals, like ordinary ethers, are stable to base.

The reactions discussed in this section are extremely important, because they are crucial to understanding the chemistry of carbohydrates (Chapter 14).

**9.8
ADDITION OF
WATER;
HYDRATION OF
ALDEHYDES AND
KETONES**

Water, like alcohols, is an oxygen nucleophile and can add reversibly to aldehydes and ketones. For example, formaldehyde in water exists mainly as the hydrate.

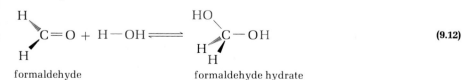

$$(9.12)$$

formaldehyde　　　　　　　　　formaldehyde hydrate

With most other aldehydes or ketones, however, the hydrates cannot be isolated because they readily lose water to re-form the carbonyl compound. An exception is trichloroacetaldehyde (chloral), which forms a stable crystalline hydrate, $CCl_3CH(OH)_2$. **Chloral hydrate** is used in medicine as a sedative and in veterinary medicine as a narcotic and anesthetic for horses, cattle, swine, and poultry. The potent drink known as a Mickey Finn is a combination of alcohol and chloral hydrate.

Problem 9.7 Hydrolysis of $CH_3CBr_2CH_3$ with sodium hydroxide does *not* give $CH_3C(OH)_2CH_3$. It gives acetone instead. Explain.

9.9 ADDITION OF GRIGNARD REAGENTS AND ACETYLIDES

Grignard reagents (Sec. 8.5) act as carbon nucleophiles toward carbonyl compounds. The R group of the Grignard reagent attacks the carbonyl carbon, forming a new carbon–carbon bond. The product is an alkoxide salt, which can then be hydrolyzed to an alcohol.

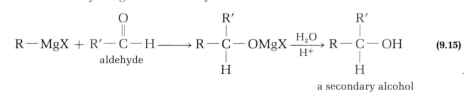

$$\mathrm{C{=}O + RMgX} \xrightarrow{\text{ether}} \mathrm{C-\overset{-}{O}\overset{+}{M}gX} \xrightarrow[\text{HCl}]{H_2O} \mathrm{C-OH + Mg^{2+}X^-Cl^-} \qquad (9.13)$$

intermediate addition product (a magnesium alkoxide)　　an alcohol

The reaction is normally carried out by slowly adding a dry ether solution of the aldehyde or ketone to the Grignard reagent. The reaction is usually exothermic and goes to completion at room temperature. After all the carbonyl compound is added, the resulting magnesium alkoxide is hydrolyzed with aqueous acid.

The reaction of a Grignard reagent with a carbonyl compound is very useful. Many alcohols can be synthesized in this way by the proper choice of reagents. The choice of carbonyl compound determines the class of alcohol produced. *Formaldehyde gives primary alcohols:*

$$R-MgX + H-\overset{\overset{\displaystyle O}{\|}}{C}-H \longrightarrow R-\overset{\overset{\displaystyle H}{|}}{\underset{\underset{\displaystyle H}{|}}{C}}-OMgX \xrightarrow[H^+]{H_2O} R-\overset{\overset{\displaystyle H}{|}}{\underset{\underset{\displaystyle H}{|}}{C}}-OH \qquad (9.14)$$

formaldehyde　　　　　　　　　　　　　　　　a primary alcohol

Other aldehydes give secondary alcohols:

$$R-MgX + R'-\overset{\overset{\displaystyle O}{\|}}{C}-H \longrightarrow R-\overset{\overset{\displaystyle R'}{|}}{\underset{\underset{\displaystyle H}{|}}{C}}-OMgX \xrightarrow[H^+]{H_2O} R-\overset{\overset{\displaystyle R'}{|}}{\underset{\underset{\displaystyle H}{|}}{C}}-OH \qquad (9.15)$$

aldehyde　　　　　　　　　　　　　　　　a secondary alcohol

Ketones give tertiary alcohols:

$$R-MgX + R'-\overset{\overset{\displaystyle O}{\|}}{C}-R'' \longrightarrow R-\overset{\overset{\displaystyle R'}{|}}{\underset{\underset{\displaystyle R''}{|}}{C}}-OMgX \xrightarrow[H^+]{H_2O} R-\overset{\overset{\displaystyle R'}{|}}{\underset{\underset{\displaystyle R''}{|}}{C}}-OH \qquad (9.16)$$

ketone　　　　　　　　　　　　　　　　a tertiary alcohol

Note that only one of the R groups (shown in black) attached to the hydroxyl-bearing carbon of the desired alcohol comes from the Grignard reagent. The rest of the alcohol skeleton comes from the carbonyl compound.

Example 9.7 **Show how the alcohol**

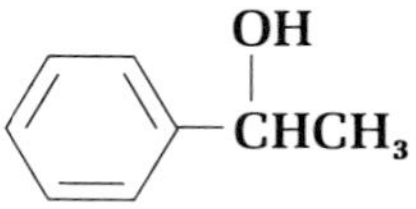

OH
|
⟨benzene⟩—CHCH₃

could be synthesized from a Grignard reagent and a carbonyl compound.

Solution The alcohol is secondary, so the carbonyl compound will be an aldehyde. There are two possibilities:

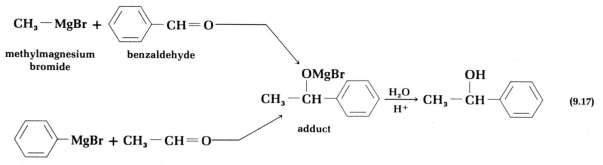

CH_3—MgBr + ⟨benzene⟩—CH=O

methylmagnesium
bromide

benzaldehyde

⟨benzene⟩—MgBr + CH_3—CH=O

phenylmagnesium
bromide

acetaldehyde

$$CH_3-\overset{\underset{\displaystyle |}{OMgBr}}{CH}-⟨benzene⟩ \xrightarrow[H^+]{H_2O} CH_3-\overset{\underset{\displaystyle |}{OH}}{CH}-⟨benzene⟩ \quad (9.17)$$

adduct

The choice between the possible sets of reactants may be based on availability or cost, or it may sometimes be made for chemical reasons (for example, the more reactive aldehyde or ketone may be selected).

Problem 9.8 Show how each of the following alcohols could be made from a Grignard reagent and a carbonyl compound.

a. 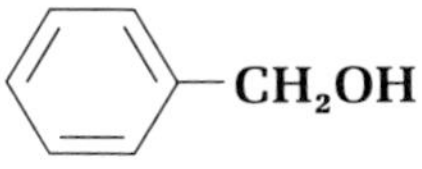 ⟨benzene⟩—CH_2OH b. 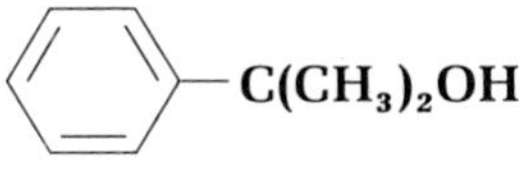 ⟨benzene⟩—$C(CH_3)_2OH$

Other organometallic reagents, such as organolithium compounds (eq. 8.6) and acetylides (eq. 6.11) react with carbonyl compounds in a manner similar to Grignard reagents. For example,

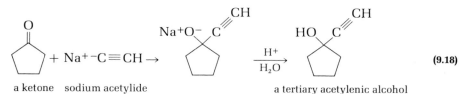

a ketone   sodium acetylide   a tertiary acetylenic alcohol

$$(9.18)$$

This type of reaction is used commercially in the final step of the synthesis of widely used oral contraceptives (see Problem 9.40).

9.10
ADDITION OF
HYDROGEN
CYANIDE;
CYANOHYDRINS

Hydrogen cyanide adds to the carbonyl group of aldehydes and ketones to form **cyanohydrins.** A basic catalyst is required.

$$\ce{>C=O + HCN ->[OH^-] >C-OH}$$ with CN attached

$$\text{(9.19)}$$

Acetone, for example, reacts as follows:

$$CH_3-\overset{\overset{\displaystyle O}{\|}}{C}-CH_3 + HCN \xrightarrow{OH^-} CH_3-\overset{\overset{\displaystyle OH}{|}}{\underset{\underset{\displaystyle CN}{|}}{C}}-CH_3 \qquad \text{(9.20)}$$

acetone   acetone cyanohydrin

Hydrogen cyanide has no unshared electron pair on carbon, so it cannot function as a carbon nucleophile. The base, however, converts some of the hydrogen cyanide to cyanide ion, which then behaves as a carbon nucleophile.

$$\ce{>C=\overset{..}{\underset{..}{O}}: + {}^{-}:C#N: <=> >C-\overset{..}{\underset{..}{O}}:^{-}}$$ with CN attached

$$\text{(9.21)}$$

$$\ce{>C-O^- + HCN <=> >C-OH + {}^-CN}$$ with CN attached

$$\text{(9.22)}$$

cyanohydrin

Because of its volatility and extreme toxicity, the hydrogen cyanide (bp 26°C) is usually generated *in situ,* by adding strong acid to a mixture of the carbonyl compound and sodium cyanide.

Cyanohydrins have an OH and a CN group attached to the same carbon. As we will see in future chapters, cyanohydrins are useful intermediates for synthesis.

Problem 9.9   **Write equations for the addition of HCN to:**
a. acetaldehyde   b. benzaldehyde

9.11
ADDITION OF
NITROGEN
NUCLEOPHILES

Ammonia, amines, and certain related compounds such as hydroxylamine ($\overset{..}{N}H_2-OH$) have an unshared electron pair on the nitrogen atom and act as nitrogen nucleophiles toward the carbonyl carbon atom. For example, primary amines react as follows:

$$\ce{>C=O + \overset{..}{N}H_2-R <=> \left[>C\overset{\displaystyle OH}{\underset{\displaystyle NHR}{<}} \right] ->[-HOH] >C=NR} \qquad \text{(9.23)}$$

primary   tetrahedral   imine
amine   addition product   (Schiff's base)

The tetrahedral addition product is similar to a hemiacetal, but it has NH in place of one of the oxygens. These addition products are normally not stable. They eliminate water to form a product with a carbon–nitrogen double bond. With primary amines, the products are called **imines,** or **Schiff's bases.** They are important intermediates in some biochemical reactions, particularly in the binding of carbonyl compounds to enzymes, which usually have free amino groups.

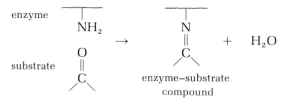

$$(9.24)$$

Example 9.8   Write the steps in the mechanism for eq. 9.23.

Solution

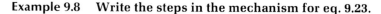
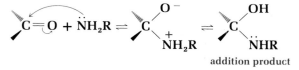

The product that is formed first is a dipolar ion. The positive nitrogen (an ammonium-type ion) loses a proton and the negative oxygen (an alkoxide ion) gains a proton, thus forming the tetrahedral addition product. A 1,2-elimination of water then gives the observed product.

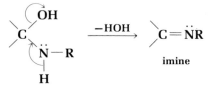

Problem 9.10   Write an equation for the reaction of benzaldehyde with aniline $(C_6H_5-NH_2)$.

Other ammonia derivatives react with carbonyl compounds in a similar manner to primary amines. Table 9.1 lists some specific examples.

Oximes and **hydrazones,** prepared from **hydroxylamine** and **hydrazines,** respectively, are usually crystalline solids whose characteristic melting points can be used to identify a particular carbonyl compound. The melting points of the solid oximes derived from three liquid carbonyl compounds are shown in eq. 9.25.

$$\underset{R}{\overset{CH_3}{\diagdown}}C{=}O + H_2N-OH \rightarrow \underset{R}{\overset{CH_3}{\diagdown}}C{=}N-OH + H_2O \qquad (9.25)$$

R = H	bp 20°C	mp 47°C
CH₃	bp 56°C	mp 80°C
C₆H₅	bp 202°C	mp 58°C

TABLE 9.1 Nitrogen derivatives of carbonyl compounds	Formula of ammonia derivative	Name	Formula of carbonyl derivative	Name
	RNH_2 or $ArNH_2$	primary amine	$\diagup C{=}NR$ or $\diagup C{=}NAr$	imine or Schiff's base
	NH_2OH	hydroxylamine	$\diagup C{=}NOH$	oxime
	NH_2NH_2	hydrazine	$\diagup C{=}NNH_2$	hydrazone
	$NH_2NHC_6H_5$	phenylhydrazine	$\diagup C{=}NNHC_6H_5$	phenylhydrazone

Problem 9.11 **Write equations for the reaction of:**
a. propanal with hydroxylamine
b. benzaldehyde with phenylhydrazine ($C_6H_5 - NHNH_2$)

**9.12
ADDITION OF
HYDROGEN**

Aldehydes and ketones are easily reduced to primary or secondary alcohols, respectively. This reduction can be accomplished in many ways, most commonly by various metal hydrides or by catalytic hydrogenation.

The most common metal hydrides used to reduce carbonyl compounds are **lithium aluminum hydride** ($LiAlH_4$) and **sodium borohydride** ($NaBH_4$). Reduction occurs by nucleophilic attack of hydride ion (H^-) on the carbonyl carbon atom. For example, with lithium aluminum hydride the reduction can be represented as follows:

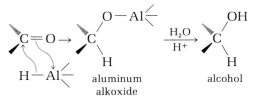

(9.26)

The initial product is an aluminum alkoxide, which is subsequently hydrolyzed by water and acid to give the alcohol. A specific example is

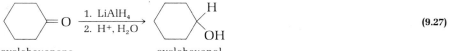

(9.27)

The net result is addition of hydrogen across the carbon–oxygen double bond.

Because the carbon–carbon double bond is not readily attacked by nucleophiles, metal hydrides can be used to reduce a carbon–oxygen double bond to the corresponding alcohol without reducing a carbon–carbon double bond present in the same compound.

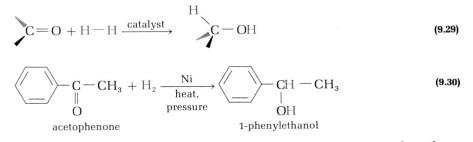

$$CH_3 — CH{=}CH — \overset{\overset{\displaystyle O}{\|}}{CH} \xrightarrow{\text{LiAlH}_4} CH_3CH{=}CH — CH_2OH \tag{9.28}$$

crotonaldehyde   crotyl alcohol
(2-butenal)   (2-buten-1-ol)

Catalytic hydrogenation of carbonyl compounds requires a catalyst, such as copper chromite or nickel.

$$\overset{\diagdown}{\underset{\diagup}{}}C{=}O + H — H \xrightarrow{\text{catalyst}} \overset{H\diagdown}{\underset{\diagup}{}}C — OH \tag{9.29}$$

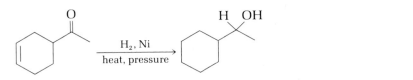

acetophenone   1-phenylethanol

The reaction conditions required are, in general, more severe than those used for the hydrogenation of a carbon–carbon double bond, so both reactions frequently occur simultaneously.

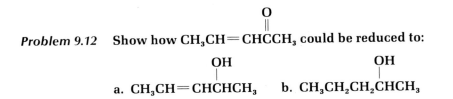

One can avoid this complication by using metal hydrides.

Problem 9.12   Show how $CH_3CH{=}CH\overset{\overset{\displaystyle O}{\|}}{C}CH_3$ could be reduced to:

a. $CH_3CH{=}CHCHCH_3$ with $\overset{\text{OH}}{|}$ above the fourth carbon

b. $CH_3CH_2CH_2CHCH_3$ with $\overset{\text{OH}}{|}$ above the fourth carbon

In Sec. 9.6 to Sec. 9.12, we described reactions of aldehydes and ketones that involve nucleophilic attack at the carbonyl carbon atom. In the remaining sections of this chapter we will describe two other types of reactions that are also characteristic of carbonyl compounds: (1) oxidation, a reaction that allows us to distinguish readily between aldehydes and ketones, and (2) reactions that involve the C—H bond on a carbon atom immediately adjacent to the carbonyl carbon.

9.13
OXIDATION OF
CARBONYL
COMPOUNDS

Aldehydes are more easily oxidized than ketones. Oxidation of an aldehyde gives an acid with the same number of carbon atoms.

$$\underset{\text{aldehyde}}{R-\overset{\displaystyle O}{\overset{\|}{C}}-H} \xrightarrow{[O]} \underset{\text{acid}}{R-\overset{\displaystyle O}{\overset{\|}{C}}-OH} \tag{9.32}$$

Several laboratory tests that distinguish aldehydes from ketones take advantage of their different susceptibilities to oxidation. One of these is **Tollens' test,** in which the silver ammonia complex ion is readily reduced by aldehydes to metallic silver.* The equation for the reaction may be written as

$$\underset{\substack{\text{alde-}\\\text{hyde}}}{RCH} + \underset{\substack{\text{silver ammonia}\\\text{complex ion}\\\text{(colorless)}}}{2\,Ag(NH_3)_2{}^+} + 3\,OH^- \rightarrow \underset{\substack{\text{acid}\\\text{anion}}}{R\overset{\displaystyle O}{\overset{\|}{C}}-O^-} + \underset{\substack{\text{silver}\\\text{mirror}}}{2\,Ag\downarrow} + 4\,NH_3\uparrow + 2\,H_2O \tag{9.33}$$

If the glass vessel in which the test is performed is thoroughly cleaned, the silver will deposit as a mirror on the glass surface. This reaction is commonly employed in silvering glass, using formaldehyde as the aldehyde because it is inexpensive.

Problem 9.13   **Write an equation for the formation of a silver mirror from formaldehyde and Tollens' reagent.**

Fehling's reagent and **Benedict's reagent,** which consist of Cu^{2+} complexed in alkaline solution with tartrate or citrate ions, respectively, react similarly with aldehydes.

$$\underset{\text{blue solution}}{R-\overset{\displaystyle O}{\overset{\|}{C}}-H + 2\,Cu^{2+}} + 5\,OH^- \rightarrow R-\overset{\displaystyle O}{\overset{\|}{C}}-O^- + \underset{\substack{\text{brick-red}\\\text{precipitate}}}{Cu_2O} + 3\,H_2O \tag{9.34}$$

The copper reagent is deep blue. When it reacts with an aldehyde, a brick-red precipitate of Cu_2O is formed.

The reaction of these reagents (Tollens', Fehling's) with aldehydes involves converting the aldehyde C—H bond to a C—O bond. The aldehyde is oxidized to a carboxylic acid having the same number of carbon atoms. Because ketones do not have a hydrogen attached to the carbonyl carbon atom, they are usually not oxidized by these reagents.

Aldehydes are so easily oxidized that samples that have been stored

---

*Silver hydroxide is insoluble in water, so the silver ion must be complexed with ammonia to keep it in solution in a basic medium.

for some time usually contain some of the corresponding acid. This is caused by air oxidation.

$$2\,RCHO + O_2 \rightarrow 2\,RCO_2H \tag{9.35}$$

Ketones can be oxidized, but more vigorous oxidizing conditions are required than for aldehydes. A bond between the carbonyl carbon and one of the adjacent carbon atoms is cleaved, leading to oxidation products with fewer carbon atoms than the starting ketone. A useful exception is the oxidation of cyclic ketones, which give products that retain all the carbon atoms. For example, cyclohexanone is oxidized commercially to **adipic acid,** an important industrial chemical used to manufacture nylon (Sec. 12.12).

one of these C—C bonds is cleaved in the oxidation

cyclohexanone adipic acid

$$\tag{9.36}$$

9.14
KETO–ENOL
TAUTOMERISM

Many aldehydes and ketones actually exist not as a single species, but as an equilibrium mixture of two forms called the **keto form** and the **enol form.** The two forms differ in the location of a proton and a double bond:

$$\tag{9.37}$$

keto form   enol form

This type of structural isomerism is called **tautomerism** (from the Greek *tauto*, the same, and *meros*, part). The two forms of the aldehyde or ketone are called **tautomers.**

Example 9.9 Write formulas for the keto and enol forms of acetone.

Solution

$$\underset{\text{keto form}}{CH_3-\overset{\overset{\displaystyle O}{\|}}{C}-CH_3} \qquad \underset{\text{enol form}}{CH_2=\overset{\overset{\displaystyle OH}{|}}{C}-CH_3}$$

Example 9.10 Are the keto and enol forms of acetone contributors to a resonance hybrid structure?

Solution Definitely not! The two forms are *structural isomers* because they differ in location of the atoms (in particular, a hydrogen atom). Although tautomers are isomers, they readily equilibrate with one another. We use the equilibrium symbol $\rightleftharpoons$ between tautomers, whereas we use the double-headed arrow $\leftrightarrow$ for resonance contributors.

Problem 9.14 **Draw structural formulas for the enol form of:**
a. cyclopentanone   b. acetaldehyde

To be capable of existing in the enol form, a carbonyl compound must have a hydrogen atom attached to the carbon atom adjacent to the carbonyl group. This hydrogen is called an **α hydrogen** and is attached to the **α carbon atom** (from the first letter of the Greek alphabet, α or alpha).

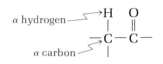

Most aldehydes and ketones exist mainly in the keto form. Acetone, for example, is 99.9997% in the keto form with only 0.0003% enol present. The main reason for the greater stability of the keto form is that the $C=O$ plus $C-H$ bond energy present in the keto form is greater than the $C=C$ plus $O-H$ bond energy of the enol form.

With some molecules, however, the enol form is more significant and may even predominate. With 2,4-pentanedione, for example, 76% of the enol form is present.

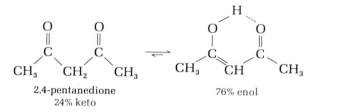

(9.38)

2,4-pentanedione
24% keto

76% enol

In this case, the enol is stabilized by an intramolecular hydrogen bond and by the presence of a conjugated system of double bonds.

The most familiar enols are the *phenols*. Here, the resonance stabilization of the aromatic ring is destroyed in the keto form.

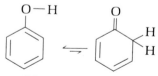

(9.39)

enol form   keto form

Carbonyl compounds that do not have an α hydrogen atom cannot form enols and must exist only in the keto form. Examples include

formaldehyde   benzaldehyde   benzophenone

a word about

18. Tautomerism and Photochromism

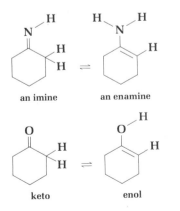

The concept of tautomerism can be expanded beyond keto and enol forms to include any pair or group of isomers that can be easily interconverted by the relocation of an atom and/or bonds. For example, imines and enamines (unsaturated amines) are tautomers whose relationship is very similar to that of keto and enol forms.

an imine   an enamine

keto   enol

Molecules can be designed so that one tautomer

can be converted to the other photochemically—that is, via the absorption of light. For example,

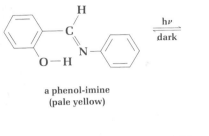

a phenol-imine
(pale yellow)

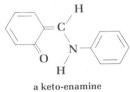

a keto-enamine
(red)

Irradiation of the pale yellow phenol-imine causes the hydrogen atom to shift from the oxygen to the nitrogen, with appropriate rebonding. If the keto-enamine product of this photochemical reaction is allowed to remain in the dark, it reverts to the more stable phenol-imine. Of what use can such a cycle of reactions, with no net reaction, possibly be?

Note that in this example, one tautomer is pale yellow, the other red. This phenomenon, in which two compounds undergo a thermally reversible photochemical color change, is called photochromism. Photochromic substances have many practical uses. One thinks immediately of glasses that, when exposed to sunlight, become darker due to being impregnated with a photochromic material. When the sunlight dims at night, or when one goes indoors, the colored photochromic substance reverts to its colorless form. Photochromic substances can also be used for data storage and display (as in digital watches), as chemical switches in computers, for micro images (microfilm and microfiche), for protection against sudden light flashes (as in nuclear explosions), for camouflage, and in many other creative ways. They have even been used frivolously, as in dolls that can become "suntanned" by a photochromic dye that turns brown on exposure to sunlight.

**9.15
ACIDITY OF
α HYDROGENS;
THE ENOLATE
ANION**

A hydrogen that is adjacent (α) to a carbonyl carbon is more acidic than ordinary C—H hydrogens for two reasons. First, the carbonyl carbon carries a partial positive charge (Sec. 9.5), thus attracting electrons from the neighboring C—H bond and increasing its acidity:

$$-\overset{H}{\underset{|}{\underset{\downarrow}{C}}} \rightarrow \overset{O^{\delta-}}{\underset{}{\overset{\|}{C}}}{}^{\delta+}$$

Bonding electrons are displaced toward the carbonyl carbon and away from the α hydrogen, making it easier for a base to remove the α hydrogen as a proton (that is, without its bonding electrons). The second and more important reason for the enhanced acidity of an α hydrogen is that the resulting anion can be stabilized by resonance.

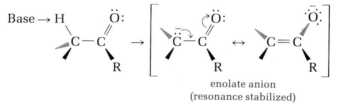

$$\text{(9.40)}$$

enolate anion
(resonance stabilized)

The anion is called an **enolate anion.** The negative charge can be distributed between the α carbon and the carbonyl oxygen atom:

$$\overset{\delta-}{\underset{}{C}}\!=\!\!=\!\!C\overset{O^{\,\delta-}}{\underset{}{}}$$

an enolate anion

Since oxygen is more electronegative than carbon, most of the charge resides on the oxygen atom.

Example 9.11   Draw the formula for the enolate anion of acetone.

Solution
$$\left[\overset{..}{\underset{..}{\underset{|}{C}}}H_2 - \overset{\overset{..}{O}:}{\underset{}{\overset{\|}{C}}} - CH_3 \leftrightarrow CH_2 = \overset{:\overset{..}{O}:^{-}}{\underset{}{\overset{|}{C}}} - CH_3\right]$$

The enolate anion is a resonance hybrid of two contributing structures that differ *only* in arrangement of the electrons.

Problem 9.15   Draw the resonance contributors to the enolate ion of:
a. cyclopentanone   b. acetaldehyde

**9.16
EQUILIBRATION
OF KETO AND
ENOL FORMS**

The equilibration of keto–enol tautomers can be catalyzed either by base or by acid. Base removes the α proton to give an enolate anion (eq. 9.40). The enolate anion can then be reprotonated either on the α carbon (to give back the original carbonyl compound) or on the oxygen (to give the enol).

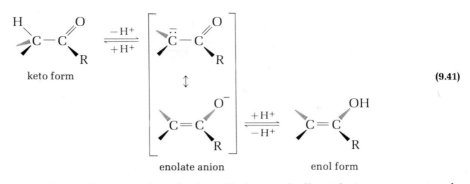

(9.41)

enol form

Ordinary bases such as hydroxide ion and alkoxide ions convert only a small fraction of a simple carbonyl compound to its enolate.

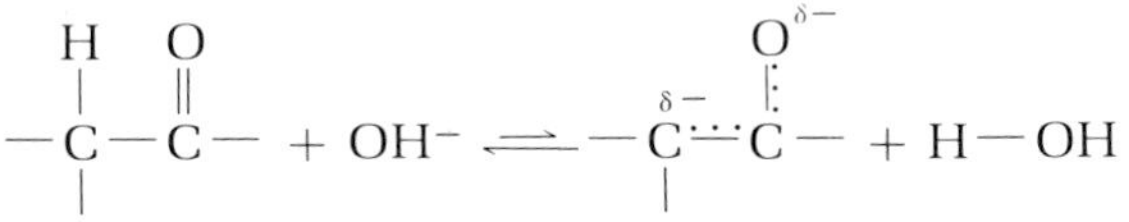

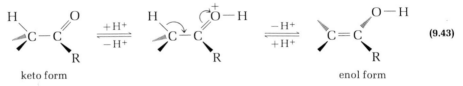

(9.42)

The equilibrium in eq. 9.42 lies mainly to the left, because carbonyl compounds are in general much weaker acids than water (or alcohols).

Acids can also catalyze the equilibration of keto–enol tautomers. They protonate the keto form, and loss of an α hydrogen then generates the enol form.

(9.43)

keto form   enol form

Problem 9.16 **Write an equation that shows how cyclohexanone and its enol can be equilibrated by:**
a. hydroxide ion   b. acid catalysis

9.17
DEUTERIUM
EXCHANGE IN
CARBONYL
COMPOUNDS
Even though the enol content of ordinary aldehydes and ketones is very low, the presence of the enol form can be demonstrated experimentally. For example, the α hydrogens can be exchanged for deuterium by placing the carbonyl compound in a solvent containing O—D bonds, such as D_2O or CH_3OD. The exchange is catalyzed by acid or base. *Only the α hydrogens exchange*, as illustrated by the following examples:

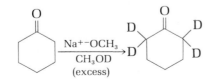

(9.44)

cyclohexanone   2,2,6,6-tetradeuteriocyclohexanone

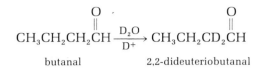

$$\text{CH}_3\text{CH}_2\text{CH}_2\overset{\overset{\displaystyle O}{\|}}{\text{CH}} \xrightarrow[\text{D+}]{\text{D}_2\text{O}} \text{CH}_3\text{CH}_2\text{CD}_2\overset{\overset{\displaystyle O}{\|}}{\text{CH}}$$ (9.45)

butanal   2,2-dideuteriobutanal

Example 9.12   **Write a mechanism for eq. 9.44.**

Solution

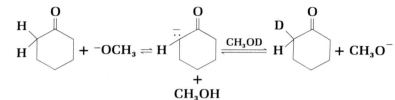

The base (methoxide ion) removes an α proton to form the enolate anion. Reprotonation, but with CH₃OD, replaces the α hydrogen with deuterium. With excess CH₃OD, all four α hydrogens will eventually be exchanged.

Problem 9.17   **Write a mechanism for eq. 9.45.**

9.18
THE ALDOL
CONDENSATION

Enolate anions can act as carbon nucleophiles and add to the carbonyl group of another aldehyde or ketone molecule. This reaction forms the basis of the **aldol condensation,** an extremely useful carbon–carbon bond-forming reaction.

The simplest aldol condensation is the combination of two acetaldehyde molecules, which occurs when a solution of the aldehyde is treated with aqueous base.

$$\overset{\overset{\displaystyle O}{\|}}{\text{CH}_3\text{CH}} + \overset{\overset{\displaystyle O}{\|}}{\text{CH}_3\text{CH}} \underset{}{\overset{\text{OH}^-}{\rightleftharpoons}} \overset{\overset{\displaystyle OH}{|}}{\text{CH}_3\text{CH}} - \text{CH}_2\overset{\overset{\displaystyle O}{\|}}{\text{CH}}$$ (9.46)

acetaldehyde   3-hydroxybutanal
                            (aldol)

The product, which has a four-carbon chain, is called **aldol** (the name comes from the fact that the product is both an *ald*ehyde and an alco*hol*).

The aldol condensation of acetaldehyde occurs according to the following three-step mechanism:

Step 1 $$\overset{\alpha}{\text{CH}_3} - \overset{\overset{\displaystyle O}{\|}}{\text{C}} - \text{H} + \text{OH}^- \rightleftharpoons \overset{..}{\text{CH}}_2 - \overset{\overset{\displaystyle O}{\|}}{\text{C}} - \text{H} + \text{HOH}$$ (9.47)

                                 enolate anion

Step 2 $$\text{CH}_3 - \overset{\overset{\displaystyle O}{\|}}{\text{CH}} + \overset{..}{\text{CH}}_2 - \overset{\overset{\displaystyle O}{\|}}{\text{CH}} \rightleftharpoons \text{CH}_3\overset{\overset{\displaystyle O^-}{|}}{\text{CH}} - \text{CH}_2\overset{\overset{\displaystyle O}{\|}}{\text{CH}}$$ (9.48)

                    nucleophile   an alkoxide ion

Step 3
$$CH_3\overset{O^-}{\underset{|}{CH}}-CH_2\overset{O}{\overset{||}{CH}} + HOH \rightleftharpoons CH_3\overset{OH}{\underset{|}{CH}}\overset{\alpha}{-}CH_2\overset{O}{\overset{||}{CH}} + OH^- \qquad (9.49)$$

In step 1, the base removes an α hydrogen to form the enolate anion. In step 2, this anion adds to the carbonyl carbon of another acetaldehyde molecule, forming a new carbon–carbon bond. Recall that ordinary bases convert only a small fraction of the carbonyl compound to its enolate (eq. 9.42), so that both species are present. In step 3, the alkoxide ion formed in step 2 accepts a proton from the solvent, thus regenerating the hydroxide ion needed for the first step.

In the aldol condensation, the α carbon of one aldehyde (or ketone) molecule becomes connected to the carbonyl carbon of another aldehyde (or ketone) molecule.

$$RCH_2\overset{O}{\overset{||}{CH}} + R\overset{\alpha}{CH_2}\overset{O}{\overset{||}{CH}} \xrightarrow[-H_2O]{OH^-} RCH_2\overset{OH}{\underset{|}{CH}}-\overset{\alpha}{\underset{\underset{R}{|}}{CH}}\overset{O}{\overset{||}{CH}} \qquad (9.50)$$

an aldol

An aldol is defined, therefore, as a 3-hydroxyaldehyde. A ketol is a 3-hydroxyketone.

Problem 9.18 Write the structure of the aldol obtained when propanal ($CH_3CH_2CH{=}O$) is treated with base.

Problem 9.19 Write out the steps in the mechanism for formation of the product in Problem 9.18.

9.19
THE MIXED ALDOL
CONDENSATION

The aldol condensation can be made very versatile by having the enolate anion of one carbonyl compound add to the carbonyl carbon of another. Consider, for example, the reaction between acetaldehyde and benzaldehyde. When treated with base, only acetaldehyde can form an enolate anion (benzaldehyde has no α hydrogen). If the enolate of acetaldehyde adds to the benzaldehyde carbonyl group, a mixed aldol condensation occurs.

$$\text{C}_6\text{H}_5-\overset{O}{\overset{||}{CH}} + \overset{\alpha}{CH_3}\overset{O}{\overset{||}{CH}} \xrightarrow{OH^-} \text{C}_6\text{H}_5-\overset{OH}{\underset{|}{CH}}-CH_2\overset{O}{\overset{||}{CH}} \xrightarrow[-H_2O]{heat} \text{C}_6\text{H}_5-CH{=}CH\overset{O}{\overset{||}{CH}} \qquad (9.51)$$

a mixed aldol   cinnamaldehyde

In this example, the resulting mixed aldol is readily dehydrated on heating to give cinnamaldehyde (the flavor constituent of cinnamon).

Problem 9.20 Write out the steps in the mechanism for eq. 9.51.

Problem 9.21   **Write the structure of the product for the mixed aldol condensation of:**
a. acetone and formaldehyde   b. propanal and benzaldehyde

The aldol condensation, and particularly the mixed aldol, is an important carbon–carbon bond-forming reaction in nature. In the cell, the reaction is catalyzed by a class of enzymes called **aldolases.** For example, the six-carbon sugar fructose (as its 1,6-diphosphate) is synthesized during the biosynthesis of glucose in plants, by the aldol condensation of two three-carbon precursors:

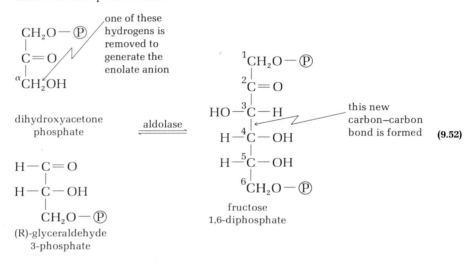

dihydroxyacetone
phosphate

(R)-glyceraldehyde
3-phosphate

fructose
1,6-diphosphate

(9.52)

The enolate ion formed by removing an α hydrogen from dihydroxyacetone phosphate adds to the aldehyde C=O of glyceraldehyde-3-phosphate. A new carbon–carbon bond is formed, which is the bond between C-3 and C-4 in the resulting fructose.

9.20
COMMERCIAL
SYNTHESES VIA
THE ALDOL
CONDENSATION

Aldols are useful in synthesis. They are more easily dehydrated than most alcohols, because the double bond in the resulting unsaturated aldehyde is conjugated with the carbonyl group. Acetaldehyde is converted to crotonaldehyde, 1-butanol, and butanal commercially by way of the aldol condensation.

$$2\ \underset{\text{acetaldehyde}}{CH_3\overset{\displaystyle O}{\overset{\|}{CH}}} \xrightarrow{\ OH^-\ } \underset{\text{aldol}}{CH_3\overset{\displaystyle OH}{\overset{|}{CH}}CH_2\overset{\displaystyle O}{\overset{\|}{CH}}} \xrightarrow[-H_2O]{\ H^+\ } \underset{\text{crotonaldehyde}}{CH_3CH=CH\overset{\displaystyle O}{\overset{\|}{CH}}} \xrightarrow[\text{catalyst}]{\ H_2\ }$$

(9.53)

$$\underset{\text{butanal}}{CH_3CH_2CH_2\overset{\displaystyle O}{\overset{\|}{CH}}} \quad \text{or} \quad \underset{\text{1-butanol}}{CH_3CH_2CH_2CH_2OH}$$

The product obtained in the final hydrogenation depends on the catalyst and reaction conditions.

Butanal can be converted in two steps, via an aldol condensation, to the well-known mosquito repellent "6-12" (2-ethylhexane-1,3-diol).

$$2 \text{ CH}_3\text{CH}_2\text{CH}_2\overset{\overset{\displaystyle O}{\|}}{\text{C}}\text{H} \xrightarrow{\text{OH}^-} \text{CH}_3\text{CH}_2\text{CH}_2\underset{\underset{\displaystyle \text{CH}_3\text{CH}_2}{|}}{\overset{\overset{\displaystyle \text{OH}}{|}}{\text{CH}}}\text{CHC}\overset{\overset{\displaystyle O}{\|}}{}\text{H} \xrightarrow[\text{Ni}]{\text{H}_2} \text{CH}_3\text{CH}_2\text{CH}_2\underset{\underset{\displaystyle \text{CH}_3\text{CH}_2}{|}}{\overset{\overset{\displaystyle \text{OH}}{|}}{\text{CH}}}\text{CHCH}_2\text{OH} \qquad (9.54)$$

butanal   butanal aldol   2-ethylhexane-1,3-diol ("6-12")

Problem 9.22   **2-Ethylhexanol has the greatest economic importance of all higher alcohols. Its esters are used as plasticizers, hydraulic oils, and synthetic lubricants. It can be manufactured from butanal via its aldol condensation product. Suggest with equations how this might be done.**

a word about _____

19. Quinones, Dyes, and Electron Transfer

Quinones constitute a unique class of carbonyl compounds. They are cyclic, conjugated diketones. The simplest example, 1,4-benzoquinone, has already been described (eq. 7.35).

All quinones are colored. Many are naturally occurring plant pigments, and they often exhibit special biological activity.

during the American Revolution were dyed with alizarin. In the mid-nineteenth century, madder was a substantial agricultural crop in Europe, with as many as 70,000 tons in annual production. But once organic chemists determined alizarin's structure, they were able to work out a commercial synthesis and reduce the price dramatically. Many alizarin-type pigments are still in use today.

Lawsone is a pigment extracted from the tropical henna shrub (*Lawsonia inermis*). It dyes wool and silk orange and can tint hair red. The Islamic prophet Muhammad is said to have dyed his beard with henna. Juglone was first isolated from walnut shells (*Juglans regia*). It occurs in the shell as the colorless hydroquinone (1,4,5-trihydroxynaphthalene) but is oxidized by the air to the quinone. Juglone stains skin brown.

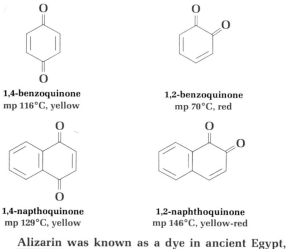

1,4-benzoquinone
mp 116°C, yellow

1,2-benzoquinone
mp 70°C, red

1,4-napthoquinone
mp 129°C, yellow

1,2-naphthoquinone
mp 146°C, yellow-red

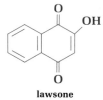

alizarin
mp 290°C, orange-red

lawsone
mp 192°C, red-brown

juglone
mp 155°C, yellow

Alizarin was known as a dye in ancient Egypt, Persia, and India. It was extracted from the madder root (an herb) and used as a mordant dye, being fixed to cloth by various metal ions (for example, aluminum). The red army coats used by the British

Perhaps the most important property of quinones is their *reversible reduction* to hydroquinones.

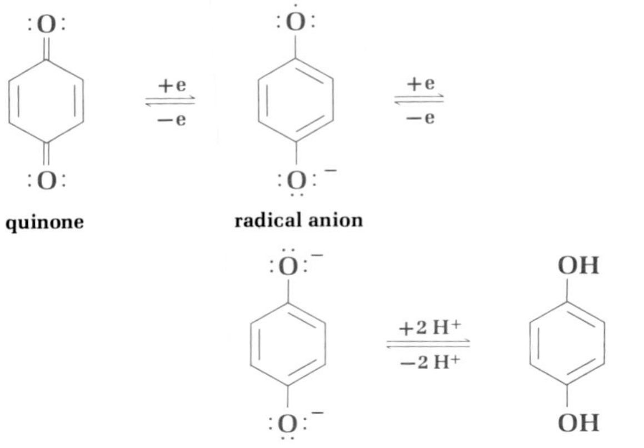

quinone — radical anion — dianion — hydroquinone

Virtually all quinones undergo this reaction. The reduction involves stepwise addition of two electrons to give a radical anion first and then a dianion. It is this property that permits quinones to play an important role in reversible biochemical oxidation–reduction (electron transport) reactions. A group of enzymes called coenzymes Q (also known as ubiquinones because of their common occurrence in animal and plant cells) participate in electron transport in mitochondria (granular bodies in the cell that are involved in the metabolism of lipids, carbohydrates, and proteins).

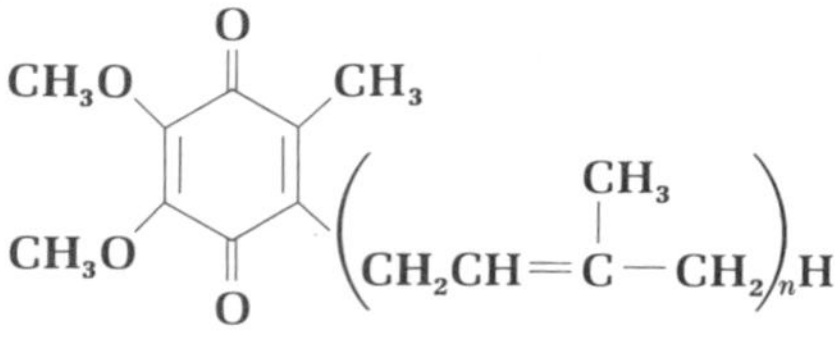

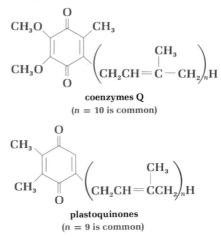

coenzymes Q
(n = 10 is common)

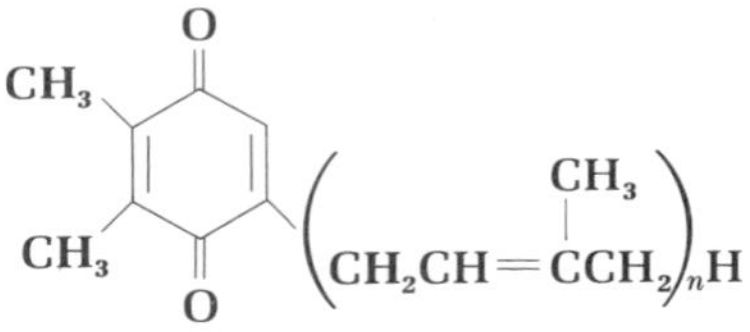

plastoquinones
(n = 9 is common)

In plant tissues the plastoquinones perform a similar function in photosynthesis. No doubt the long isoprenoid carbon chain in these quinones is necessary to promote fat solubility of these coenzymes.

The quinone known as vitamin K (Figure 9.1) is required for the normal clotting of blood, where it plays an essential role in the biosynthesis of the enzyme prothrombin.

ADDITIONAL PROBLEMS

9.23. Name each of the following compounds.

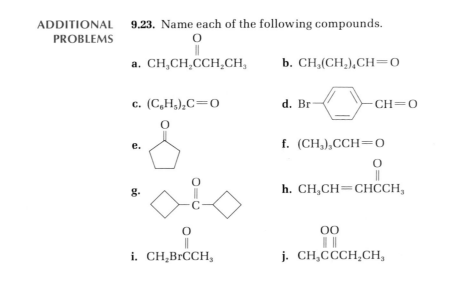

a. $CH_3CH_2C(=O)CH_2CH_3$

b. $CH_3(CH_2)_4CH{=}O$

c. $(C_6H_5)_2C{=}O$

d. 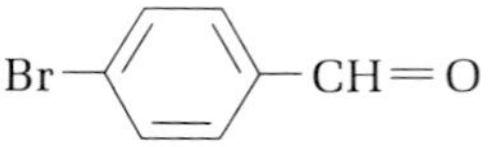

e. 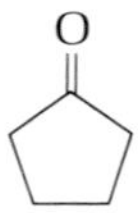

f. $(CH_3)_3CCH{=}O$

g. 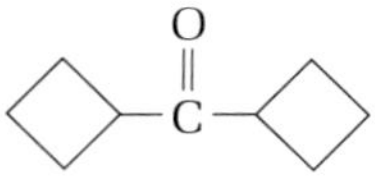

h. $CH_3CH{=}CHC(=O)CH_3$

i. $CH_2BrC(=O)CH_3$

j. $CH_3C(=O)C(=O)CH_2CH_3$

9.24. Write structural formulas for each of the following.
a. 2-octanone   **b.** 4-methylpentanal
c. m-chlorobenzaldehyde   **d.** 3-methylcyclohexanone
e. 2-butenal   **f.** benzyl phenyl ketone
g. p-tolualdehyde   **h.** p-benzoquinone
i. 2,2-dibromohexanal   **j.** 1-phenyl-2-butanone

9.25. Give an example of each of the following.
a. acetal   **b.** hemiacetal
c. ketal   **d.** hemiketal
e. cyanohydrin   **f.** imine
g. oxime   **h.** phenylhydrazone
i. enol   **j.** aldehyde with no α hydrogen

9.26. Write an equation for the reaction, if any, of p-bromobenzaldehyde with each of the following, and name the organic product.
a. Tollens' reagent   **b.** hydroxylamine
c. H_2, nickel   **d.** ethylmagnesium bromide, then H_3O^+

e. phenylhydrazine   **f.** aniline (⟨ ⟩ —NH_2)

g. cyanide ion   **h.** excess methanol, dry HCl
i. ethylene glycol, H^+ **j.** lithium aluminum hydride

9.27. What simple chemical test can distinguish between the members of the following pairs of compounds?
a. hexanal and 2-hexanone
b. benzyl alcohol and benzaldehyde
c. cyclopentanone and 2-cyclopentenone

9.28. Use the structures shown in Figure 9.1 to write an equation for the following reactions of natural products.
a. cinnamaldehyde + Tollens' reagent
b. vanillin + hydroxylamine
c. carvone + sodium borohydride
d. camphor + (1) methylmagnesium bromide and (2) H_3O^+

9.29. The boiling points of the isomeric carbonyl compounds heptanal, 4-heptanone, and 2,4-dimethyl-3-pentanone are 155°C, 144°C, and 124°C, respectively. Suggest a possible explanation for the observed order of boiling points.

9.30. Write out all the steps in the formation of a cyclic ketal from acetone and ethylene glycol (eq. 9.9).

9.31. Complete each of the following equations.
a. butanal + excess ethanol, $H^+ \rightarrow$ **b.** $CH_3CH(OCH_3)_2 + H_2O$, $H^+ \rightarrow$

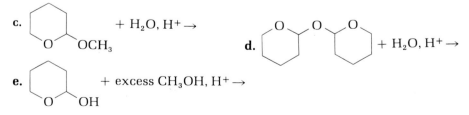

c. [structure] + H_2O, $H^+ \rightarrow$ **d.** [structure] + H_2O, $H^+ \rightarrow$

e. [structure] + excess CH_3OH, $H^+ \rightarrow$

9.32. Write an equation for the reaction of each of the following with methyl-magnesium bromide, followed by hydrolysis with aqueous acid.

a. acetaldehyde   **b.** acetophenone
c. formaldehyde   **d.** cyclohexanone

9.33. Using a Grignard reagent and the appropriate aldehyde or ketone, show how each of the following could be prepared.

a. 1-pentanol   **b.** 3-pentanol
c. 2-methyl-2-butanol   **d.** 1-cyclopentylcyclopentanol
e. 1-phenyl-1-propanol   **f.** 3-butene-2-ol

9.34. Write an equation for the reaction of:

a. cyclohexanone + Na$^+$$^-$C≡CH → $\xrightarrow[\text{H}^+]{\text{H}_2\text{O}}$

b. cyclopentanone + HCN →

c. 2-butanone + NH$_2$OH $\xrightarrow{\text{H}^+}$
d. p-tolualdehyde + benzylamine →
e. propanal + phenylhydrazine →

9.35. Write out each step in the mechanism for:

a. ⬡—CH=O + ⬡—NH$_2$ → ⬡—CH=N—⬡ + H$_2$O

b. (CH$_3$)$_2$C=O + NH$_2$OH → (CH$_3$)$_2$C=NOH + H$_2$O

9.36. Give the structure of the product.

a. CH$_3$C(=O)—⬡ $\xrightarrow[\text{2. H}_2\text{O, H}^+]{\text{1. LiAlH}_4}$

b. CH$_3$CH$_2$C(=O)—⬡ $\xrightarrow[\text{CH}_3\text{OD (excess)}]{\text{CH}_3\text{O}^-\text{Na}^+}$

c. ⬡—CH=CH—CH=O $\xrightarrow[\text{Ni, heat}]{\text{excess H}_2}$

d. CH$_2$=CH—⬡—CH=O $\xrightarrow[\text{2. H}_2\text{O, H}^+]{\text{1. NaBH}_4}$

9.37. Write the formula for all possible enols of:
a. 2-butanone   **b.** phenylacetaldehyde   **c.** 2,4-pentanedione

9.38. How many hydrogens are replaced by deuterium when each of the following compounds is treated with NaOD in D$_2$O?
a. 3-methylcyclopentanone   **b.** 2-methylbutanal

9.39. Write out the steps in the mechanism for the aldol condensation of butanal (the first step in the synthesis of the mosquito repellent "6-12," eq. 9.54).

9.40. The final steps in the synthesis of two oral contraceptives, Enovid and Norlutin, are shown. For each step, supply the missing reagent and tell what general type of reaction is involved.

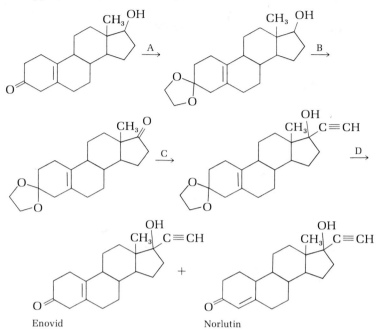

Enovid   +   Norlutin

ten

Carboxylic Acids and Their Derivatives

The most important organic acids are the **carboxylic acids.** Their functional group is the **carboxyl group,** the name being a contraction of the names of its two component parts, the *carbonyl* group and the *hydroxyl* group. The general formula for a carboxylic acid can be written in expanded or abbreviated form:

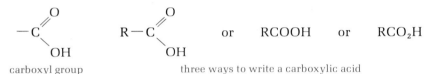

carboxyl group   three ways to write a carboxylic acid

In this chapter we will describe the structure, acidity, preparation, and reactions of acids. We will also discuss some common related classes of compounds in which the hydroxyl group of an acid is replaced by other functions (for example, OR, halogen, and others).

Because of their abundance in nature, carboxylic acids were among the earliest classes of compounds to be studied by organic chemists. It is not surprising, then, that many of them have common names. These names are usually derived from some Latin or Greek word that indicates the original source of the acid. Table 10.1 lists the first ten unbranched carboxylic acids, together with their common and IUPAC names. Many of these acids were first isolated from fats, so they are sometimes referred to as **fatty acids** (we will consider the structures of fats in detail in the next chapter). To obtain the IUPAC name of a carboxylic acid (final column, Table 10.1), replace the final *e* in the name of the corresponding alkane by the suffix *oic,* and add the word *acid.*

Substituted acids are named in two ways. In the IUPAC system, *the chain is numbered beginning with the carboxyl carbon atom* and substituents are located in the usual way. If the common name of the acid is used, substituents are located with Greek letters, beginning with the α-carbon atom.

Carbon atoms	Formula	Source	Common name	IUPAC name
1	HCOOH	ants (Latin, *formica*)	formic acid	methanoic acid
2	CH$_3$COOH	vinegar (Latin, *acetum*)	acetic acid	ethanoic acid
3	CH$_3$CH$_2$COOH	milk (Greek, *protos pion*, first fat)	propionic acid	propanoic acid
4	CH$_3$(CH$_2$)$_2$COOH	butter (Latin, *butyrum*)	butyric acid	butanoic acid
5	CH$_3$(CH$_2$)$_3$COOH	valerian root (Latin, *valere*, to be strong)	valeric acid	pentanoic acid
6	CH$_3$(CH$_2$)$_4$COOH	goats (Latin, *caper*)	caproic acid	hexanoic acid
7	CH$_3$(CH$_2$)$_5$COOH	vine blossom (Greek, *oenanthe*)	enanthic acid	heptanoic acid
8	CH$_3$(CH$_2$)$_6$COOH	goats (Latin, *caper*)	caprylic acid	octanoic acid
9	CH$_3$(CH$_2$)$_7$COOH	pelargonium (an herb with stork-shaped capsules; Greek, *pelargos*, stork)	pelargonic acid	nonanoic acid
10	CH$_3$(CH$_2$)$_8$COOH	goats (Latin, *caper*)	capric acid	decanoic acid

TABLE 10.1   Aliphatic carboxylic acids

$$\overset{\beta}{\underset{3}{CH_3}} - \overset{\alpha}{\underset{2}{CH}} - \overset{}{\underset{1}{COOH}} \qquad HO - \overset{\delta}{\underset{5}{CH_2}} - \overset{\gamma}{\underset{4}{CH_2}} - \overset{\beta}{\underset{3}{CH_2}} - \overset{\alpha}{\underset{2}{CH_2}} - \overset{}{\underset{1}{COOH}}$$

$$\underset{Br}{|}$$

IUPAC   2-bromopropanoic acid   5-hydroxypentanoic acid
Common   α-bromopropionic acid   δ-hydroxyvaleric acid
                                 (δ = delta, γ = gamma, β = beta, α = alpha)

Problem 10.1   **Write the structure for:**
a. 3-chlorobutanoic acid   b. γ-hydroxybutyric acid

Problem 10.2   **Give the IUPAC name and the common name for:**

a. ⬡—CH$_2$CH$_2$COOH   **b. CCl$_3$COOH**

When the carboxyl group is connected to a ring, the ending -*carboxylic acid* is added to the name of the parent cycloalkane.

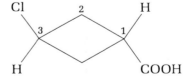

cyclopentanecarboxylic acid   *trans*-3-chlorocyclobutanecarboxylic acid

Aromatic acids are named by attaching the suffix *-oic* or *-ic acid* to an appropriate prefix derived from the aromatic hydrocarbon. A few examples are:

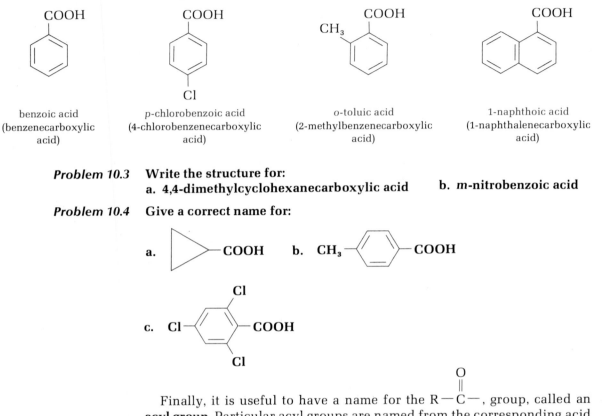

benzoic acid
(benzenecarboxylic acid)

p-chlorobenzoic acid
(4-chlorobenzenecarboxylic acid)

o-toluic acid
(2-methylbenzenecarboxylic acid)

1-naphthoic acid
(1-naphthalenecarboxylic acid)

Problem 10.3   **Write the structure for:**
a. 4,4-dimethylcyclohexanecarboxylic acid   **b. m-nitrobenzoic acid**

Problem 10.4   **Give a correct name for:**

a. △—COOH   b. CH₃—⬡—COOH

c. Cl—⬡(Cl)(Cl)—COOH

Finally, it is useful to have a name for the $R-\overset{O}{\underset{||}{C}}-$, group, called an **acyl group.** Particular acyl groups are named from the corresponding acid by changing the *-ic* ending to *-yl*.

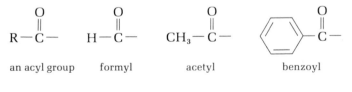

R—C—   H—C—   CH₃—C—   ⬡—C—

an acyl group   formyl   acetyl   benzoyl

Problem 10.5   **Write the formula for:**
a. 4-acetylbenzoic acid   **b. benzoyl bromide**
c. butanoyl chloride   **d. hexanoyl fluoride**

**10.3
PHYSICAL
PROPERTIES
OF ACIDS**   The first members of the carboxylic acid series are colorless liquids with rather sharp or unpleasant odors. Acetic acid, which constitutes about 4 or 5% of vinegar, gives vinegar its characteristic odor and flavor. Butyric acid gives rancid butter its disagreeable odor, and the goat acids (caproic, caprylic, and capric) smell like goats.

TABLE 10.2 Physical properties of some carboxylic acids	Name	bp, °C	mp °C	Solubility, g/100 g H_2O at 25°C
	formic	101	8 ⎫	
	acetic	118	17 ⎪	
	propanoic	141	−22 ⎬	miscible (∞)
	butanoic	164	−8 ⎭	
	caproic	205	−1.5	1.0
	caprylic	240	17	0.06
	capric	270	31	0.01
	benzoic	249	122	0.4 (but 6.8 at 95°C)

As their structures would suggest, carboxylic acids are polar. Like alcohols, they form hydrogen bonds with themselves or with other molecules. Consequently, they have rather high boiling points for their molecular weights—higher even than comparable alcohols. For example, acetic acid and n-propyl alcohol have the same formula weights (60) and boil at 118°C and 97°C, respectively. Determination of molecular weights shows that formic and acetic acids are dimers in nonpolar solvents and even in the vapor state. Two units are firmly held together by hydrogen bonds:

$$R-C\overset{\displaystyle O\cdots H-O}{\underset{\displaystyle O-H\cdots O}{}}C-R$$

Hydrogen bonding also explains the water-solubility of the lower-molecular-weight carboxylic acids. Table 10.2 lists some physical properties of carboxylic acids.

10.4 ACIDITY AND ACIDITY CONSTANTS

Carboxylic acids dissociate in water to a **carboxylate anion** and a hydronium ion.

$$R-C\overset{\displaystyle O}{\underset{\displaystyle OH}{}} + H\ddot{O}H \rightleftharpoons R-C\overset{\displaystyle O}{\underset{\displaystyle O^-}{}} + H-\overset{H}{\underset{}{\overset{+}{O}}}-H \qquad \textbf{(10.1)}$$

carboxylate ion   hydronium ion

Acidity is measured quantitatively by the **acidity or ionization constant**, K_a. The K_a of an acid is given by the expression

$$K_a = \frac{[RCO_2^-][H_3O^+]}{[RCO_2H]} \qquad \textbf{(10.2)}$$

which is the equilibrium constant for the reaction in eq. 10.1, omitting the concentration of water from the denominator. (Water is the solvent for

TABLE 10.3	Name	Formula	K_a	pK_a
The ionization constants of some acids	formic	HCOOH	2.1×10^{-4}	3.68
	acetic	CH_3COOH	1.8×10^{-5}	4.74
	propanoic	CH_3CH_2COOH	1.4×10^{-5}	4.85
	butanoic	$CH_3CH_2CH_2COOH$	1.6×10^{-5}	4.80
	chloroacetic	$ClCH_2COOH$	1.5×10^{-3}	2.82
	dichloroacetic	$Cl_2CHCOOH$	5.0×10^{-2}	1.30
	trichloroacetic	CCl_3COOH	2.0×10^{-1}	0.70
	2-chlorobutanoic	$CH_3CH_2CHClCOOH$	1.4×10^{-3}	2.85
	3-chlorobutanoic	$CH_3CHClCH_2COOH$	8.9×10^{-5}	4.05
	benzoic acid	C_6H_5COOH	6.6×10^{-5}	4.18
	o-chlorobenzoic	$o\text{-}Cl\!-\!C_6H_4COOH$	12.5×10^{-4}	2.90
	m-chlorobenzoic	$m\text{-}Cl\!-\!C_6H_4COOH$	1.6×10^{-4}	3.80
	p-chlorobenzoic	$p\text{-}Cl\!-\!C_6H_4COOH$	1.0×10^{-4}	4.00
	p-nitrobenzoic	$p\text{-}NO_2\!-\!C_6H_4COOH$	4.0×10^{-4}	3.40
	phenol	C_6H_5OH	1.0×10^{-10}	10.0
	ethanol	CH_3CH_2OH	1.0×10^{-16}	16.0

these measurements and is present in such large excess that its concentration remains constant and unaffected by the small amount of acid dissolved in it.) The larger the value of K_a, the stronger the acid (the larger the K_a, the larger the numerator in eq. 10.2, or the larger the concentration of H_3O^+). An acid with a K_a of 10^{-4} is 10 times stronger than an acid with a K_a of 10^{-5}.

Because K_a is usually an exponential number, acidity is sometimes expressed by pK_a, the negative logarithm of K_a ($pK_a = -\log K_a$). For example, if $K_a = 10^{-5}$, then $pK_a = 5$. Table 10.3 lists the acidity constants for some carboxylic acids and other acids. In comparing data in this table, remember that the larger the value of K_a, or the smaller the value of pK_a, the stronger the acid.

Example 10.1 Which is the stronger acid, formic or acetic, and by how much?

Solution Formic acid is stronger; it has the larger K_a. The ratio of acidities is

$$\frac{2.1 \times 10^{-4}}{1.8 \times 10^{-5}} = 1.16 \times 10^1 = 11.6$$

Problem 10.6 Using the data given in Table 10.3, determine which is the stronger acid, acetic or chloroacetic, and by how much.

Before we can explain the acidity differences in Table 10.3, we must examine the structural features that make carboxylic acids acidic.

**10.5
RESONANCE IN
THE CARBOXYLATE
ION**

You might wonder why carboxylic acids are so much more acidic than alcohols or phenols, inasmuch as all three classes ionize by losing H^+ from a hydroxyl group. The answer lies mainly in a comparison of the possibilities for charge delocalization through resonance in the resulting anions. (We have already made this comparison for alcohols and phenols in Sec. 7.6.) Let us use a specific example.

From Table 10.3 we see that acetic acid is approximately 10^{11}, or one hundred thousand million, times stronger an acid than ethanol.

$$CH_3CH_2OH \rightleftharpoons CH_3CH_2O^- + H^+ \qquad K_a = 10^{-16} \tag{10.3}$$
$$\text{ethoxide ion}$$

$$\underset{\text{acetate ion}}{CH_3\overset{\overset{O}{\|}}{C}-OH} \rightleftharpoons CH_3\overset{\overset{O}{\|}}{C}-O^- + H^+ \qquad K_a = 10^{-5} \tag{10.4}$$

In ethoxide ion the negative charge is localized on a single oxygen atom. In acetate ion, on the other hand, the negative charge can be delocalized through resonance.

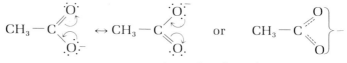

resonance in a carboxylate anion

The negative charge is spread equally over the two oxygens, so each oxygen in the carboxylate ion carries only half of the negative charge. Hence the acetate ion is stabilized by resonance, compared to the ethoxide ion. This stabilization drives the equilibrium more to the right in eq. 10.4 than in eq. 10.3. Consequently, more H^+ is formed from acetic acid than from ethanol. That is, acetic acid is a stronger acid than ethanol.

Example 10.2 **Phenoxide ions, like carboxylate ions, are stabilized by resonance (Sec. 7.6). Why aren't phenols just as strong acids as carboxylic acids?**

Solution **Charge delocalization is not as great in phenoxide ions as in carboxylate ions because some contributors to the resonance hybrid put negative charge on carbon instead of on oxygen, thus disrupting aromaticity. In the carboxylate ion, both contributors are identical and have the negative charge on oxygen, a more electronegative atom than carbon.**

Physical evidence supports the importance of resonance in carboxylate ions. In formic acid the two carbon–oxygen bonds have different lengths. But in the salt sodium formate, both carbon–oxgyen bonds of the formate ion are identical, and their length is between those of normal double and single carbon–oxygen bonds.

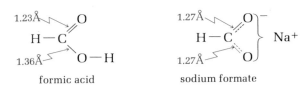

formic acid   sodium formate

**10.6
EFFECT OF
STRUCTURE ON
ACIDITY. THE
INDUCTIVE EFFECT**
The data in Table 10.3 show that even among the carboxylic acids (wherein the ionizing functional group is kept constant), acidities can vary appreciably depending on what other groups are attached to the carboxyl group. Compare, for example, the K_a of acetic acid with those of mono-, di-, and trichloroacetic acids, and note that the acidity varies over 10,000-fold.

The most important factor affecting acidity is the **inductive effect** of the groups close to the carboxyl group. *Electron-withdrawing groups enhance acidity, and electron-releasing groups reduce acidity.*

Let us examine the carboxylate ions formed when acetic acid and its chloro derivatives ionize:

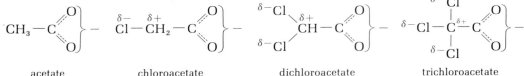

acetate   chloroacetate   dichloroacetate   trichloroacetate

Because chlorine is more electronegative than carbon, the C—Cl bond is polarized (Sec. 1.5) with the chlorine partially negative and the carbon partially positive. Thus electrons are pulled away from the carboxylate ion and toward the chlorine. This effect tends to spread or disperse the negative charge over more atoms than in acetate ion itself and thus stabilizes the ion. The more chlorines, the greater the effect, and the greater the strength of the acid.

Example 10.3   Explain the acidity order in Table 10.3 for butanoic acid and its 2- and 3-chloro derivatives.

Solution   The 2-chloro substituent increases the acidity of butanoic acid substantially, due to its inductive effect. In fact, the effect is about the same as for chloroacetic and acetic acids. The 3-chloro substituent exerts a similar *but much smaller* effect, because the C—Cl bond is now further away from the carboxylate group. *Inductive effects fall off rapidly with distance.*

Problem 10.7   Account for the relative acidities of benzoic acid and its *o*-, *m*-, and *p*-chloro derivatives (Table 10.3).

Formic acid is 11.6 times stronger than acetic acid. This fact suggests that CH_3 is more electron-releasing than hydrogen. This observation is consistent with what we observed earlier (Sec. 3.14) about carbocation stabilities—that methyl groups are more effective than hydrogen atoms in donating electrons to, and therefore stabilizing, a positive carbon atom.

**10.7
NEUTRALIZATION
OF ACIDS TO
FORM SALTS**

Carboxylic acids can be neutralized by bases to form salts. For example,

$$\tag{10.5}$$

carboxylic acid   strong base   a sodium salt

The salts can be isolated by evaporating the water. Salts are named as shown in the following examples:

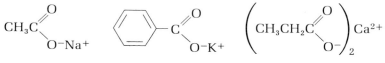

sodium acetate   potassium benzoate   calcium propanoate

The cation is named first, followed by the name of the carboxylate ion, which is obtained by changing the *-ic* ending of the acid to *-ate*.

Example 10.4   **Name the salt $CH_3CH_2CH_2C$**

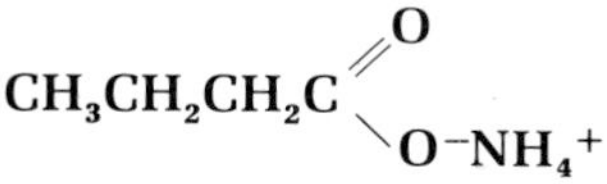

Solution   **Ammonium butanoate (IUPAC), or ammonium butyrate (common).**

Problem 10.8   **Write an equation, analogous to eq. 10.5, for the preparation of potassium 3-bromopropanoate from the corresponding acid.**

As we shall see later (Chapter 11), salts of certain acids are useful as soaps and detergents.

**10.8
PREPARATION
OF ACIDS**

Organic acids can be prepared in many ways, four of which will be described here: (1) oxidation of primary alcohols or aldehydes, (2) oxidation of alkyl side chains on aromatic rings, (3) reaction of Grignard reagents with carbon dioxide, and (4) hydrolysis of alkyl cyanides (or nitriles).

10.8a Oxidation of Primary Alcohols and Aldehydes The oxidation of primary alcohols (eq. 7.26) and aldehydes (eq. 9.32) to carboxylic acids has already been mentioned. It is easy to see that these are oxidation reactions because, as we go from an alcohol to an aldehyde to an acid, we replace C—H bonds by C—O bonds. In doing so we make the carbon atom more oxidized.

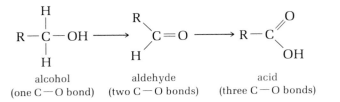

alcohol   aldehyde   acid

(one C—O bond) (two C—O bonds) (three C—O bonds)

(10.6)

The most commonly used oxidizing agents for these purposes are potassium permanganate ($KMnO_4$), chromic acid (CrO_3), nitric acid, and, with aldehydes only, silver oxide (Ag_2O).

10.8b Oxidation of Aromatic Side Chains Aromatic acids can be prepared by oxidizing an alkyl side chain on an aromatic ring.

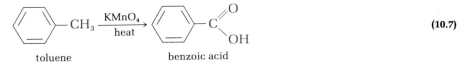

(10.7)

toluene   benzoic acid

This reaction emphasizes once again the striking stability of aromatic rings, because it is the alkane-like methyl group, and not the aromatic ring, that is oxidized. This reaction is commercially important. For example, **terephthalic acid,** one of the two raw materials needed to manufacture Dacron (called Terylene in England) is produced in this way, using a cobalt catalyst for the air oxidation.

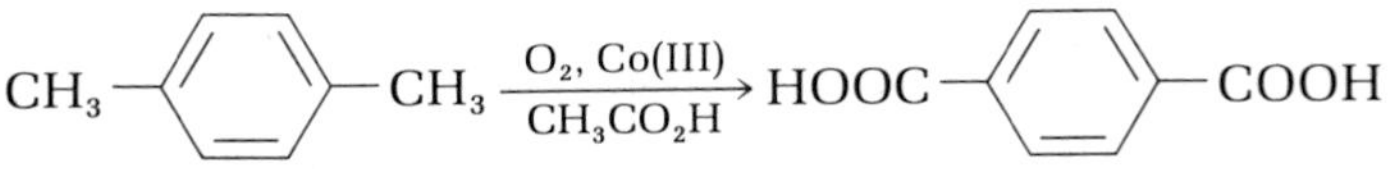

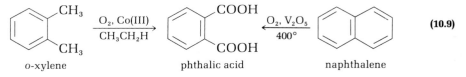

(10.8)

p-xylene   terephthalic acid

Phthalic acid, useful for making plasticizers, resins, and dyestuffs, is manufactured by a similar oxidation of either o-xylene or naphthalene.

![chemical reaction scheme](o-xylene with CH3 groups →[O2, Co(III) / CH3CH2H] phthalic acid with COOH groups ←[O2, V2O5 / 400°] naphthalene)

(10.9)

o-xylene   phthalic acid   naphthalene

10.8c From Grignard Reagents and Carbon Dioxide As we saw in Sec. 9.9, Grignard reagents add to the carbonyl groups of aldehydes or ketones to give alcohols. In a similar manner, they add to a carbonyl group in carbon dioxide to give acids.

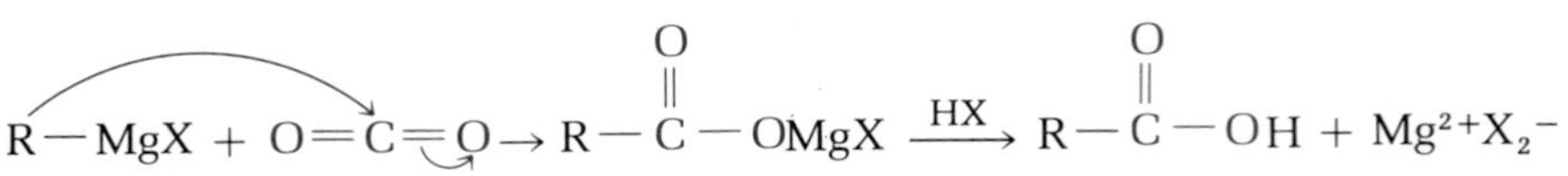

$$R\text{—MgX} + O{=}C{=}O \rightarrow R\text{—}\underset{\displaystyle \overset{\displaystyle O}{\|}}{C}\text{—OMgX} \xrightarrow{\text{HX}} R\text{—}\underset{\displaystyle \overset{\displaystyle O}{\|}}{C}\text{—OH} + Mg^{2+}X_2^{-}$$

(10.10)

The reaction gives good yields and is an excellent laboratory method for making both aliphatic and aromatic acids. Note that the acid obtained has one more carbon atom than the alkyl or aryl halide from which the Grignard reagent is prepared.

Problem 10.9 **Write an equation for the preparation of benzoic acid from bromobenzene via a Grignard reagent.**

10.8d Hydrolysis of Organic Cyanides (Nitriles) The carbon–nitrogen triple bond of organic cyanides can be hydrolyzed to give a carboxylic acid. The overall reaction is

$$R-C\equiv N + 2\ H_2O \xrightarrow[OH^-]{H^+ \text{ or}} R-C\overset{O}{\underset{OH}{\diagdown}} + NH_3 \tag{10.11}$$

a cyanide or an acid
nitrile

The reaction requires an acid or base as a catalyst, and the nitrogen atom of the cyanide is lost as ammonia. The required alkyl cyanides are generally made from the corresponding alkyl halide (usually primary) and sodium cyanide by an S_N2 displacement (Table 6.1, item 15). For example,

$$CH_3CH_2CH_2Br \xrightarrow{NaCN} CH_3CH_2CH_2CN \xrightarrow[H^+]{H_2O} CH_3CH_2CH_2CO_2H + NH_4^+ \tag{10.12}$$

n-propyl bromide   butyronitrile   butyric acid
(butanenitrile)   (butanoic acid)

Organic cyanides are customarily named after the corresponding acid, by changing the -oic suffix to onitrile (hence butyronitrile in eq. 10.12). In the IUPAC system, the suffix -nitrile is added to the name of the hydrocarbon with the same number of carbon atoms (hence butanenitrile in eq. 10.12). Sometimes organic cyanides may be named as alkyl cyanides.

Example 10.5 **Name CH_3CN in three ways.**

Solution **Acetonitrile (it gives acetic acid on hydrolysis), ethanenitrile (IUPAC), or methyl cyanide. The first of these names sees the greatest usage.**

Note that in nitrile hydrolysis, as with the Grignard method, the acid obtained has one more carbon atom than the alkyl halide from which the cyanide is prepared. Consequently, both methods provide a way of increasing the length of a carbon chain.

Problem 10.10 **Write equations for synthesizing phenylacetic acid from benzyl bromide by two routes.**

**10.9
DERIVATIVES
OF CARBOXYLIC
ACIDS**

Carboxylic acids can be converted to a number of structurally related derivatives, compounds in which the hydroxyl part of the carboxyl group is replaced by various other groups. In the remainder of this chapter we will consider the preparation and reactions of the more important of these acid derivatives. Their general formulas are

$$
\underset{\text{an ester}}{R-\overset{\overset{\displaystyle O}{\|}}{C}-OR'} \qquad \underset{\text{a primary amide}}{R-\overset{\overset{\displaystyle O}{\|}}{C}-NH_2} \qquad \underset{\text{an acyl halide}}{R-\overset{\overset{\displaystyle O}{\|}}{C}-X} \begin{pmatrix}\text{X is usually}\\ \text{Cl or Br}\end{pmatrix} \qquad \underset{\text{an acid anhydride}}{R-\overset{\overset{\displaystyle O}{\|}}{C}-O-\overset{\overset{\displaystyle O}{\|}}{C}-R}
$$

Esters and amides occur widely in nature. But anhydrides of carboxylic acids are uncommon in nature and acyl halides are solely creatures of the laboratory.

**10.10
ESTERS**

Esters are derived from acids by replacing the OH group by an OR group. Esters are named in a manner analogous to salts.

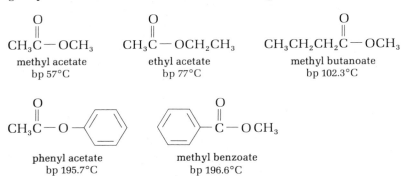

$$
\underset{\substack{\text{methyl acetate}\\ \text{bp 57°C}}}{CH_3\overset{\overset{\displaystyle O}{\|}}{C}-OCH_3} \qquad \underset{\substack{\text{ethyl acetate}\\ \text{bp 77°C}}}{CH_3\overset{\overset{\displaystyle O}{\|}}{C}-OCH_2CH_3} \qquad \underset{\substack{\text{methyl butanoate}\\ \text{bp 102.3°C}}}{CH_3CH_2CH_2\overset{\overset{\displaystyle O}{\|}}{C}-OCH_3}
$$

phenyl acetate
bp 195.7°C

methyl benzoate
bp 196.6°C

Notice that the R part of the OR group is named first, followed by the name of the acid, with the *-ic* ending changed to *-ate*.

Problem 10.11   **Write the name for:**

$$
\textbf{a. } H-\overset{\overset{\displaystyle O}{\|}}{C}-OCH_3 \qquad \textbf{b. } CH_3CH_2\overset{\overset{\displaystyle O}{\|}}{C}-OCH_2CH_2CH_3
$$

Esters are usually rather pleasant-smelling substances that are responsible for the flavor and fragrance of many fruits and flowers. Some of the more common are *n*-pentyl acetate (bananas), octyl acetate (oranges), ethyl butyrate (pineapples), and pentyl butyrate (apricots). Natural flavors can be exceedingly complex. For example, no fewer than 53 esters have been identified among the volatile constituents of Bartlett pears! Mixtures of esters are also used as artificial flavors.

Problem 10.12 **Write the formulas for the four specific esters named in the previous paragraph.**

**10.11
PREPARATION OF
ESTERS; FISCHER
ESTERIFICATION**

When a carboxylic acid and an alcohol are heated in the presence of an acid catalyst (usually HCl or H_2SO_4), an equilibrium is established with the ester and water.

$$\underset{\text{an acid}}{R-\overset{\overset{\text{O}}{\|}}{C}-OH} + \underset{\text{an alcohol}}{HO-R'} \underset{}{\overset{H^+}{\rightleftharpoons}} \underset{\text{an ester}}{R-\overset{\overset{\text{O}}{\|}}{C}-OR'} + H_2O \qquad \textbf{(10.13)}$$

The process is called **Fischer esterification** after Emil Fischer, a nineteenth-century organic chemist who developed the method. Although the reaction is an equilibrium, it can be used to make esters in high yield by shifting the equilibrium to the right. This can be accomplished by several techniques. If either the alcohol or the acid is inexpensive, a large excess can be used. Alternatively, the ester and/or water may be removed as formed (by distillation, for example), thus driving the reaction forward.

Problem 10.13 **Using eq. 10.13 as a guide, write an equation for the preparation of n-butyl acetate from the corresponding acid and alcohol.**

**10.12
THE MECHANISM
OF ACID-
CATALYZED
ESTERIFICATION**

In considering eq. 10.13, one might ask the mechanistic question: is the water molecule formed from the H of the acid and the OH of the alcohol, or from the OH of the acid and the H of the alcohol? Putting it another way, does the O of the water molecule come from the acid or from the alcohol? This question may seem rather trivial. But in answering it, we will find a key to understanding much of the chemistry of acids, esters, and their derivatives.

This question has been settled through the use of isotopic labeling. For example, Fischer esterification of benzoic acid with methanol enriched in the ^{18}O isotope of oxygen gives labeled methyl benzoate.

$$\langle\!\!\!\!\bigcirc\!\!\!\!\rangle\!-\overset{\overset{\text{O}}{\|}}{C}-OH + H^{18}OCH_3 \xrightarrow{H^+} \underset{\text{methyl benzoate}}{\langle\!\!\!\!\bigcirc\!\!\!\!\rangle\!-\overset{\overset{\text{O}}{\|}}{C}-{}^{18}OCH_3} + HOH \qquad \textbf{(10.14)}$$

None of the ^{18}O appears in the water. Thus it is clear that the water is formed using the *OH group of the acid*, not the OH of the alcohol.

How can we explain this experimental fact? A mechanism consistent with this result is as follows (the O of the alcohol is shown in color so that its path can be traced):

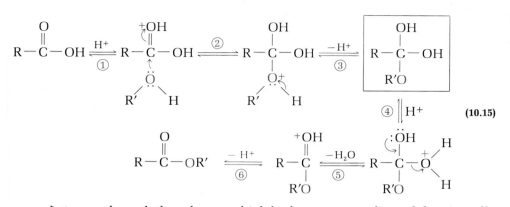

(10.15)

Let us go through the scheme, which looks more complicated than it really is, one step at a time. *Step 1 explains the acid catalysis.* In step 1, the carbonyl group of the acid is protonated. As with aldehydes and ketones (eq. 9.4), protonation increases the positive charge on the carboxyl carbon atom and enhances its susceptibility to nucleophilic attack. *Step 2 is crucial.* It involves nucleophilic addition of the alcohol to the protonated acid. In this step, the new carbon–oxygen bond (the ester bond) forms. Steps 3 and 4 are equilibria in which various oxgyens lose or gain a proton. Such equilibria are reversible, extremely rapid, and constantly going on in any acidic solution of any oxygen-containing compound. It does not matter, in step 4, which of the two hydroxyl groups is protonated, because they are equivalent. Step 5 involves the breaking of a carbon–oxygen bond and the loss of water. This step is the reverse of step 2. For it to occur, the hydroxyl group must be protonated to improve its leaving-group capability. Finally, in step 6, the protonated ester loses its proton. This is the reverse of step 1.

Some further features of eq. 10.15 are worth examining. We begin with an acid. The carboxyl carbon is trigonal and sp^2-hybridized. We end with an ester, and here too the ester carbon is trigonal and sp^2-hybridized. However, the reaction proceeds through a neutral intermediate (shown in color) in which the carbon atom has four groups attached to it and is thus sp^3-hybridized and tetrahedral. If we omit all of the proton-transfer steps in eq. 10.15 we can focus on this feature of the reaction.

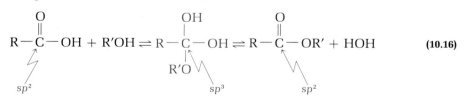

(10.16)

The net result of this process is a substitution of the OR' group of the alcohol for the OH group of the acid. It has been shown, however, that the reaction is not a direct substitution. Instead, it occurs in two steps: nucleophilic addition followed by elimination. We will see in the next and subsequent sections of this chapter that this is a general mechanism for nucleophilic substitutions at unsaturated carbon atoms.

Problem 10.14 **Using eq. 10.15 as a guide, write out the steps in the mechanism for the acid-catalyzed preparation of ethyl acetate from ethanol and acetic acid. In the United States, this method is used commercially to produce over 100 million pounds of ethyl acetate annually, mainly for use as a solvent in the paint industry. Ethyl acetate is also used as a solvent for nail polish and various glues.**

All of the steps in eq. 10.15 are reversible, so it is possible to hydrolyze an ester to the corresponding acid and alcohol by using an acid catalyst and a large excess of water. However, esters are more commonly hydrolyzed with the aid of a base, as described in the following section.

10.13 SAPONIFICATION AND AMMONOLYSIS OF ESTERS

Alkaline hydrolysis of esters is called **saponification** (from the Latin *sapon*, soap), because the same type of reaction is used to make soaps from fats (Chapter 11). The general reaction is

$$R-C\underset{OR'}{\overset{O}{\big|}}\quad Na^+OH^- \xrightarrow{heat} R-C\underset{O^-Na^+}{\overset{O}{\big|}}\quad + R'OH \qquad (10.17)$$

ester   alkali   salt of an acid   alcohol

The mechanism involves nucleophilic attack by hydroxide ion on the carbonyl carbon of the ester:

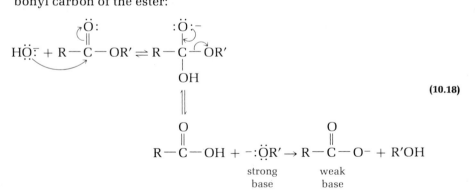

$$(10.18)$$

strong base   weak base

Note the similarity of this mechanism to that of Fischer esterification. Once again, the key step is nucleophilic addition to the carbonyl group. The reaction proceeds via a tetrahedral intermediate, though the reactant and the product are trigonal. *Saponification is not reversible,* because in the final step the strongly basic alkoxide ion removes a proton from the acid to form a carboxylate ion and an alcohol molecule.

Problem 10.15 **Write an equation for the saponification of methyl benzoate.**

Saponification is especially useful as a means of breaking down some unknown ester, perhaps isolated from nature, into its component parts for structure determination.

Ammonia converts esters to amides.

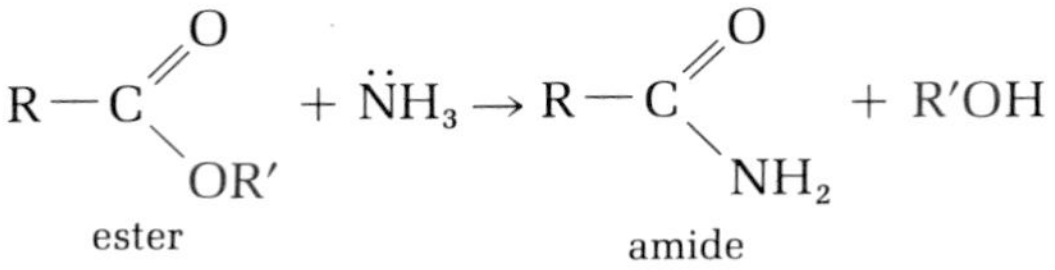

$$R-C\overset{O}{\underset{OR'}{\big\langle}} + \ddot{N}H_3 \rightarrow R-C\overset{O}{\underset{NH_2}{\big\langle}} + R'OH \qquad (10.19)$$

ester   amide

The reaction mechanism is very much like that of saponification. The unshared electron pair on the nitrogen atom of ammonia initiates nucleophilic attack on the ester carbonyl group.

Problem 10.16 **Write an equation for the reaction of ethyl acetate with ammonia.**

Problem 10.17 **Write out the steps in the mechanism of eq. 10.19.**

**10.14
REACTION OF
ESTERS WITH
GRIGNARD
REAGENTS**

Esters react with two equivalents of a Grignard reagent to give a tertiary alcohol. The reaction proceeds by nucleophilic attack on the ester carbonyl group. The initial product, a ketone, reacts further in the usual way (eq. 9.16) to give the tertiary alcohol.

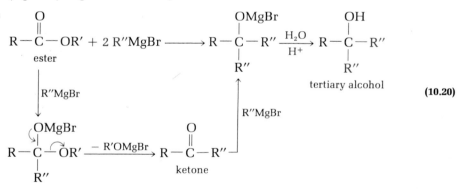

This method is particularly useful for making tertiary alcohols in which at least two of three alkyl groups attached to the hydroxyl-bearing carbon atom are identical.

Problem 10.18 **Write the structure of the tertiary alcohol that would be obtained from**

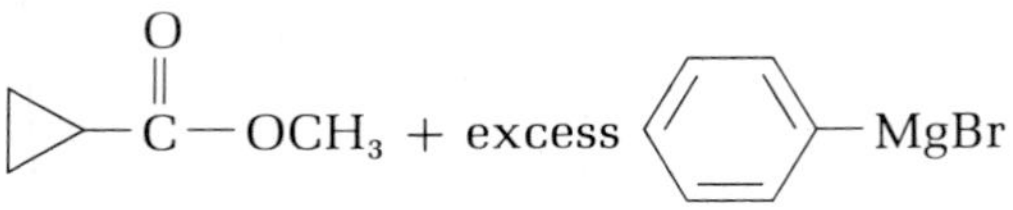

$$\triangleright\!\!-\!\overset{O}{\underset{}{\overset{\|}{C}}}\!-\!OCH_3 + \text{excess} \quad \bigcirc\!\!-\!MgBr$$

10.15
REDUCTION
OF ESTERS

Esters can be reduced to alcohols by lithium aluminum hydride.

$$R-\overset{\overset{\displaystyle O}{\|}}{C}-OR' \xrightarrow[\text{ether}]{\text{LiAlH}_4} RCH_2OH + R'OH \tag{10.21}$$

ester   primary alcohol

The mechanism is similar to the hydride reduction of other carbonyl compounds (eq. 9.26) and to the Grignard addition described in Sec. 10.14.

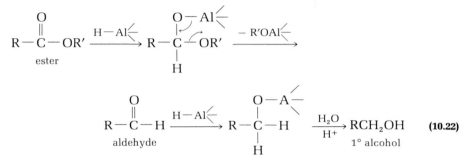

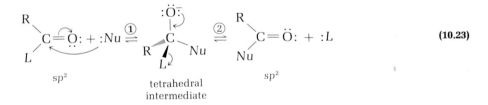

10.16
THE NEED FOR
ACTIVATED ACYL
COMPOUNDS

Most reactions of carboxylic acids, esters, and related compounds involve nucleophilic attack on the carbonyl carbon atom. We saw examples in Fischer esterification, in saponification and ammonolysis of esters, and in the first stage of the reactions of esters with Grignard reagents or lithium aluminum hydride. All of these reactions can be summarized in a single mechanistic equation:

The carbonyl carbon, initially trigonal, is attacked by a nucleophile Nu: to form a **tetrahedral intermediate** (step 1). Loss of some leaving group L (step 2) then regenerates the carbonyl group with its trigonal carbon atom. The net result is the replacement of L by Nu.

Biochemists look at eq. 10.23 in a slightly different way. They refer to the overall reaction as **acyl transfer.** The acyl group is transferred from L to Nu.

Regardless of how we consider the reaction, one important feature that can affect the rate of both steps is the nature of group L. The rates of both steps are enhanced by increasing the electron-withdrawing properties of L. Step 1 is favored because the carbonyl carbon is more positive and therefore more susceptible to nucleophilic attack. And step 2 is also facilitated because the more electronegative we make *L*, the better leaving group it becomes.

In general, esters are *less* reactive toward nucleophiles than are aldehydes or ketones. The reason is that in esters, the positive charge on the carbonyl carbon can be delocalized to the oxygen atom.

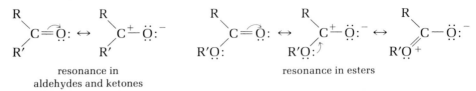

resonance in   resonance in esters
aldehydes and ketones

Consequently, the carbonyl carbon is less positive in esters than in aldehydes or ketones and less susceptible to nucleophilic attack.

Now let us examine some of the ways in which the carboxyl group can be modified to enhance its reactivity toward nucleophiles.

10.17   **Acyl halides** are among the most reactive of carboxylic acid derivatives.
ACYL HALIDES   *Acyl chlorides* are more common and less expensive than other acyl halides. They are usually prepared from acids by reaction with thionyl chloride or phosphorus pentachloride (compare with Sec. 7.10).

$$R-C\overset{O}{\underset{OH}{\big\langle}} \;+\; SOCl_2 \;\longrightarrow\; R-C\overset{O}{\underset{Cl}{\big\langle}} \;+\; HCl\uparrow \;+\; SO_2\uparrow \qquad \textbf{(10.24)}$$

thionyl
chloride

$$R-C\overset{O}{\underset{OH}{\big\langle}} \;+\; PCl_5 \;\longrightarrow\; R-C\overset{O}{\underset{Cl}{\big\langle}} \;+\; HCl \;+\; POCl_3 \qquad \textbf{(10.25)}$$

phosphorus   phosphorus
pentachloride   oxychloride

Acyl halides have irritating odors. Benzoyl chloride, for example, is a lachrymator (tear gas).

Problem 10.19   **Write an equation for the preparation of benzoyl chloride, using eq. 10.24 as a guide.**

Acyl halides react rapidly with most nucleophiles. Since they react with water, they fume in air.

$$\underset{\text{acetyl chloride}}{CH_3-\overset{\overset{O}{\|}}{C}-Cl} + HOH \xrightarrow{\text{rapid}} \underset{\text{acetic acid}}{CH_3-\overset{\overset{O}{\|}}{C}-OH} + \underset{\text{(fumes)}}{HCl} \qquad \textbf{(10.26)}$$

Problem 10.20   **Explain why acyl halides may be irritating to the nose.**

Acyl halides react rapidly with alcohols to form esters.

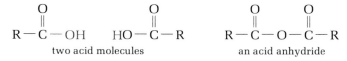

$$\text{benzoyl chloride} \qquad + \text{CH}_3\text{OH} \xrightarrow[\text{temp.}]{\text{room}} \qquad \text{methyl benzoate} + \text{HCl} \qquad \textbf{(10.27)}$$

Indeed, the most common way to prepare an ester in the laboratory is to convert an acid to its acid chloride and then react the latter with an alcohol. Although two steps are necessary (compared with one step for Fischer esterification), the method may be preferable, especially if either the acid or the alcohol is expensive. (Recall that the Fischer method is an equilibrium reaction and is often carried out with a large excess of one of the reactants.)

Acyl halides react rapidly with ammonia to form amides.

$$\underset{\text{acetyl chloride}}{\overset{\overset{\text{O}}{\parallel}}{\text{CH}_3\text{C}}-\text{Cl}} + 2\,\text{NH}_3 \rightarrow \underset{\text{acetamide}}{\overset{\overset{\text{O}}{\parallel}}{\text{CH}_3\text{C}}-\text{NH}_2} + \text{NH}_4{}^+\text{Cl}^- \qquad \textbf{(10.28)}$$

The reaction is much more rapid than the ammonolysis of esters (eq. 10.19). Note, however, that two equivalents of ammonia are needed, one to form the amide and one to neutralize the hydrogen chloride.

Example 10.6   **Explain why acyl chlorides are more reactive than esters toward nucleophiles.**

Solution   **The electronegativity order is Cl > OR. Therefore, the carbonyl carbon is more positive in acyl halides than in esters and more reactive toward nucleophiles. Also, Cl⁻ is a better leaving group than OR⁻.**

10.18   Acid anhydrides are related to acids by removing water from two carboxyl
ACID ANHYDRIDES   groups.

$$\underset{\text{two acid molecules}}{\overset{\overset{\text{O}}{\parallel}}{\text{R}-\text{C}}-\text{OH} \qquad \text{HO}-\overset{\overset{\text{O}}{\parallel}}{\text{C}}-\text{R}} \qquad \underset{\text{an acid anhydride}}{\overset{\overset{\text{O}}{\parallel}}{\text{R}-\text{C}}-\text{O}-\overset{\overset{\text{O}}{\parallel}}{\text{C}}-\text{R}}$$

The most important aliphatic anhydride commercially is **acetic anhydride**. About a million tons are manufactured annually, mainly to react with alcohols to form acetates. The two most common uses are in making cellulose acetate (Sec. 14.12b) and aspirin (Sec. 11.7).

Acid anhydrides are more reactive than esters, but less reactive than acyl halides, toward nucleophiles. Some typical reactions of acetic anhydride are:

$$
CH_3-\overset{\overset{\displaystyle O}{\|}}{C}-O-\overset{\overset{\displaystyle O}{\|}}{C}-CH_3
\begin{cases}
\xrightarrow{HO-H} CH_3\overset{\overset{\displaystyle O}{\|}}{C}-OH + CH_3\overset{\overset{\displaystyle O}{\|}}{C}-OH \\
\xrightarrow{RO-H} CH_3\overset{\overset{\displaystyle O}{\|}}{C}-OR + CH_3\overset{\overset{\displaystyle O}{\|}}{C}-OH \\
\xrightarrow{NH_2-H} CH_3\overset{\overset{\displaystyle O}{\|}}{C}-NH_2 + CH_3\overset{\overset{\displaystyle O}{\|}}{C}-OH
\end{cases}
\tag{10.29}
$$

acetic anhydride
bp 139.5°C

Water hydrolyzes anhydrides back to the acid. Alcohols give esters and ammonia gives amides. In each case, one equivalent of the acid is also produced.

Problem 10.21   **Write the formula for:**
a. butanoic anhydride   **b. benzoic anhydride**

Problem 10.22   **Write an equation for the reaction of acetic anhydride with 1-butanol.**

a word about ————————————————————————

20. Nature's Acyl-Activating Groups

Acyl transfer (eq. 10.23) plays an important role in many biochemical processes. However, acyl halides and acid anhydrides are far too corrosive to be cell constituents—they are hydrolyzed quite rapidly by water and are therefore incompatible with cellular fluid. Most ordinary esters, on the other hand, react too slowly with nucleophiles to carry out acyl transfer efficiently at body temperatures. Consequently, other groups have evolved to activate acyl groups in the cell. The most important of these is coenzyme A (the A stands for acetylation, one of the functions of this enzyme). Coenzyme A is a complex *thiol* (Figure 10.1). It is usually abbreviated by the symbol CoA—SH. Though its structure is made up of three parts—adenosine diphosphate (ADP), pantothenic acid (a vitamin), and 2-aminoethanethiol—it is the thiol group that gives coenzyme A its most important functions.

Coenzyme A forms thioesters, and these thioesters are the active acyl-transfer agents. Of the thioesters that coenzyme A forms, the acetyl ester, called acetyl–coenzyme A and abbreviated

$$CH_3\overset{\overset{\displaystyle O}{\|}}{C}-S-CoA,$$ is the most important. Acetyl–CoA reacts with many nucleophiles to transfer the acetyl group.

$$
CH_3\overset{\overset{\displaystyle O}{\|}}{C}-S-CoA + Nu\colon \xrightarrow[\text{enzyme}]{H_2O}
$$

acetyl–CoA

$$
CH_3\overset{\overset{\displaystyle O}{\|}}{C}-Nu + CoA-SH
$$

The reactions are usually enzyme-mediated and occur rapidly at ordinary cell temperatures.

One might ask why thioesters are superior to ordinary esters as acyl-transfer agents. The reason is that resonance is *less* important in thioesters than in ordinary oxyesters.

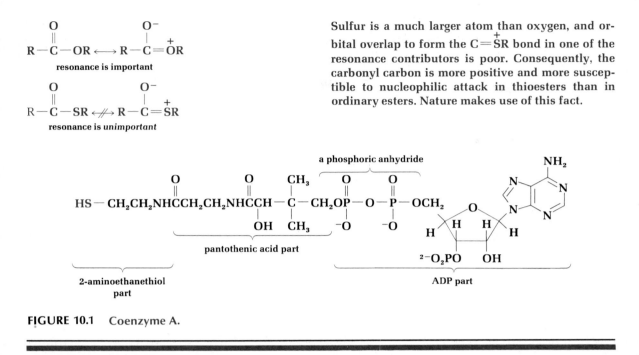

$$R-\overset{\overset{\text{O}}{\|}}{C}-OR \longleftrightarrow R-\overset{\overset{\text{O}^-}{|}}{C}=\overset{+}{OR}$$

resonance is important

$$R-\overset{\overset{\text{O}}{\|}}{C}-SR \overset{}{\longleftrightarrow\!\!\!/} R-\overset{\overset{\text{O}^-}{|}}{C}=\overset{+}{SR}$$

resonance is *unimportant*

Sulfur is a much larger atom than oxygen, and orbital overlap to form the $C=\overset{+}{S}R$ bond in one of the resonance contributors is poor. Consequently, the carbonyl carbon is more positive and more susceptible to nucleophilic attack in thioesters than in ordinary esters. Nature makes use of this fact.

FIGURE 10.1 Coenzyme A.

10.19 AMIDES

Amides are the least reactive of the common carboxylic acid derivatives. As a consequence, they occur widely in nature. The most important amides are the proteins. We shall devote a separate chapter (Chapter 15) to their chemistry. Here we will concentrate on just a few properties of simple amides.

Simple amides, which have the general formula $RCONH_2$, can be prepared via the reaction of ammonia with esters (eq. 10.19), with acyl halides (eq. 10.28), or with acid anhydrides (eq. 10.29). Amides can also be prepared by heating the ammonium salts of acids.

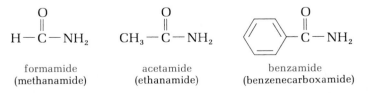

$$R-\overset{\overset{\text{O}}{\|}}{C}-OH + NH_3 \rightarrow R-\overset{\overset{\text{O}}{\|}}{C}-O^-NH_4{}^+ \overset{\text{heat}}{\longrightarrow} R-\overset{\overset{\text{O}}{\|}}{C}-NH_2 + H_2O \quad \textbf{(10.30)}$$

ammonium salt   amide

Amides are commonly named by replacing the *-ic* or *-oic* acid ending by the *-amide* ending. The IUPAC names are shown in parentheses.

$$H-\overset{\overset{\text{O}}{\|}}{C}-NH_2 \qquad CH_3-\overset{\overset{\text{O}}{\|}}{C}-NH_2 \qquad \underset{\text{benzamide}}{\text{⬡}}-\overset{\overset{\text{O}}{\|}}{C}-NH_2$$

formamide   acetamide   benzamide
(methanamide)   (ethanamide)   (benzenecarboxamide)

Amides in which one or both of the hydrogens on the nitrogen atom are replaced by organic groups are also known (Sec. 12.8, Sec. 12.12 and Chapter 15).

Amides have a planar geometry. Even though the carbon–nitrogen bond is normally written as a single bond, rotation around that bond is quite restricted. The reason is that resonance of the type

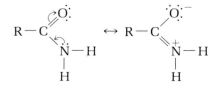

is very important in amides. The dipolar contributor is so important that the carbon–nitrogen bond behaves a lot like a double bond. Consequently, the structure is planar and rotation at the C—N bond is restricted. Indeed, the C—N bond in amides is only 1.32 Å long–much shorter than the usual carbon–nitrogen single bond (about 1.47 Å).

As the dipolar resonance contributor suggests, amides are highly polar and readily form hydrogen bonds.

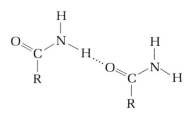

They have high boiling points for their molecular weights, although alkyl substitution on the nitrogen tends to lower their boiling and melting points by decreasing the hydrogen-bonding possibilities.

	$\overset{\displaystyle O}{\overset{\|}{H-C-NH_2}}$	$\overset{\displaystyle O}{\overset{\|}{H-C-N(CH_3)_2}}$	$\overset{\displaystyle O}{\overset{\|}{CH_3C-NH_2}}$	$\overset{\displaystyle O}{\overset{\|}{CH_3C-N(CH_3)_2}}$
	formamide	N,N-dimethylformamide	acetamide	N,N-dimethylacetamide
bp	210°C	153°C	222°C	165°C
mp	2.5°C	−60.5°C	81°C	−20°C

Amides react with nucleophiles. For example, they can be hydrolyzed by water.

$$\underset{\text{amide}}{R-\overset{\displaystyle O}{\overset{\|}{C}}-NH_2} + H-OH \xrightarrow[\text{OH}^-]{\text{H}^+ \text{ or}} \underset{\text{acid}}{R-\overset{\displaystyle O}{\overset{\|}{C}}-OH} + NH_3 \qquad \textbf{(10.31)}$$

Prolonged heating and acid or base catalysis are usually needed.

Problem 10.23   Write an equation for the hydrolysis of acetamide.

Amides can be reduced by lithium aluminum hydride to give amines.

$$R-\overset{\overset{\displaystyle O}{\|}}{C}-NH_2 \xrightarrow[\text{ether}]{\text{LiAlH}_4} RCH_2NH_2 \qquad\qquad (10.32)$$

amide   amine

The reaction is similar to the reduction of esters (eq. 10.22). Amides may also be dehydrated to the corresponding cyanides (nitriles).

$$R-\overset{\overset{\displaystyle O}{\|}}{C}-NH_2 \xrightarrow[\text{heat}]{\text{P}_2\text{O}_5} R-C\equiv N + H_2O \qquad\qquad (10.33)$$

a word about ————————————————————————————

21. Urea

Urea is a rather special amide, the diamide of carbonic acid.

$$HO-\overset{\overset{\displaystyle O}{\|}}{C}-OH \qquad H_2N-\overset{\overset{\displaystyle O}{\|}}{C}-NH_2$$

carbonic acid   urea
                                   mp 133°C

A colorless, water-soluble crystalline solid, urea is the normal end product of the metabolism of proteins. An average adult excretes approximately 30 g of urea in the urine daily.

Urea is produced commercially, mainly for use as a fertilizer (it contains 40% nitrogen by weight). Over 2 billion pounds are manufactured annually via the reaction of ammonia with carbon dioxide.

$$CO_2 + 2\,NH_3 \xrightarrow[\text{pressure}]{150-200°C} H_2N-\overset{\overset{\displaystyle O}{\|}}{C}-NH_2 + H_2O$$

Urea is also used as a raw material for certain drugs (eq. 12.28) and plastics.

**10.20
A SUMMARY OF
CARBOXYLIC ACID
DERIVATIVES**

Although we have described many reactions in this chapter, we can summarize in a single chart the reactions of acyl halides, anhydrides, esters, and amides (whose reactivity toward nucleophiles decreases in that order) with some of the more common nucleophiles (water, alcohols, and ammonia).

$$R - \overset{\overset{\displaystyle O}{\|}}{C} - Z + Nu - H \rightarrow R - \overset{\overset{\displaystyle O}{\|}}{C} - Nu + HZ \qquad \text{(10.34)}$$

Nu	Reaction type	Product type $\overset{O}{\underset{\|}{RC}} - Nu$	Z	Acid derivative	By-product HZ
HO—	hydrolysis	acid	—Cl	acid chloride	HCl
RO—	alcoholysis	ester	$-O\overset{\overset{\displaystyle O}{\|}}{C}R$	acid anhydride	RCO_2H
NH_2—	ammonolysis	amide	—OR′	ester	R′OH
			$-NH_2$	amide	NH_3

All of these reactions are of a single mechanistic type, as described in Sec. 10.16. The particular choice of Z for a given Nu depends on the nature of R, the desired reaction conditions, and other factors, such as the reactivity of the reactants, expense, scale for carrying out the reaction, and so on.

ADDITIONAL PROBLEMS

10.24. Write a structural formula for each of the following acids.
a. 3-methylpentanoic acid
b. 2,2-dichlorobutanoic acid
c. γ-hydroxyvaleric acid
d. p-toluic acid
e. cyclobutanecarboxylic acid
f. 2-propanoylbenzoic acid
g. phenylacetic acid
h. 2-naphthoic acid

10.25. Name each of the following acids.
a. $(CH_3)_2CHCH_2CH_2COOH$
b. $CH_3CHBrCH(CH_3)COOH$

c.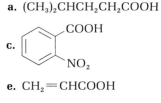

d. $CH_3CH(C_6H_5)COOH$

e. $CH_2{=}CHCOOH$
f.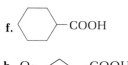

g. CH_3CF_2COOH
h. 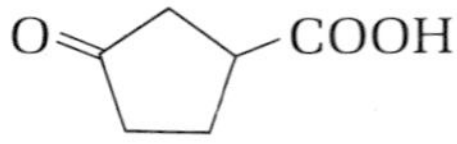 O$=$⬠—COOH

i. 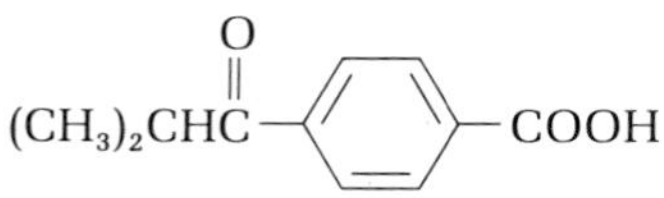 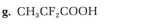 $(CH_3)_2CH\overset{\overset{\displaystyle O}{\|}}{C}$—⬡—COOH
j. $CH_3C{\equiv}CCOOH$

10.26. Which will have the higher boiling point? Explain your reasoning.
a. CH_3CH_2COOH or $CH_3CH_2CH_2CH_2OH$?
b. $CH_3CH_2CH_2CH_2COOH$ or $(CH_3)_3CCOOH$

10.27. In each of the following pairs of acids, which would be expected to be the stronger, and why?
a. CH_2ClCO_2H and CH_2BrCO_2H
b. $o\text{-}BrC_6H_4CO_2H$ and $m\text{-}BrC_6H_4CO_2H$
c. CCl_3CO_2H and CF_3CO_2H
d. $C_6H_5CO_2H$ and $p\text{-}CH_3OC_6H_4CO_2H$
e. $ClCH_2CH_2CO_2H$ and $CH_3CHClCO_2H$

10.28. Write a balanced equation for the neutralization of:
a. chloroacetic acid with potassium hydroxide
b. decanoic acid with calcium hydroxide

10.29. Give equations for the synthesis of:
a. $CH_3CH_2CH_2CO_2H$ from $CH_3CH_2CH_2CH_2OH$
b. $CH_3CH_2CH_2CO_2H$ from $CH_3CH_2CH_2OH$ (two ways)

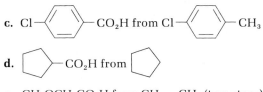

c. Cl—⟨⟩—CO_2H from Cl—⟨⟩—CH_3

d. ⟨⟩—CO_2H from ⟨⟩

e. $CH_3OCH_2CO_2H$ from CH_2—CH_2 (two steps) with O bridge

f. ⟨⟩—CO_2H from ⟨⟩—Br

10.30. The acid-catalyzed hydrolysis of an alkyl cyanide (eq. 10.11) involves, as the first step, nucleophilic attack of water on the protonated cyanide. Write out the steps in a plausible mechanism for the reaction.

10.31. The Grignard route for the synthesis of $(CH_3)_3CCOOH$ from $(CH_3)_3CBr$ is far superior to the cyanide (nitrile) route. Explain why.

10.32. Write a structure for each of the following compounds.
a. sodium 2-chloropropanoate   **b.** calcium acetate
c. isopropyl acetate   **d.** ethyl formate
e. phenyl benzoate   **f.** benzonitrile
g. propionic anhydride   **h.** m-toluamide
i. 4-chlorobutanoyl chloride   **j.** formyl fluoride

10.33. Name each of the following compounds.

a. Cl—⟨⟩—$COO^-NH_4{}^+$ **b.** $[CH_3(CH_2)_2CO_2{}^-]_2Ca^{2+}$

c. $(CH_3)_2CHCOOC_6H_5$ **d.** $CF_3CO_2CH_3$
e. CH_3COSH **f.** CH_3COSCH_3

g. $HCONH_2$ **h.** ▷—$\overset{O}{\overset{\|}{C}}$—$O$—$\overset{O}{\overset{\|}{C}}$—◁

10.34. Write out each step in the Fischer esterification of benzoic acid with methanol (if necessary, use eq. 10.15 as a guide).

10.35. Write an equation for the reaction of ethyl benzoate with:
a. hot aqueous sodium hydroxide
b. ammonia, heat
c. n-propylmagnesium bromide (two equivalents), then H_3O^+
d. lithium aluminum hydride (two equivalents), then H_3O^+

10.36. Write out all the steps in the mechanism for the:
a. saponification of $CH_3CH_2CO_2CH_3$
b. ammonolysis of $CH_3CH_2CO_2CH_3$

10.37. Explain the following differences in reactivity toward nucleophiles.
a. Esters are less reactive than ketones.
b. A given acid chloride is more reactive than the anhydride of the same acid.
c. Benzoyl chloride is less reactive than cyclohexanecarbonyl chloride.

10.38. Tell what Grignard reagent and what ester could be used to prepare:

$$\text{a. } CH_3CH_2CH_2 - \overset{\overset{\displaystyle OH}{|}}{\underset{\underset{\displaystyle C_6H_5}{|}}{C}} - CH_2CH_2CH_3 \qquad \text{b. } CH_3CH_2CH_2C(C_6H_5)_2OH$$

10.39. Write an equation for the:
a. hydrolysis of acetyl chloride
b. reaction of benzoyl chloride with methanol
c. esterification of 1-pentanol with acetic anhydride
d. ammonolysis of 4-bromobutanoyl bromide
e. Fischer esterification of valeric acid with ethanol

10.40. Write an equation for each of the following reactions.
a. $CH_3CH_2CH_2CO_2H + PCl_5 \rightarrow$
b. $CH_3(CH_2)_8CO_2H + SOCl_2 \rightarrow$

c. 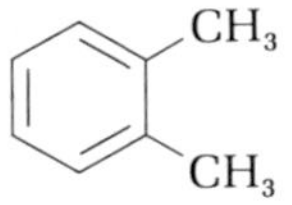 $+ KMnO_4 \rightarrow$

d.  $-CO_2^-NH_4^+ + heat \rightarrow$

e. $CH_3(CH_2)_5CONH_2 + LiAlH_4 \rightarrow$

f.  $-CO_2CH_2CH_3 + LiAlH_4 \rightarrow$

10.41. Write the important resonance contributors to the structure of propanoamide, and tell which atoms will lie in a single plane.

10.42. Mandelic acid, which has the formula $C_6H_5CH(OH)COOH$, can be isolated from bitter almonds (called *mandel* in German). It is sometimes used in medicine to treat urinary infections. Suggest a two-step synthesis of mandelic acid from benzaldehyde, using its cyanohydrin as an intermediate.

eleven

Difunctional Acids, Fats, and Detergents

With few exceptions, we have to this point described the chemistry of compounds with a single functional group. But of course many organic compounds contain several functional groups in the same molecule. One might at first think that the chemistry of such molecules could be described as the sum of the chemistries of each of the functional groups, and to a first approximation this is correct. It is particularly true if the functional groups are far apart from one another on the carbon framework so that, in a sense, one group doesn't "know" that the other group is there.

But if two groups are close to one another in a molecule, each group can "feel" the other's presence and influence its chemistry. In this chapter we will discuss a number of difunctional (and in some cases polyfunctional) molecules. We will start with dicarboxylic acids, wherein both functional groups are identical. Then we will move on to molecules *with two different functions,* such as hydroxy or keto acids.* In all cases we shall see how the proximity of groups influences their mutual behavior.

Most simple **dicarboxylic acids** occur in nature and have common names derived from their source. **Oxalic acid,** for example, whose structure is simply two connected carboxyl groups, $HOOC-COOH$, occurs in many plants and vegetables of the *oxalis* family, such as garden sorrel. Its IUPAC name is *ethanedioic acid* (the suffix indicates the presence of two carboxyl groups). Table 11.1 lists some of the common aliphatic dicarboxylic acids.

The principle that one functional group will affect the chemistry of a nearby functional group in the same molecule can be clearly seen in the acidity constants of the dicarboxylic acids. Dicarboxylic acids have two ionization constants:

$$HOOC(CH_2)_nCOOH \overset{K_1}{\rightleftharpoons} {}^-OOC(CH_2)_nCOOH \overset{K_2}{\rightleftharpoons} {}^-OOC(CH_2)_nCOO^- \quad \text{(11.1)}$$

Note that K_1 for the dicarboxylic acids (Table 11.1) is *greater* than K_a of acetic acid (1.8×10^{-5}). This is because the carboxyl group is an electron-withdrawing group and hence stabilizes the negative charge on the monocarboxylate ion:

*Because of their special importance, we will leave amino acids for another chapter (Chapter 15).

Formula	Common name	IUPAC name	Acidity constants K_1	K_2
HOOC—COOH	oxalic	ethanedioic	5.4×10^{-2}	5.4×10^{-5}
HOOC—CH$_2$—COOH	malonic	propanedioic	1.4×10^{-3}	0.2×10^{-5}
HOOC—(CH$_2$)$_2$—COOH	succinic	butanedioic	6.2×10^{-5}	0.2×10^{-5}
HOOC—(CH$_2$)$_3$—COOH	glutaric	pentanedioic	4.6×10^{-5}	0.4×10^{-5}
HOOC—(CH$_2$)$_4$—COOH	adipic	hexanedioic	3.7×10^{-5}	0.4×10^{-5}

TABLE 11.1   Aliphatic dicarboxylic acids

$$\overset{\longleftarrow \quad +}{\text{HOOC} \sim\!\!\sim\!\!\sim \text{CO}_2{}^-}$$

The closer the two carboxyl groups, the stronger the effect is. Thus oxalic acid has the largest K_1. The value decreases and approaches that of acetic acid as the number of CH$_2$ groups between the carboxyls increases (Table 11.1). On the other hand, K_2 is *less* than K_a of acetic acid because repulsion between the two negative charges destabilizes the dicarboxylate dianion. (K_2 of oxalic acid is *less* than K_a of formic acid, a better reference compound for it.)

   Oxalic acid occurs in the cell sap of many plants, such as rhubarb and spinach. Though toxic, it decomposes and is rendered harmless during cooking. **Succinic acid** was first isolated from the distillate of amber (in Latin, *succinum*). **Glutaric acid** and **adipic acid** are both found in sugar beets. Commercially, adipic acid is the most important aliphatic dicarboxylic acid, being manufactured on a very large scale for use in making nylon (Sec. 12.12).

   The three benzenedicarboxylic acids are generally known by their common names:

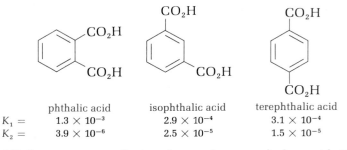

	phthalic acid	isophthalic acid	terephthalic acid
$K_1 =$	1.3×10^{-3}	2.9×10^{-4}	3.1×10^{-4}
$K_2 =$	3.9×10^{-6}	2.5×10^{-5}	1.5×10^{-5}

All three are manufactured on a large scale by oxidation of the corresponding xylenes (Sec. 10.8b).

Problem 11.1   **Account for the large K_1/K_2 ratio for phthalic acid compared to the corresponding ratios for isophthalic and terephthalic acids.**

The effect of heat on the various dicarboxylic acids depends on their structures—and in particular on the distance between the carboxyl groups. Oxalic acid decomposes to CO_2 and formic acid, which breaks down further to CO and water.

$$HOOC-COOH \xrightarrow{200°C} CO_2 + HCOOH \xrightarrow{200°C} CO + H_2O \qquad (11.2)$$
$$\text{oxalic acid} \qquad\qquad \text{formic acid}$$

Malonic acid loses CO_2 even more readily.

$$HOOC-CH_2-COOH \xrightarrow{135°C} CO_2 + CH_3COOH \qquad (11.3)$$
$$\text{malonic acid} \qquad\qquad \text{acetic acid}$$

This reaction probably proceeds by a cyclic transition state to give the enol of acetic acid, which then tautomerizes.

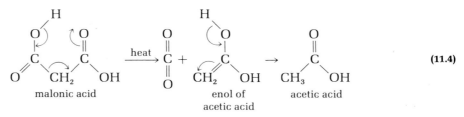

(11.4)

malonic acid   enol of   acetic acid
                      acetic acid

A similar reaction occurs on heating most dicarboxylic acids with two carboxyl groups attached to the same carbon atom.

Example 11.1   **What product is obtained when $CH_3CH_2CH(COOH)_2$ is heated?**

Solution   **$CH_3CH_2CH_2COOH$, by a mechanism exactly analogous to eq. 11.4.**

Problem 11.2   **Predict the product obtained on heating:**

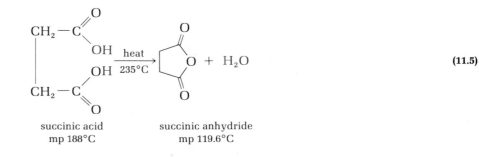

Dicarboxylic acids with two or three carbon atoms between the two carboxyl groups lose water on heating, to form cyclic five- or six-membered ring anhydrides.

$$(11.5)$$

succinic acid   succinic anhydride
mp 188°C   mp 119.6°C

However, such dehydrations are most often brought about by heating the dicarboxylic acid with a dehydrating agent, such as acetic anhydride.

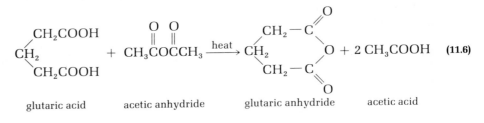

glutaric acid   acetic anhydride   glutaric anhydride   acetic acid

Problem 11.3   **Predict the products of the following reactions:**

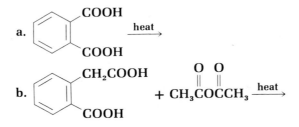

11.4
ESTERS OF
DICARBOXYLIC
ACIDS

Dicarboxylic acids can, of course, be converted to esters, and some of these esters have practical importance. Phthalic anhydride is converted to a variety of dialkyl phthalates.

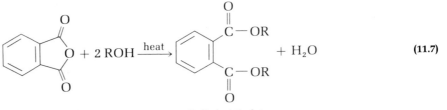

phthalic anhydride   a dialkyl phthalate

Phthalates with R = butyl or 2-ethylhexyl, for example, are used as plasticizers. When added to various polymers, such as polyvinyl chloride, they increase the polymers' flexibility and make them less brittle.

Difunctional acids react with difunctional alcohols to form **polyesters.** Although many polyesters are known, the most common example is **Dacron,** the polyester of terephthalic acid (Sec. 10.8 and 11.2) and ethylene glycol (Sec. 7.14 and 8.8).

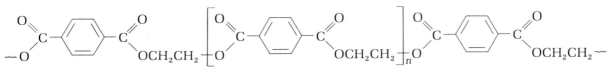

the polyester Dacron,
poly(ethylene terephthalate)

FIGURE 11.1

The *Gossamer Albatross*, its wings and pilot's compartment covered by Mylar. (Courtesy of DuPont Magazine.)

The value of n is about 100 ± 20. The crude polyester can be spun into fibers for use in textiles. The fibers are highly resistant to wrinkling. The polyester can also be fabricated into a particularly strong film called **Mylar.** Mylar is so strong, yet light, that it was selected for use in the human-powered *Gossamer Albatross*, which flew across the English channel on June 12, 1979. The polyester film, which covered the 93-foot wingspread as well as the pilot's compartment beneath it, weighed only about 2 lb (0.9 kg) (Figure 11.1). In the United States, production of polyester fibers exceeds 4 billion pounds per year.

Problem 11.4 **Write out the steps in making two units of a Dacron polyester from terephthalic acid and ethylene glycol.**

Problem 11.5 **Kodel is a polyester with the following structure:**

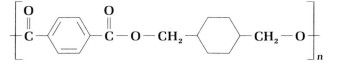

From what two "monomers" do you think it is made?

**11.5
UNSATURATED
ACIDS**

The simplest unsaturated acid is propenoic acid, commonly called **acrylic acid.** Several of its derivatives are commercial chemicals used mainly to make polymers. The most important are **acrylonitrile**, used to make Orlon or Acrilan, and the ester **methyl methacrylate,** which, when polymerized, produces Lucite or Plexiglas (review Table 3.3).

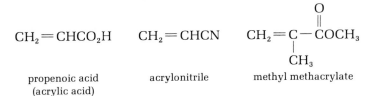

$$CH_2=CHCO_2H \qquad CH_2=CHCN$$

propenoic acid   acrylonitrile   methyl methacrylate
(acrylic acid)

The free-radical polymerization of acrylonitrile gives a polymer that can be spun into acrylic fibers.

$$CH_2=CH \xrightarrow{\text{catalyst}} \left(CH_2CH \right)_n \qquad \qquad \text{(11.8)}$$
$$\quad\; |\qquad\qquad\qquad\qquad\; | $$
$$\quad CN \qquad\qquad\qquad\quad CN$$

Orlon, Acrilan

Acrylonitrile is also copolymerized with butadiene to make synthetic rubber. Annual world production of acrylonitrile is over 2 million tons.

Methyl methacrylate is made from acetone cyanohydrin (eq. 9.20), via methanolysis and dehydration.

$$\begin{array}{cc} CH_3 & OH \\ & C \\ CH_3 & CN \end{array} + CH_3OH + H_2SO_4 \rightarrow \begin{array}{cc} CH_2 & O \\ & \| \\ C-COCH_3 \\ CH_3 \end{array} + NH_4^+{}^-HSO_4 \quad \text{(11.9)}$$

acetone cyanohydrin   methyl methacrylate

When polymerized, methyl methacrylate gives **Plexiglas,** a crystal-clear plastic that is hard and resistant to fracturing. Methyl methacrylate is also used in acrylic paints.

Problem 11.6 **Using eq. 11.8 as a guide, write a formula for Plexiglas.**

Many long-chain unsaturated carboxylic acids are present in fats and oils, whose chemistry we will discuss later in this chapter.

Two unsaturated acids played an important role in the early study of *cis-trans* (geometric) isomerism (Sec. 3.5). They are both butenedioic acids but have common names, **maleic acid** and **fumaric acid.**

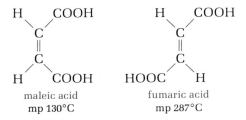

maleic acid   fumaric acid
mp 130°C   mp 287°C

The effect of geometry on the behavior of these acids is striking. On heating just above its melting point (in a vacuum, to remove the water), maleic acid loses water to form the anhydride.

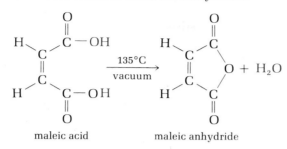

maleic acid   maleic anhydride

(11.10)

Fumaric acid cannot form an analogous cyclic anhydride, because the carboxyl groups are too far apart.

In general, the chemistry of unsaturated acids is characteristic of the chemistry of the two functional groups.

Example 11.2 **Predict the reaction of acrylic acid with bromine.**

Solution $CH_2 = CHCOOH + Br_2 \rightarrow CH_2 - CHCOOH$
$\qquad\qquad\qquad\qquad\qquad\qquad\quad |\qquad |$
$\qquad\qquad\qquad\qquad\qquad\qquad\ \ Br\quad Br$

Problem 11.7 **Predict the reaction of acrylic acid with thionyl chloride (eq. 10.24).**

**11.6
HYDROXY ACIDS** Many hydroxy acids occur in nature, are biologically important, and have common names. Here are a few examples:

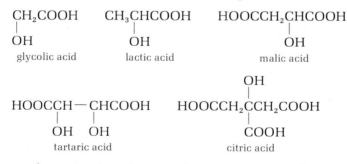

Lactic acid is found in sour milk as a result of lactose fermentation. It is also produced during muscular activity. **Tartaric acid** is prepared from grape juice ferments. It is used in carbonated beverages and effervescent tablets. The monopotassium salt of tartaric acid is **cream of tartar,** which is used in medicine as a laxative and in baking powder as a source of acidity. **Glycolic acid** is found in cane-sugar juice, but it is manufactured from chloroacetic acid and sodium hydroxide and sees many industrial uses as an inexpensive acid.

Problem 11.8   **Write an equation for the manufacture of glycolic acid.**

Malic acid was first isolated from unripe apples (Latin *malum*, apple) and is present in many fruit juices. So too is **citric acid,** which constitutes about 6% of lemon juice. Citric acid is added to many soft drinks and candies for flavor. It is also an important intermediate in carbohydrate metabolism and is a normal constituent of blood and urine.

Note that these five common hydroxy acids have the hydroxyl and carboxyl groups attached to the same carbon atom. They are all α-hydroxy acids. There are some examples of hydroxy acids in nature for which this is not so. For example, 3-hydroxybutanoic acid [$CH_3CH(OH)CH_2CO_2H$] is an important intermediate in fat metabolism and a normal component of human blood plasma.

Now let us consider a few simple reactions of hydroxy acids. Hydroxy acids are simultaneously alcohols and carboxylic acids. In the last chapter we learned that these two functional groups react with one another to form esters (Sec. 10.11). We might well ask, then, how both groups can exist side by side in the same molecule. The answer lies in their relative positions. Hydroxy acids in which the two functions are separated by *three* or *four* carbon atoms *do* react *intra*molecularly to form *cyclic* esters called **lactones.** For example,

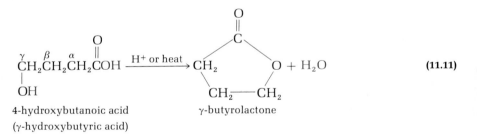

$$\text{(11.11)}$$

4-hydroxybutanoic acid
(γ-hydroxybutyric acid) γ-butyrolactone

Often the lactones form spontaneously, especially from γ-hydroxy acids. The most common lactones have five- or six-membered rings, though lactones with smaller or larger rings are known. Two examples of six-membered lactones that occur in nature are **coumarin,** which is responsible for the pleasant odor of newly mown hay, and **nepetalactone,** the compound in catnip that excites cats.

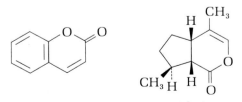

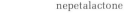

coumarin   nepetalactone

If the hydroxyl and carboxyl groups in a hydroxy acid are separated by fewer than three carbons, the lactones are more difficult to form, because the rings would be four- or three-membered and hence strained. Consequently β- and α-hydroxy acids do not readily form lactones. Instead, β-hydroxy acids readily lose water, on heating, to give unsaturated acids.

$$\overset{\beta}{R}\overset{\alpha}{CH}CH_2COOH \xrightarrow{\text{heat}} RCH{=}CHCOOH + HOH \qquad (11.12)$$
$$\underset{OH}{\big|}$$

a β-hydroxy acid

α-Hydroxy acids react *intermolecularly* to give cyclic diesters called **lactides**.

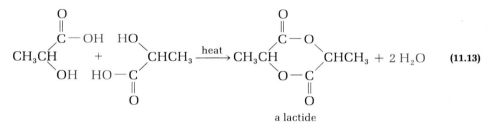

a lactide

We see, then, that the chemistry of hydroxy acids depends very much on the location of the two functions with respect to each other.

Problem 11.9 **Predict the effect of heat or H⁺ on each of the following hydroxy acids.**
a. $CH_2CH_2CH_2CH_2COOH$
  $\underset{OH}{\big|}$
b. $CH_3CHCH_2CH_2COOH$
  $\underset{OH}{\big|}$
c. $CH_3CH_2CHCH_2COOH$
  $\underset{OH}{\big|}$
d. $CH_3CH_2CH_2CHCOOH$
  $\underset{OH}{\big|}$

11.7
PHENOLIC ACIDS;
ASPIRIN

Phenolic acids are a special class of hydroxy acids. The most important commercial phenolic acid is **salicylic acid** (*o*-hydroxybenzoic acid). It is manufactured by heating sodium phenoxide with carbon dioxide, under pressure.

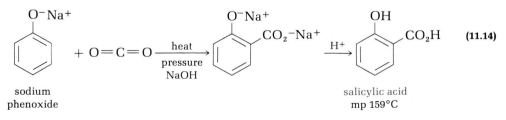

sodium
phenoxide

salicylic acid
mp 159°C

The methyl ester, **methyl salicylate,** is the chief component of oil of wintergreen. Methyl salicylate is used to flavor gum or candy. It is also used in rubbing linaments, where its mild irritating action on skin provides a counter-irritant for sore muscles.

methyl salicylate
bp 223°C

Problem 11.10 **Write an equation for the preparation of methyl salicylate from methanol and salicylic acid.**

The main use of salicylic acid is to prepare aspirin. Reaction with acetic anhydride converts the phenolic hydroxyl group of salicylic acid to its acetyl ester, which is aspirin:

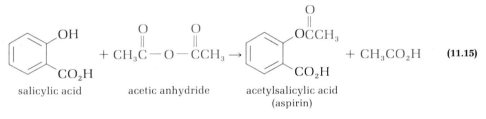

salicylic acid   acetic anhydride   acetylsalicylic acid
                                                          (aspirin)

Annual aspirin production in the United States is over 50 million pounds, enough to produce over 50 billion standard 5-grain tablets. Aspirin is widely used, either by itself or mixed with other drugs, as an analgesic and antipyretic. It is not without its dangers, however. Repeated use may cause gastrointestinal bleeding, and large single doses (10–20 g) can cause death.

**11.8
KETO ACIDS**

Keto acids, especially those with the keto group either α or β to the carboxyl group, are important intermediates in biological oxidations and reductions.

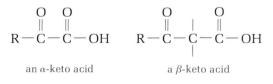

an α-keto acid   a β-keto acid

The best known of the α-keto acids is **pyruvic acid,** which plays an important role in carbohydrate metabolism. It is a source of acetyl groups for acetyl–coenzyme A (see A Word About Nature's Acyl-Activating Groups, page 254).

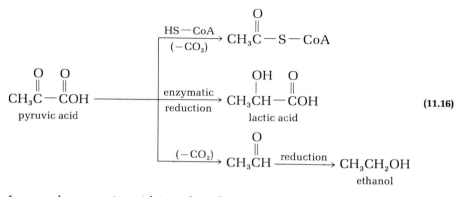

(11.16)

In muscle, pyruvic acid is reduced enzymatically to lactic acid. Pyruvic acid is also a key intermediate in the fermentation of glucose to ethanol. It is converted to ethanol through enzymatic decarboxylation and reduction (eq. 11.16).

β-Keto acids are important intermediates in fat metabolism. A key reaction here is their facile decarboxylation. For example, the β-keto acid **acetoacetic acid** readily loses CO_2 on heating.

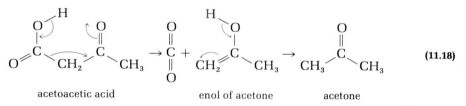

(11.17)

The reaction is similar to the decarboxylation of malonic acid (eq. 11.3), and the mechanism is similar (compare with eq. 11.4).

(11.18)

| acetoacetic acid | enol of acetone | acetone |

Problem 11.11 **Write an equation for the expected reaction when** [structure] **is heated.**

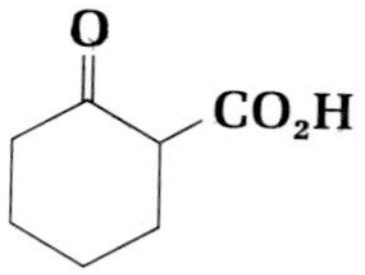

Because of the importance of β-keto acids, several methods have been developed for synthesizing them. The most important of these methods is described in the next section.

Being adjacent to a carbonyl group, the α-hydrogens of an ester can be removed by a strong base to form an **ester enolate.**

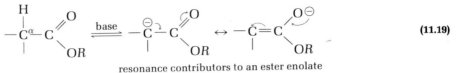

resonance contributors to an ester enolate   **(11.19)**

Bases commonly used for this purpose are sodium alkoxides or sodium hydride. The ester enolate can then act as a carbon nucleophile and add to the carbonyl group of another ester molecule. The reaction is very similar to the aldol condensation (Sec. 9.18) and is called the **Claisen condensation.** It is a way of making **β-ketoesters.** Let us use ethyl acetate as an example of how the reaction works.

Treatment of ethyl acetate with sodium ethoxide in ethanol produces the β-keto ester **ethyl acetoacetate.** The overall reaction is

$$CH_3\overset{O}{\overset{||}{C}}-OCH_2CH_3 + H-CH_2\overset{O}{\overset{||}{C}}-OCH_2CH_3 \xrightarrow[\text{in ethanol}]{NaOCH_2CH_3}$$

ethyl acetate   ethyl acetate   **(11.20)**

$$CH_3\overset{O}{\overset{||}{C}}-CH_2\overset{O}{\overset{||}{C}}-OCH_2CH_3 + CH_3CH_2OH$$

ethyl acetoacetate

The reaction consists of three steps:

Step 1 $CH_3\overset{O}{\overset{||}{C}}-OCH_2CH_3 + Na^{+-}OCH_2CH_3 \rightleftharpoons Na^{+-}CH_2\overset{O}{\overset{||}{C}}OCH_2CH_3 + CH_3CH_2OH$ **(11.21)**

                     sodium ethoxide   ester enolate

Step 2 $CH_3\overset{O}{\overset{(||)}{C}}-OCH_2CH_3 + {}^-CH_2\overset{O}{\overset{||}{C}}OCH_2CH_3 \rightleftharpoons$

$$CH_3\overset{{}^-O}{\underset{|}{\overset{|}{C}}}-OCH_2CH_3 \rightleftharpoons CH_3\overset{O}{\overset{||}{C}}CH_2\overset{O}{\overset{||}{C}}OCH_2CH_3 + {}^-OCH_2CH_3 \quad \textbf{(11.22)}$$

$$\underset{\overset{|}{\underset{O}{\overset{||}{C}}-OCH_2CH_3}}{CH_2}$$

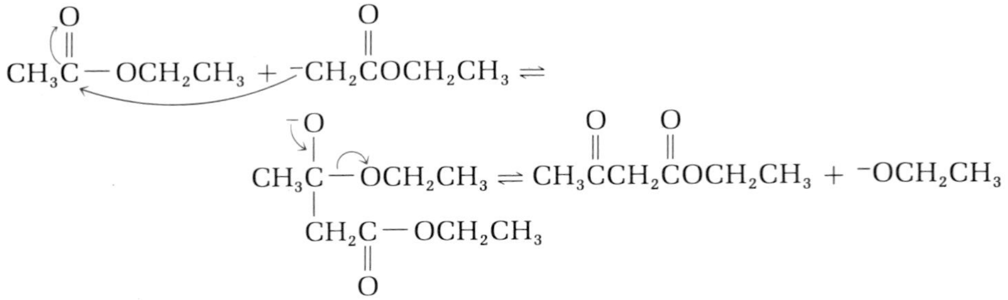

Step 3 $CH_3\overset{O}{\overset{||}{C}}CH_2\overset{O}{\overset{||}{C}}OCH_2CH_3 + {}^-OCH_2CH_3 \rightarrow CH_3\overset{O}{\overset{||}{C}}\!=\!\!\overset{\ominus}{C}H\!=\!\!\overset{O}{\overset{||}{C}}OCH_2CH_3 + CH_3CH_2OH$ **(11.23)**

                                           enolate ion of a β-keto ester

In the first step, the base (sodium ethoxide) removes an α-hydrogen from the ester to form an ester enolate (eq. 11.21). The ester enolate then adds to the carbonyl group of a second ester molecule, displacing ethoxide ion (eq. 11.22). These first two steps of the reaction are completely reversible. It is step 3 that drives the entire set of equilibria forward. In this step the β-keto ester is converted to its enolate anion. The methylene (CH_2) hydrogens in ethyl acetoacetate are α to *two* carbonyl groups and hence are appreciably more acidic than ordinary α-hydrogens. They are easily removed by the base (ethoxide ion) to form the resonance-stabilized enolate ion shown in eq. 11.23. To complete the Claisen condensation, the solution is acidified, to regenerate the β-ketoester from its enolate anion.

Example 11.3   **Identify the product of the Claisen condensation of ethyl propanoate.**

$$\underset{\displaystyle CH_3CH_2\overset{\textstyle O}{\overset{\|}{C}}-OCH_2CH_3}{}$$

Solution   **The product is**

$$CH_3CH_2\overset{\overset{\textstyle O}{\|}}{\underset{\beta}{C}}-\underset{\underset{\textstyle CH_3}{|}}{\overset{\alpha}{CH}}-\overset{\overset{\textstyle O}{\|}}{C}OCH_2CH_3$$

Note that it is the α-carbon of one ester molecule that becomes joined to the carbonyl carbon of the other ester. The product is always a β-keto ester (compare with eq. 9.53 for the aldol condensation).

Problem 11.12   **Using eqs. 11.21–11.23 as a guide, write out the steps in the mechanism for the Claisen condensation of ethyl propanoate.**

The Claisen condensation is a useful reaction for making new carbon–carbon bonds in the laboratory. It also plays a role in the metabolism of fats. In the remainder of this chapter we will turn our attention to fats and oils and to the soaps and detergents that can be made from them.

11.10
FATS AND OILS;
TRIESTERS OF
GLYCEROL

Fats and oils are familiar parts of daily life. Examples of common fats include butter, lard, and the fatty portions of meat. Oils come mainly from vegetable sources and include corn oil, cottonseed oil, olive oil, soybean oil, and many others. Although fats are solids and oils are liquids, they have the same basic organic structure. *Fats and oils are triesters of glycerol and are called* **triglycerides.** When we boil a fat or oil with alkali, as in the saponification of an ester (eq. 10.17), and acidify the resulting solution, we obtain glycerol and a mixture of **fatty acids.**

	Common name	Number of carbons	Structural formula	mp, °C
Saturated	lauric	12	$CH_3(CH_2)_{10}COOH$	44
	myristic	14	$CH_3(CH_2)_{12}COOH$	58
	palmitic	16	$CH_3(CH_2)_{14}COOH$	63
	stearic	18	$CH_3(CH_2)_{16}COOH$	70
	arachidic	20	$CH_3(CH_2)_{18}COOH$	77
Unsaturated	oleic	18	$CH_3(CH_2)_7CH{=}CH(CH_2)_7COOH$	13
	linoleic	18	$CH_3(CH_2)_4CH{=}CHCH_2CH{=}CH(CH_2)_7COOH$	-5
	linolenic	18	$CH_3CH_2CH{=}CHCH_2CH{=}CHCH_2CH{=}CH(CH_2)_7COOH$	-11

TABLE 11.2   Common acids obtained from fats

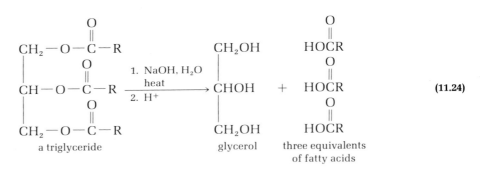

(11.24)

a triglyceride   glycerol   three equivalents
of fatty acids

The most common saturated and unsaturated fatty acids obtained in this way are listed in Table 11.2. Although exceptions are known, *the most common fatty acids have unbranched chains and an even number of carbon atoms.* If double bonds are present, they usually have the *cis* (or Z) configuration and are not conjugated.

Problem 11.13   **Write out the complete structure of linolenic acid, showing the *cis* geometry at each double bond.**

There are two types of triglycerides, **simple** and **mixed**.

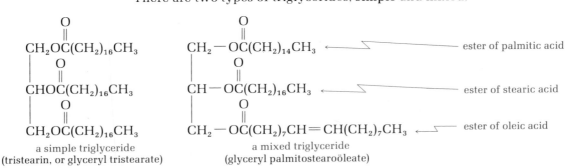

a simple triglyceride   a mixed triglyceride
(tristearin, or glyceryl tristearate)   (glyceryl palmitostearoöleate)

Problem 11.14   **Draw the structures for:**
a. tripalmitin   b. glyceryl stearopalmitoöleate

Simple triglycerides are relatively rare in nature. In general a particular fat or oil does not consist of a single triglyceride but of a complex mixture of triglycerides. For this reason, the composition of a fat or oil is usually expressed in terms of the percentages of the various acids that can be obtained by its hydrolysis. Some fats and oils give mainly 1 or 2 acids, with only minor amounts of other acids. Olive oil, for example, gives 83% oleic acid. Palm oil gives 43% palmitic acid and 43% oleic acid, with lesser amounts of stearic and linoleic acids. Butter fat, on the other hand, gives at least 14 different acids on hydrolysis and is somewhat exceptional in that about 9% of these are acids with fewer than 10 carbon atoms.

**11.11
HYDROGENATION
OF VEGETABLE
OILS**

What is it, in terms of structure, that makes some triglycerides solids (fats) and others liquids (oils)? The distinction is clear from their composition. *Oils contain a much higher percentage of unsaturated fatty acids than do fats.* For example, most vegetable oils (such as corn or soybean oil) give about 80% unsaturated acids on hydrolysis. For fats (such as beef tallow) the figure is much lower, just a little over 50%. You will note in Table 11.2 that the melting points of unsaturated fatty acids are, in general, appreciably lower than those of saturated acids. Compare, for example, the melting points of stearic and oleic acids, which differ in structure by only one double bond. The same difference applies to triglycerides: the more double bonds in the fatty acid portion of the triester, the lower the melting point.

The reason for the effect of saturation or unsaturation on melting point becomes apparent when we examine space-filling models. Figure 11.2 shows a model of a fully saturated triglyceride, tripalmitin. Note that the long, saturated chains have fully extended, staggered conformations. In this way the chains, and indeed the molecules, can pack together fairly

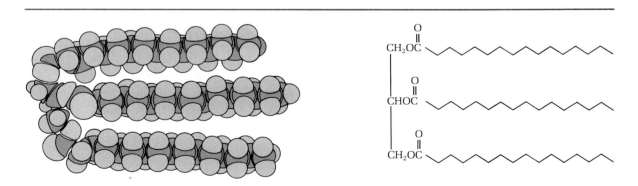

FIGURE 11.2   Space-filling and schematic model of tripalmitin.

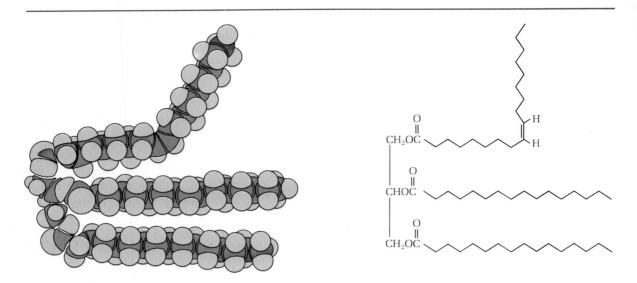

FIGURE 11.3   Space-filling and schematic model of glyceryl dipalmitoöleate.

regularly, as in a crystal. Consequently, they usually form solids at room temperature.

The result of introducing just one *cis* double bond in one of the chains is shown in Figure 11.3. Clearly the chains in a single molecule (and the molecules themselves if many are brought close to one another) cannot align nicely in a crystalline array, and the substance remains a liquid. The more double bonds, the more disorderly the structure and the lower the melting point.

We can convert vegetable oils, which are highly unsaturated, into vegetable fats such as Crisco or Spry by catalytically hydrogenating some or all of the double bonds (Sec. 3.9). This process, called **hardening,** is illustrated by the hydrogenation of triolein to tristearin.

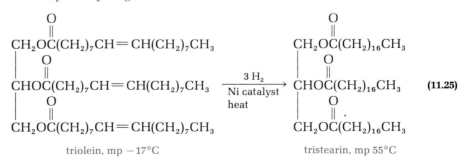

$$\begin{array}{l} O \\ \| \\ CH_2OC(CH_2)_7CH{=}CH(CH_2)_7CH_3 \\ \quad O \\ \quad \| \\ CHOC(CH_2)_7CH{=}CH(CH_2)_7CH_3 \\ \quad O \\ \quad \| \\ CH_2OC(CH_2)_7CH{=}CH(CH_2)_7CH_3 \end{array} \xrightarrow[\substack{\text{Ni catalyst} \\ \text{heat}}]{3\ H_2} \begin{array}{l} O \\ \| \\ CH_2OC(CH_2)_{16}CH_3 \\ \quad O \\ \quad \| \\ CHOC(CH_2)_{16}CH_3 \\ \quad O \\ \quad \| \\ CH_2OC(CH_2)_{16}CH_3 \end{array}$$

(11.25)

triolein, mp −17°C   tristearin, mp 55°C

Oleomargarine is made by hydrogenating cottonseed, soybean, peanut, or corn oil until the desired butterlike consistency is obtained. The product may be churned with milk and artificially colored to mimic butter's flavor and appearance.

**11.12
SAPONIFICATION
OF FATS AND
OILS; SOAP**

When a fat or oil is heated with alkali, the ester linkages are broken to give glycerol and salts of the fatty acids (Sec. 10.13). The reaction is illustrated here with the saponification of tripalmitin.

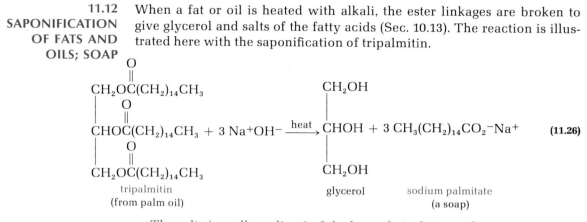

$$\underset{\substack{\text{tripalmitin}\\\text{(from palm oil)}}}{\begin{array}{c}\text{CH}_2\text{OC(CH}_2)_{14}\text{CH}_3\\ |\quad\overset{\text{O}}{\overset{\|}{}}\\ \text{CHOC(CH}_2)_{14}\text{CH}_3\\ |\quad\overset{\text{O}}{\overset{\|}{}}\\ \text{CH}_2\text{OC(CH}_2)_{14}\text{CH}_3\end{array}} + 3\,\text{Na}^+\text{OH}^- \xrightarrow{\text{heat}} \underset{\text{glycerol}}{\begin{array}{c}\text{CH}_2\text{OH}\\ |\\ \text{CHOH}\\ |\\ \text{CH}_2\text{OH}\end{array}} + \underset{\substack{\text{sodium palmitate}\\\text{(a soap)}}}{3\,\text{CH}_3(\text{CH}_2)_{14}\text{CO}_2^-\text{Na}^+} \qquad \textbf{(11.26)}$$

The salts (usually sodium) of the long-chain fatty acids are **soaps**.

a word about _____

22. Soaps and How They Work

The conversion of animal fats (for example, goat tallow) into soap by heating them with wood ashes (which are alkaline) is one of the oldest of chemical processes. Soap has been produced for at least 2300 years, having been known to the ancient Celts and Romans. Yet as recently as the sixteenth and seventeenth centuries, soap was a rather rare substance in central Europe, being used mainly in medicine. When, in 1672, an admirer sent a gift of Italian soap to a female member of the German aristocracy, it had to be accompanied by instructions for its use. But by the nineteenth century, soap had come into such widespread use that the German organic chemist Justus von Liebig remarked that the quantity of soap consumed by a nation was an accurate measure of its wealth and civilization. At present, annual

world production of ordinary soaps (not including synthetic detergents) is well over 6 million tons.

Soaps are made by either a batch process or a continuous process. In the batch process, the fat or oil is heated with a slight excess of alkali (NaOH) in an open kettle. When saponification is complete, salt is added to precipitate the soap as thick curds. The water layer, which contains salt, glycerol, and excess alkali, is drawn off and the glycerol is recovered by distillation. The crude soap curds, which contain some salt, alkali, and glycerol as impurities, are purified by boiling with water and reprecipitating with salt several times. Finally, the curds are boiled with enough water to form a smooth mixture that, on standing, gives a homogeneous upper layer of soap. This soap may be sold without further processing, as a cheap industrial soap. Various fillers such as sand or pumice may be added, to make scouring soaps. Other treatments transform the crude soap to toilet soaps, powdered or flaked soaps, medicated or perfumed soaps, laundry soaps, liquid soaps, or (by blowing air in) floating soaps.

In the continuous process, which is more common today, the fat or oil is hydrolyzed by water at high temperatures and pressures in the presence of a catalyst, usually a zinc soap. The fat or oil and the water are introduced continuously into opposite ends of a large reactor, and the fatty acids and glycerol are removed as formed, by distillation. The acids are then carefully neutralized with an appropriate amount of alkali to make soap.

How do soaps work? Most dirt sticks to clothes or the body by a thin film of oil. If the oil film can be removed, the dirt particles can be washed away. Soap molecules consist of a long, hydrocarbon-like chain of carbon atoms with a highly polar or ionic group at one end (Figure 11.4). The carbon chain is lipophilic (attracted to or soluble in fats and oils) and the polar end is hydrophilic (attracted to or soluble in water). In a sense, soap molecules are schizophrenic, having two different "personalities." Let us see what happens when we add soap to water.

When soap is shaken with water, it forms a colloidal dispersion—not a true solution. These soap solutions contain aggregates of soap molecules called micelles. The nonpolar or lipophilic carbon chains are directed toward the center of the micelle. The polar or hydrophilic ends of the molecule form the "surface" of the micelle that is presented to the water (Figure 11.5). In ordinary soaps, the outer part of each micelle is negatively charged, and the posi-

tive sodium ions congregate near the micelles.

When acting to remove dirt, soap molecules surround and emulsify the droplets of oil or grease. The lipophilic "tails" of the soap molecules dissolve in the oil. The hydrophilic ends extend out of the oil droplet toward the water. In this way, the oil droplets are stabilized in the water solution, because the negative surface charge of the droplets prevents their coalescence. This is shown in Figure 11.6.

Another striking property of soap solutions is their unusually low surface tension, which gives a soap solution more "wetting" power than plain water has. It is a combination of the emulsifying power and the surface action of soap solutions that enables them to detach dirt, grease, and oil particles from the surface to be cleaned and to emulsify them so that they can be washed away. The same principles of cleansing action apply to synthetic detergents.

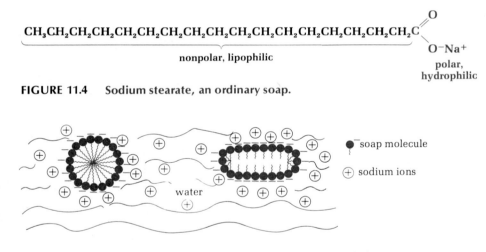

$$CH_3CH_2CH_2CH_2CH_2CH_2CH_2CH_2CH_2CH_2CH_2CH_2CH_2CH_2CH_2CH_2CH_2C$$

$$\underbrace{}_{\text{nonpolar, lipophilic}}$$

O

O$^-$Na$^+$

polar, hydrophilic

FIGURE 11.4   Sodium stearate, an ordinary soap.

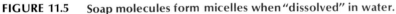

● $^-$soap molecule

⊕ sodium ions

water

FIGURE 11.5   Soap molecules form micelles when "dissolved" in water.

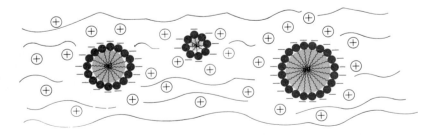

FIGURE 11.6   Oil droplets become emulsified by soap molecules.

11.13
SYNTHETIC
DETERGENTS
(SYNDETS)

A few years ago the annual world production of synthetic detergents (often called **syndets**) first exceeded that of ordinary soaps, and this trend seems likely to continue. The need for syndets arose because the use of ordinary unimproved soaps raises two problems. First, being salts of weak acids, their solutions are rather alkaline due to partial hydrolysis.

$$\underset{\text{soap}}{R-\overset{\displaystyle O}{\overset{\|}{C}}-O^-Na^+} + H-OH \rightleftharpoons R-\overset{\displaystyle O}{\overset{\|}{C}}-OH + \underset{\text{alkali}}{Na^+OH^-} \tag{11.27}$$

The alkali can be harmful to certain fabrics. Yet ordinary soaps cannot function in solutions of low pH (that is, in acid) because the long-chain fatty acid will precipitate from solution as a scum. For example, sodium stearate, a typical soap, is destroyed by conversion to stearic acid on acidification.

$$\underset{\text{sodium stearate}}{C_{17}H_{35}C\overset{\displaystyle O}{\underset{O^-Na^+}{\diagup}}} + H^+Cl^- \rightarrow \underset{\text{stearic acid}}{C_{17}H_{35}C\overset{\displaystyle O}{\underset{OH}{\diagup}}} \downarrow + Na^+Cl^- \tag{11.28}$$

The second problem with ordinary soaps is that they form insoluble salts with the calcium, magnesium, or ferric ions present in "hard" water.

$$\underset{\substack{\text{sodium stearate} \\ \text{(soluble)}}}{2C_{17}H_{35}C\overset{\displaystyle O}{\underset{O^-Na^+}{\diagup}}} + Ca^{++} \rightarrow \underset{\substack{\text{calcium stearate} \\ \text{(insoluble)}}}{(C_{17}H_{35}COO)_2{}^-Ca^{++}}\downarrow + 2Na^+ \tag{11.29}$$

The insoluble salts are responsible for "rings" around the bathtub or collar and for films that dull the look of clothes and hair.

These problems with ordinary soaps have been solved or diminished in several ways. For example, water can be "softened," either municipally or in individual households, to remove the offending calcium or magnesium ions. In softened water these ions are replaced by sodium ions. If this water is also used for drinking, however, it may cause health problems, especially for older people who have to limit their sodium intake.

Phosphates can also be added to soaps (and to syndets). Phosphates form soluble complexes with metal ions, thus keeping these ions from forming insoluble salts with the soap. But widespread use of phosphates in the past has created problems. As a result of their use in detergents, tremendous quantities of phosphates have eventually found their way into lakes, rivers, and streams. Since they are fertilizers, phosphates have been known to stimulate plant growth to such an extent that the plants exhaust the dissolved oxygen in the water, in turn causing fish to die. Phosphates are still used in detergents, but their use is limited by law to levels that are unlikely to be harmful.

Another way to correct the problems of ordinary soaps is to design and synthesize inexpensive but more effective detergents. These syndets must have several design features, including a long lipophilic chain and a polar or ionic hydrophilic end. Also, the polar end should not form insoluble salts with the ions present in hard water and should not affect the acidity of water. Finally, the carboxylate group of an ordinary soap should be replaced.

The first syndets, designed in the 1940s, were the sodium salts of alkyl hydrogen sulfates. Long-chain alcohols were prepared by the hydrogenolysis of fats and oils.

$$
\begin{array}{c}
CH_3(CH_2)_{10}\overset{\overset{\displaystyle O}{\|}}{C}\!+\!OCH_2 \\
CH_3(CH_2)_{10}\overset{\overset{\displaystyle O}{\|}}{C}\!+\!OCH \\
CH_3(CH_2)_{10}\overset{\overset{\displaystyle O}{\|}}{C}\!+\!OCH_2
\end{array}
+ 6\,H_2 \xrightarrow[\text{heat, pressure}]{\text{copper chromite}} 3CH_3(CH_2)_{10}CH_2OH +
\begin{array}{c}
HOCH_2 \\
HOCH \\
HOCH_2
\end{array}
\qquad (11.30)
$$

1-dodecanol
(lauryl alcohol)

The long-chain alcohol was treated with sulfuric acid to make the alkyl hydrogen sulfate (eq. 7.25), which was then neutralized with base.

$$
\underset{\text{lauryl alcohol}}{CH_3(CH_2)_{10}CH_2OH} + \underset{\text{sulfuric acid}}{HOSO_2OH} \rightarrow \underset{\text{lauryl hydrogen sulfate}}{CH_3(CH_2)_{10}CH_2OSO_2OH} + H_2O
$$

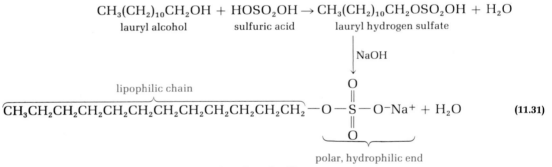

$$
\overset{\text{lipophilic chain}}{\overbrace{CH_3CH_2CH_2CH_2CH_2CH_2CH_2CH_2CH_2CH_2CH_2CH_2}}\!-\!O\!-\!\underset{\underset{\displaystyle O}{\|}}{\overset{\overset{\displaystyle O}{\|}}{S}}\!-\!O^-Na^+ + H_2O \qquad (11.31)
$$

polar, hydrophilic end

sodium lauryl sulfate

Sodium lauryl sulfate is an excellent detergent. Because it is a salt of a strong acid, its solutions are nearly neutral. Its calcium and magnesium salts do not precipitate from solution, so it is effective in hard as well as soft water. But its supply is too limited to meet the demand, and thus the need for other syndets arose.

At present, the most widely used syndets are straight-chain alkylbenzenesulfonates. They are made in three steps. Straight-chain alkenes with 10–14 carbons are treated with benzene and a Friedel–Crafts catalyst (AlCl$_3$ or HF, eq. 4.14) to form an alkylbenzene. Sulfonation and neutralization with base complete the process.

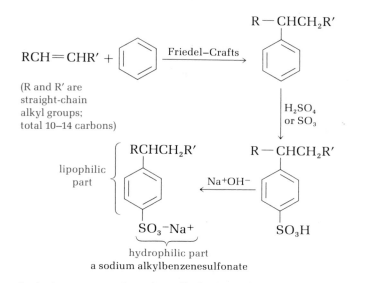

(11.32)

a sodium alkylbenzenesulfonate

It is important that the alkyl chain have no branches. The first alkylbenzenesulfonates had branched side chains and were not biodegradable. They created severe pollution problems in the 1950s, causing foaming in sewage-treatment plants and in lakes and rivers. But since about 1965 alkylbenzenesulfonates with unbranched side chains have been used. They are fully biodegradable by microorganisms and do not accumulate in the environment.

Problem 11.15 **Propylene tetramer, which can be made by polymerizing propylene, has the formula**

$$\begin{matrix} CH_3 & CH_3 & CH_3 \\ | & | & | \\ CH_3CHCH_2CHCH_2CHCH=CHCH_3 \end{matrix}$$

Show how it can be converted to a branched-chain alkylbenzenesulfonate detergent by steps analogous to those in eq. 11.32.

a word about ⸺⸺⸺⸺⸺⸺⸺⸺⸺⸺⸺⸺⸺⸺

23. The Metabolism of Fats

We have just seen that some detergents, those with unbranched carbon chains, are biodegradable. That is, they can be "digested" by microorganisms and broken down into smaller and smaller carbon chains. Hence they do not survive as such but are rendered harmless in the environment. To under-

stand why unbranched carbon chains are biodegradable but branched chains are not, we must understand how the straight-chain fatty acids of fats are metabolized.

When a fat or oil is ingested, it is hydrolyzed by enzymes called <u>lipases</u> into fatty acids and glycerol (eq. 11.24). This hydrolysis occurs mainly in the small intestines, where the fatty acids are absorbed and transported to other organs for further metabolism. Ultimately, the fatty acids are oxidized to carbon dioxide and water to furnish energy. How does this oxidation of the long carbon chains occur?

First, the carboxyl group of the fatty acid is activated by conversion to a thioester, through reaction with coenzyme A (see A Word About Nature's Acyl-Activating Groups, page 254).

$$RCH_2CH_2COOH + CoA\!-\!SH \underset{ATP}{\overset{enzyme}{\rightleftharpoons}}$$
fatty acid

$$RCH_2CH_2\overset{\overset{O}{\|}}{C}\!-\!SCoA + H\!-\!OH \qquad (11.33)$$
fatty acyl-CoA

The activation requires an enzyme and adenosine triphosphate (ATP). The alkyl chain is then shortened *2 carbons at a time* by a four-step process summarized in Figure 11.7. This figure shows how stearoyl–CoA, which has an 18-carbon chain, is broken into palmitoyl–CoA (16 carbons) and acetyl–CoA (2 carbons). This process is repeated with palmitoyl–CoA to give myristoyl–CoA, and so on until the carbon chain is chopped down, 2 carbons at a time, to acetyl–CoA. Let us look briefly at each of the steps shown in Figure 11.7.

In the first step an enzyme and flavin adenine dinucleotide (FAD) dehydrogenate the thioester to give an unsaturated ester with a double bond between carbon-2 and carbon-3 in the chain. In the next step, the double bond is hydrated (Sec. 3.10) to give a 3-hydroxy thioester. In step 3 the alcohol function of the hydroxyester is oxidized to a keto function, with nicotinamide adenine dinucleotide (NAD$^+$) as the oxidizing agent (see p. 180). Finally, the β-ketothioester is cleaved by coenzyme A. This is the key step in the process, and it is the reverse of a Claisen ester condensation (Sec. 11.9). The mechanism can be pictured as shown in Figure 11.8.

In the biodegradation of syndets such as the alkylbenzenesulfonates, the benzene ring is oxidized first, leading to a long-chain fatty acid, which is then degraded according to eq. 11.33 and Figure 11.7. If the alkyl chain is branched, the process at some stage (depending on where the branches are) will give a hydroxyester (step 2) in which the alcohol function is tertiary instead of secondary. This blocks the oxidation step and prevents further degradation of the carbon chain. For this reason branched-chain alkylbenzene sulfonates are not fully biodegradable.

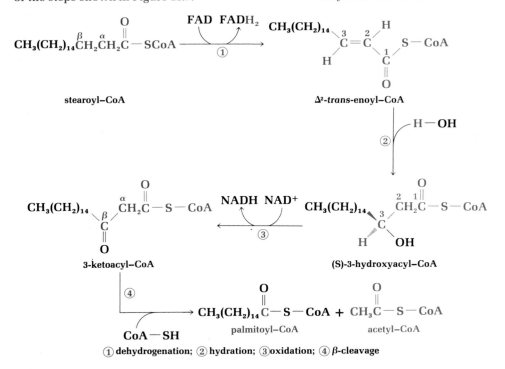

① dehydrogenation; ② hydration; ③ oxidation; ④ β-cleavage

FIGURE 11.7   Steps in saturated-fatty-acid oxidation.

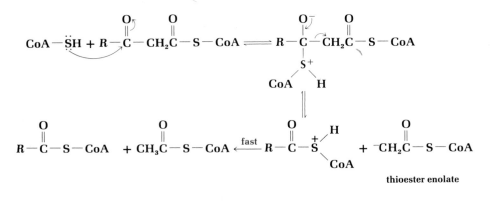

FIGURE 11.8

11.14 OTHER LIPIDS

Fats and oils represent only one type of a more general class of substances called **lipids** (from the Greek *lipos*, fat). Lipids are constituents of plants or animals that are distinguished by their solubility properties. Lipids are soluble in ether and other nonpolar solvents and are insoluble in water. This solubility property distinguishes lipids from several other major classes of natural products such as proteins, carbohydrates, and nucleic acids which, in general, are *not* soluble in nonpolar organic solvents. In this final section of the chapter, we will briefly describe a few other types of lipids.

Phospholipids constitute about 40% of cell membranes, the remaining 60% being proteins. Phospholipids are related in structure to fats and oils, except that one of the three ester groups is replaced by a phosphatidylamine.

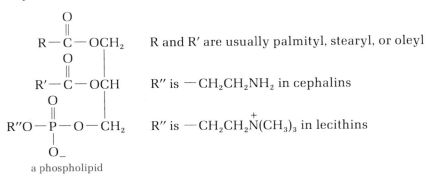

R and R′ are usually palmityl, stearyl, or oleyl

R″ is $-CH_2CH_2NH_2$ in cephalins

R″ is $-CH_2CH_2\overset{+}{N}(CH_3)_3$ in lecithins

a phospholipid

Phospholipids arrange themselves in bilayers in membranes, the hydrocarbon "tails" pointing in and the polar phosphatidylamine ends constituting the membrane surface:

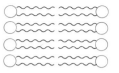

arrangements of phospholipids in a bilayer membrane

Waxes are lipids that differ from fats and oils in being simple mono-esters. The acid and alcohol portions both have long saturated carbon chains. A typical example is *myricyl palmitate,* the main component of beeswax.

$$CH_3(CH_2)_{13}CH_2\overset{\displaystyle O}{\overset{\|}{C}}-O-(CH_2)_{29}CH_3$$

myricyl palmitate (beeswax)

Problem 11.16   **When a fat or oil is boiled with alkali, it "dissolves." The same is not true for a wax. Explain.**

Some waxes are simply hydrocarbons (Sec. 2.7).

Waxes are more brittle, usually harder, and less greasy than fats. They are used to make polishes, cosmetics, ointments, and other pharmaceutical preparations, as well as candles and phonograph records. In nature, waxes coat the leaves and stems of plants that grow in arid regions, thus reducing evaporation. Similarly, insects with a high surface-area-to-volume ratio are often coated with a natural protective wax.

Steroids form another major class of lipids. The common structural feature of steroids is a system of four fused rings. The A, B, and C rings are six-membered, whereas the D-ring is five-membered.

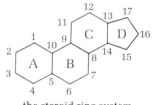

the steroid ring system, showing the numbering

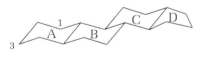

steroid shape, with chair cyclohexanes

In most steroids, the rings are not aromatic. Usually there are methyl substituents attached to C-10 and C-13 (called "angular" methyl groups) and some sort of side chain attached to C-17. See, for example, the formula for **cholesterol** on page 187.

Steroids perform many biological functions. Among the more important steroids are the **sex hormones,** compounds produced in the gonads (ovaries and testes) that control reproductive physiology and secondary sex characteristics. The female sex hormones are of two types. The **estrogens,** the most common of which is **estradiol,** are essential for changes during the menstrual cycle and for development of female secondary sex characteristics. **Progesterone,** which prepares the uterus for implantation of the fertilized egg, also maintains the pregnancy and prevents further ovulation during pregnancy. Progesterone is used clinically to prevent abortion in certain difficult pregnancies. It differs structurally from estrogens such as estradiol in that the A ring is not aromatic. Oral contraceptives ("the pill"), such as **Enovid,** have structures similar to progesterone.

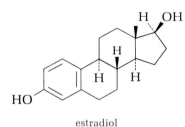

estradiol

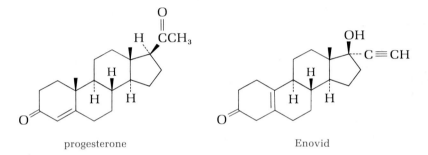

progesterone   Enovid

The male sex hormones are called **androgens.** They regulate the development of male reproductive organs and secondary sex characteristics such as facial and body hair, deep voice, and male musculature. Two important androgens are **testosterone** and **androsterone.**

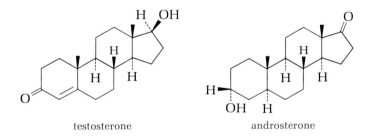

testosterone   androsterone

Note that the only structural difference between the male hormone testosterone and the female hormone progesterone is the replacement of a hydroxyl group by an acetyl group at C-17 in the D ring.

All steroidal hormones are biosynthesized from cholesterol. Among the most striking features of the steroidal hormones, the great changes in bioactivity that result from seemingly minor changes in structure testify to the extreme specificity of biochemical reactions.

a word about _____

24. Prostaglandins

The prostaglandins are a group of compounds structurally and biosynthetically related to the unsaturated fatty acids. They were discovered in the 1930s, when it was found that human semen contained substances that could stimulate smooth muscle tissue, such as uterine muscle, to contract. On the assumption that these substances came from the prostate gland, they were named <u>prostaglandins</u>. We now know that these substances are widely distributed in almost all human tissues, that they are biologically active in minute concentration, and that they have various effects on fat metabolism, heart rate, and blood pressure. Because they are present in such low concentrations, it was not until the 1960s that sufficient quantities were isolated to determine their structures. Well over a dozen naturally occurring prostaglandins have now been identified.

The prostaglandins have 20 carbon atoms. Experiments with radioactive tracers have shown that they are synthesized in the body via oxidation and cyclization of the C-20 unsaturated fatty acids, such as arachidonic acid. Carbon-8 through carbon-12 of the chain are looped to form a cyclopentane ring, and an oxygen function (carbonyl or hydroxyl

group) is present at C-9. Various numbers of double bonds or hydroxyl groups may also be present.

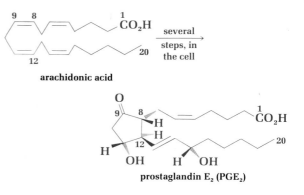

prostaglandin E₂ (PGE₂)

Since 1968 organic chemists have worked out many syntheses of prostaglandins in the laboratory. Some of these are efficient enough to be used commercially, and they have supplied biologists and physicians with enough material to explore clinical uses. It seems likely that prostaglandins will now find important medical uses, especially in the treatment of inflammatory diseases such as asthma and rheumatoid arthritis, the control of hypertension, regulation of blood pressure and metabolism, and inducing labor and bringing about therapeutic abortions.

ADDITIONAL PROBLEMS

11.17. Using the common names in Table 11.1 as a guide, write structural formulas for:

a. dimethyl oxalate   **b.** adipoyl chloride
c. glutaric anhydride   **d.** dimethyl diethylmalonate
e. hydroxysuccinic acid   **f.** 2,3-dimethylbutanedioic acid

11.18. From the data in Table 11.1, we see that the ratio of K_1/K_2 for malonic acid is about 700, whereas the same ratio for adipic acid is only 9. Explain.

11.19. 1,1-, *cis*-1,2- and *trans*-1,2-cyclopropanedicarboxylic acids behave differently when heated. Write equations showing what happens in each case.

11.20. Complete the following equations:

a.   ⬡—COOH   →(heat)
           —COOH

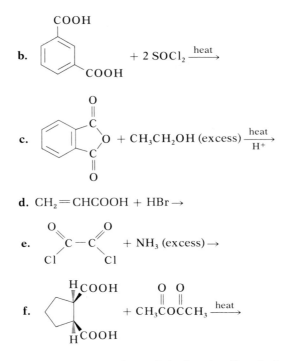

b. (benzene-1,3-dicarboxylic acid) COOH / COOH + 2 SOCl₂ $\xrightarrow{\text{heat}}$

c. (phthalic anhydride structure) + CH₃CH₂OH (excess) $\xrightarrow[\text{H}^+]{\text{heat}}$

d. $CH_2 = CHCOOH + HBr \rightarrow$

e. (Cl–CO–CO–Cl, oxalyl chloride) + NH₃ (excess) →

f. (cyclopentane with two COOH groups, H stereochemistry) + CH₃COCCH₃ (O O) $\xrightarrow{\text{heat}}$

11.21. Dacron can be made by heating dimethyl terephthalate with ethylene glycol in the presence of an acid catalyst. Write out the steps in the process.

11.22. Explain what might happen to the shape of a Dacron molecule 100 units long if the terephthalic acid used in its manufacture were only 98% pure and contained 2% of isophthalic acid as an impurity.

11.23. Write out the repeating unit in the polyester that would be obtained from:
a. isophthalic acid and ethylene glycol
b. adipic acid and 1,3-propanediol

11.24. Addition of bromine to maleic acid gives a dibromosuccinic acid. Addition of bromine to fumaric acid gives a *different* dibromosuccinic acid. Recalling the mechanism of bromine addition to double bonds (Sec. 3.15), write equations for each reaction and explain the structural difference between the two dibromo-succinic acids.

11.25. Write structural formulas for each of the following:
a. diethyl tartrate
b. maleic anhydride
c. potassium hydrogen tartrate (cream of tartar)
d. phenyl salicylate (*salol*—used as an analgesic and also in suntan oils)
e. 3-butenoic acid
f. α-methyl-γ-butyrolactone
g. ethyl acetoacetate
h. pyruvic acid

11.26. Consider the structure of the catnip ingredient nepatalactone (Sec. 11.6).
a.· Draw the structure and show, by dotted lines, that it is composed of two iso-prene units.
b. Circle the chiral centers and determine their configurations (R or S).

11.27. A phenolic acid $C_9H_8O_3(\mathbf{A})$ exists in two isomeric forms. Both rapidly decolorize permanganate and, on moderate oxidation, yield salicylic and oxalic acids as the only organic products. One isomer of **A** easily loses water, when heated, to yield $C_9H_6O_2$. The other fails to dehydrate under the same conditions. Suggest structural formulas for the isomers.

11.28. Write the structure of the Claisen condensation product of ethyl phenylacetate, and show the steps in its formation.

11.29. When heated with sodium ethoxide in ethanol, diethyl adipate undergoes an *intra*molecular Claisen condensation. What is the structure of the product and how is it formed?

11.30. When treated with sodium ethoxide in ethanol, a mixture of ethyl benzoate and ethyl acetate gives a *crossed* Claisen condensation product. What is its structure and how is it formed?

11.31. Using Table 11.2 as a guide, write the structural formulas for:
a. potassium palmitate   **b.** magnesium oleate
c. trilaurin   **d.** glyceryl butyropalmitoöleate
e. myristyl linoleate   **f.** ethyl arachidate

11.32. Write equations for the (a) saponification, (b) hydrogenation, and (c) hydrogenolysis of glyceryl trilinolenate.

11.33. Write an equation for the following reactions.

$$\overset{O}{\overset{\|}{}}$$
a. $C_{15}H_{31}CO^-Na^+ + HCl \rightarrow$

$$\overset{O}{\overset{\|}{}}$$
b. $C_{15}H_{31}CO^-Na^+ + Mg^{2+} \rightarrow$

11.34. Write equations for the preparation of an alkylbenzenesulfonate synthetic detergent, starting with 1-decene and benzene.

11.35. A synthetic detergent widely used in dishwashing liquids has the structure $CH_3(CH_2)_{11}(OCH_2CH_2)_3OSO_3^-Na^+$. Write a series of equations showing how this detergent could be synthesized from $CH_3(CH_2)_{10}CH_2OH$ and ethylene oxide (review Sec. 8.9).

11.36. In 1904 the German biochemist F. Knoop fed a series of acids with the general formula

$$\text{C}_6\text{H}_5-(CH_2)_nCH_2CO_2H$$

to rabbits. He observed that if *n* was 2, 4, 6, . . . , the rabbit excreted phenylacetic acid in the urine, but if *n* was 3, 5, 7, . . . , the rabbit excreted benzoic acid. What conclusions regarding fatty acid oxidation could he draw from these results? Are they consistent with present theory?

11.37. Using Figure 11.7 as a guide, write out the steps in the biological oxidation of butanoic acid to acetic acid.

11.38. Write the structure for:
a. a fat   **b.** a vegetable oil

c. a wax **d.** an ordinary soap
e. a synthetic detergent **f.** a steroid
g. a phospholipid **h.** an α-ketoacid

11.39. Predict the products of the following reactions. (Consult the text for the structures of the starting materials.)
a. estradiol + acetic anhydride
b. progesterone + LiAlH₄
c. testosterone + peracetic acid
d. androsterone + chromic acid

11.40. Answer the following questions regarding cortisone, a drug used in the treatment of arthritis.

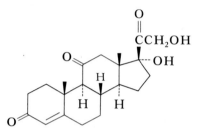

cortisone

a. Draw the structure and number all the carbon atoms.
b. How many chiral centers are present, and what is the configuration (R or S) of each?

twelve

Amines and Related Nitrogen Compounds

We come now to the last of the major functional groups that we shall survey in this text, the **amines.** Amines can be considered as organic relatives of ammonia, derived by replacing one, two, or all three hydrogens of ammonia by organic groups. Like ammonia, amines are bases. In fact, amines are the most important type of organic base that occurs in nature. In this chapter, we will first describe the structure, preparation, chemical properties, and industrial uses of some simple amines. Toward the end of the chapter we will discuss a few natural and synthetic amines with important biological activity and applications.

**12.2
CLASSIFICATION
AND STRUCTURE
OF AMINES**

The relationship between ammonia and the amines is illustrated in the following structures:

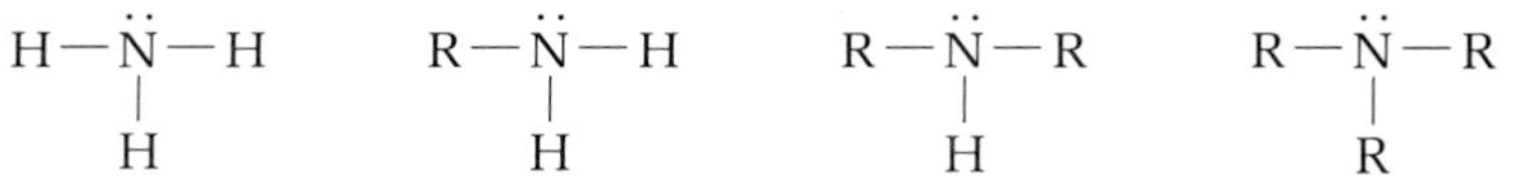

$$H-\overset{\overset{..}{|}}{\underset{H}{N}}-H \qquad R-\overset{\overset{..}{|}}{\underset{H}{N}}-H \qquad R-\overset{\overset{..}{|}}{\underset{H}{N}}-R \qquad R-\overset{\overset{..}{|}}{\underset{R}{N}}-R$$

    ammonia   primary amine   secondary amine   tertiary amine

For convenience amines are classified as **primary amines, secondary amines,** or **tertiary amines,** depending on whether one, two, or three organic groups are attached to the nitrogen. The R groups in these structures may be alkyl or aryl and, when two or more R groups are attached to the nitrogen, they may be identical or different from one another. We will see that in some secondary or tertiary amines the nitrogen may be part of a ring.

Problem 12.1 **Classify the following amines as primary, secondary, or tertiary.**

a. $(CH_3)_3CNH_2$

b.

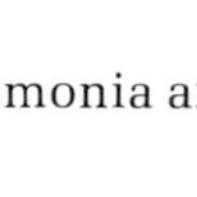

$$\text{[cyclopentane ring with N—H]}$$

c.  CH_3—⟨benzene ring⟩—NH_2

d. 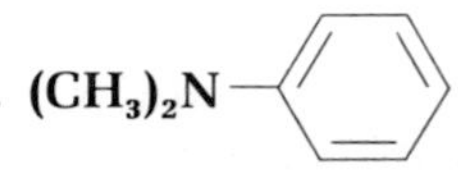 $(CH_3)_2N$—⟨benzene ring⟩

FIGURE 12.1

Amines have a
pyramidal structure, as
seen in these models
of trimethylamine.
The space-filling
model (*b*) is a "top"
view, and the ball at
the center of that
model represents the
orbital that contains
the unshared electron
pair.

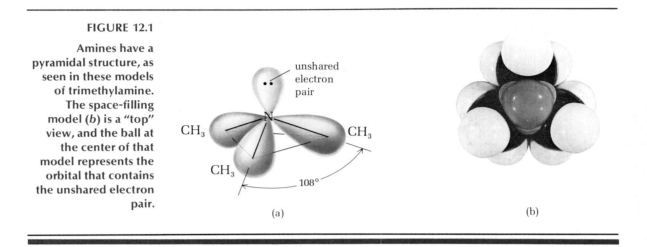

(a)   (b)

The nitrogen atom in an amine is trivalent, and it also carries an un-shared electron pair. Thus the nitrogen orbitals are sp^3-hybridized. The overall geometry is pyramidal (nearly tetrahedral), as shown for trimethyl-amine in Figure 12.1. From this geometry, one might think that a tertiary amine with three different groups attached to the nitrogen should be chiral and separable into enantiomers. This is true in principle, but in practice the two enantiomers usually interconvert rapidly at room temperature (and often even at quite low temperatures) through an "umbrella-in-the-wind" type of process:

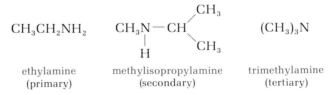

$$\text{planar transition state}$$

(12.1)

12.3 NOMENCLATURE OF AMINES

Simple amines are named by naming the alkyl groups attached to the nitro-gen and adding the suffix -*amine*.

$$CH_3CH_2NH_2 \qquad CH_3N{-}CH\begin{smallmatrix}CH_3\\ \\CH_3\end{smallmatrix} \qquad (CH_3)_3N$$

$$\underset{|}{}$$

ethylamine   methylisopropylamine   trimethylamine
(primary)   (secondary)   (tertiary)

The $-NH_2$ or *amino group* may sometimes be named as a substituent, as in the following examples:

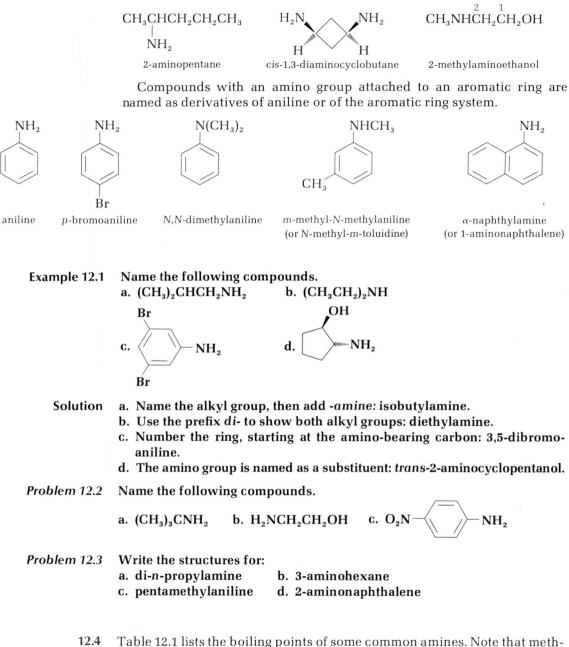

aniline   p-bromoaniline   N,N-dimethylaniline   m-methyl-N-methylaniline   α-naphthylamine
                                                    (or N-methyl-m-toluidine)   (or 1-aminonaphthalene)

Example 12.1   **Name the following compounds.**
a. (CH₃)₂CHCH₂NH₂   b. (CH₃CH₂)₂NH

Solution   a. **Name the alkyl group, then add -amine: isobutylamine.**
b. **Use the prefix di- to show both alkyl groups: diethylamine.**
c. **Number the ring, starting at the amino-bearing carbon: 3,5-dibromo-aniline.**
d. **The amino group is named as a substituent: trans-2-aminocyclopentanol.**

Problem 12.2   **Name the following compounds.**

a. (CH₃)₃CNH₂   b. H₂NCH₂CH₂OH   c. O₂N—⟨ ⟩—NH₂

Problem 12.3   **Write the structures for:**
a. **di-n-propylamine**   b. **3-aminohexane**
c. **pentamethylaniline**   d. **2-aminonaphthalene**

**12.4
PHYSICAL
PROPERTIES
OF AMINES**   Table 12.1 lists the boiling points of some common amines. Note that methylamine and ethylamine are gases, with boiling points below room temperature. Although these boiling points are well above those of alkanes with comparable molecular weights, they are much below those of the comparable alcohols, methanol and ethanol, as shown in Table 12.2. We conclude from these data that, although intermolecular N—H····N hydrogen bonds are important and raise the boiling points of primary and

Name	Formula	bp, °C	Dissociation constant, K_b	pK_b
ammonia	NH_3	−33.4	2.0×10^{-5}	−4.70
methylamine	CH_3NH_2	−6.3	44×10^{-5}	−3.36
dimethylamine	$(CH_3)_2NH$	7.4	51×10^{-5}	−3.29
trimethylamine	$(CH_3)_3N$	2.9	5.9×10^{-5}	−4.23
ethylamine	$CH_3CH_2NH_2$	16.6	47×10^{-5}	−3.33
n-propylamine	$CH_3CH_2CH_2NH_2$	48.7	38×10^{-5}	−3.42
n-butylamine	$CH_3CH_2CH_2CH_2NH_2$	77.8	40×10^{-5}	−3.40
aniline	$C_6H_5NH_2$	184.0	4.2×10^{-10}	−9.38
N-methylaniline	$C_6H_5NHCH_3$	195.7	7.1×10^{-10}	−9.15
N,N-dimethylaniline	$C_6H_5N(CH_3)_2$	193.5	11×10^{-10}	−8.96
ethylenediamine	$H_2NCH_2CH_2NH_2$	116.5	8.5×10^{-5}	−4.07
hexamethylenediamine	$H_2NCH_2CH_2CH_2CH_2CH_2CH_2NH_2$	204.5	85×10^{-5}	−3.07
pyridine	C_5H_5N	115.3	23×10^{-10}	−8.64

TABLE 12.1   Properties of some common amines

TABLE 12.2

A comparison of alkane, amine, and alcohol boiling points*

alkane	CH_3CH_3 (30) bp −88.6°C	$CH_3CH_2CH_3$ (44) bp −42.1°C
amine	CH_3NH_2 (31) bp −6.3°C	$CH_3CH_2NH_2$ (45) bp +16.6°C
alcohol	CH_3OH (32) bp +65.0°C	CH_3CH_2OH (46) bp +78.5°C

*Molecular weights are shown in parentheses.

secondary amines over those of the corresponding alkanes, they are not so strong as the intermolecular O—H····O bonds in alcohols. The reason is that nitrogen is not so electronegative as oxygen.

Problem 12.4   **Explain why the difference between the boiling points of isobutane (2-methylpropane; bp −10.2°C) and trimethylamine (bp +2.9°C) is much smaller than the difference between the boiling points of n-butane (bp −0.5°C) and n-propylamine (bp 48.7°C). All four compounds have nearly identical formula weights.**

All three classes of amines can form hydrogen bonds with the —OH group of water (that is, O—H····N). Thus, most simple amines containing up to about five or six carbon atoms are either completely or appreciably soluble in water.

**12.5
PREPARATION
OF AMINES**

Before we discuss other properties of amines, we will briefly review several ways of making them.

12.5a Alkylation of Ammonia and Amines This method was discussed in Sec. 6.5. Ammonia or amines react with primary or secondary alkyl halides by an S_N2 displacement process. A new bond is formed between the nitrogen atom and the alkyl group. Hence the reaction is called **alkylation.** For example, the nitrogen in aniline can be methylated once or twice.

(12.2)

aniline   N-methylaniline   N,N-dimethylaniline

The alkylation can be intramolecular, as in the following final step in a laboratory synthesis of nicotine:

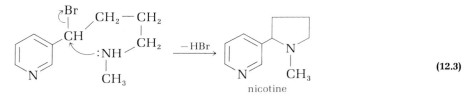

(12.3)

nicotine

12.5b Reduction of Aromatic Nitro Compounds Most aromatic amines are prepared by reducing the corresponding nitro compound, these being available through electrophilic aromatic nitration (see eq. 4.11 and especially Sec. 4.12). The nitro group is very easily reduced, and the reduction can be accomplished either catalytically or by various chemical reducing agents such as metals (iron, tin, zinc) and acid.

$$CH_3-\!\!\left\langle\;\right\rangle\!\!-NO_2 \xrightarrow[\text{or } SnCl_2,\text{ HCl or LiAlH}_4]{H_2,\text{ Ni, heat, pressure}} CH_3-\!\!\left\langle\;\right\rangle\!\!-NH_2$$ (12.4)

p-nitrotoluene   p-toluidine

12.5c Reduction of Amides and Nitriles Many nitrogen-containing functional groups can be reduced to amines. Among the most useful of these reactions is the reduction of amides (eq. 10.32) and nitriles. In the laboratory, the reducing agent most commonly used for this purpose is lithium aluminum hydride.

$$R-\overset{\overset{\displaystyle O}{\|}}{C}-N\overset{R'}{\underset{R''}{\big\backslash}} \xrightarrow{\text{LiAlH}_4} RCH_2N\overset{R'}{\underset{R''}{\big\backslash}} \qquad \text{(R' and R'' may be H} \atop \text{or organic groups.)}$$ (12.5)

$$R-C\!\equiv\!N \xrightarrow[H_2,\text{ Ni}]{\text{LiAlH}_4} RCH_2NH_2$$ (12.6)

In the reduction of amides, the carbonyl group is converted to a CH_2 group. The reduction of nitriles, which can be accomplished either with $LiAlH_4$ or catalytically, gives primary amines, whereas the reduction of amides can give any class of amine, depending on the structure of the amide.

Example 12.2 **Give the structure of the main organic product in each of the following reactions.**

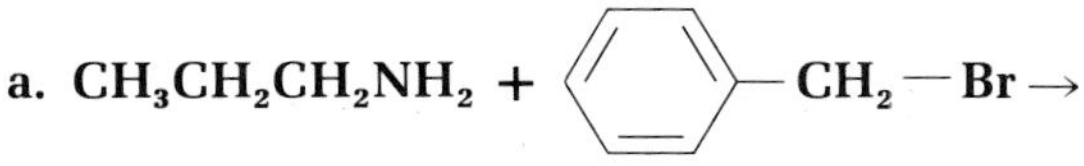

a. $CH_3CH_2CH_2NH_2$ + CH_2—Br →

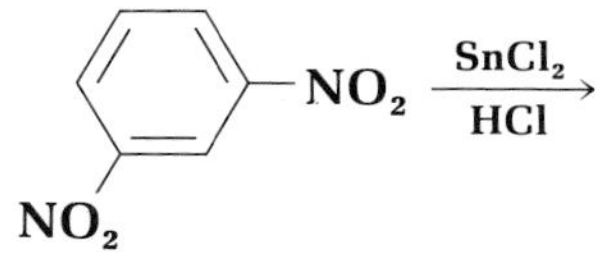

b. —NO_2 $\xrightarrow[\text{HCl}]{\text{SnCl}_2}$

c. $CH_3\overset{\overset{\displaystyle O}{||}}{C}NHCH_2CH_3$ $\xrightarrow{\text{LiAlH}_4}$

d. $NCCH_2CH_2CH_2CH_2CN$ $\xrightarrow[\text{catalyst}]{\text{excess } H_2}$

Solution a. S_N2 displacement occurs readily with benzyl halides:

$CH_3CH_2CH_2NHCH_2$

b. Both nitro groups are reduced:

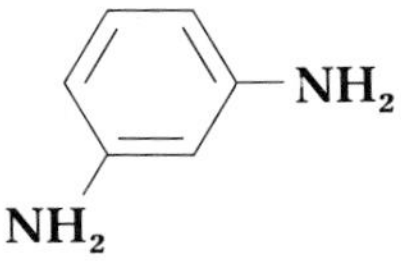

—NH_2

NH_2

c. The C=O is reduced to CH_2: $CH_3CH_2NHCH_2CH_3$.
d. Both CN groups are reduced: $H_2N(CH_2)_6NH_2$. The raw material for nylon (Sec. 12.12) is manufactured commercially in this way.

Problem 12.5 Give syntheses of each of the following amines.

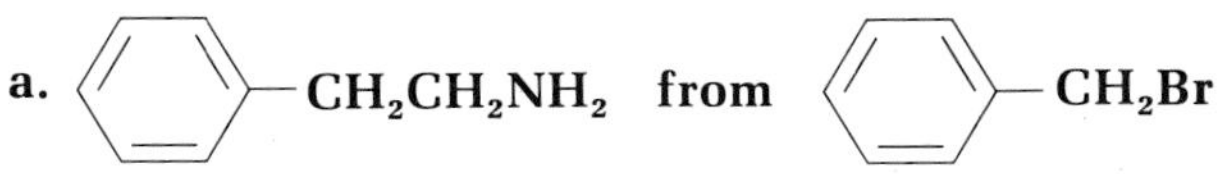

a. —$CH_2CH_2NH_2$ from —CH_2Br

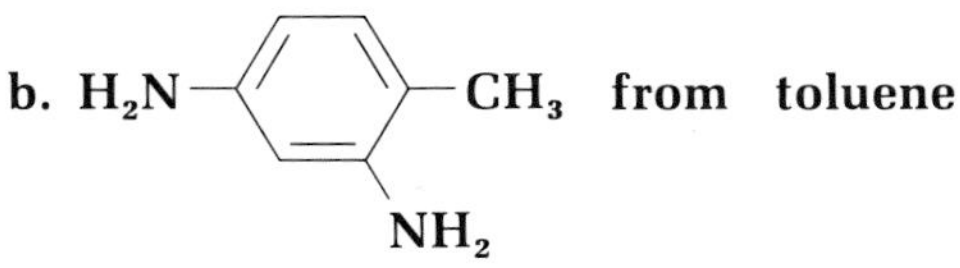

b. H_2N— —CH_3 from toluene

NH_2

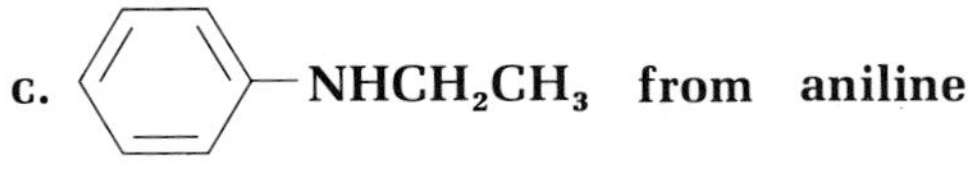

c. —$NHCH_2CH_3$ from aniline

12.6
THE BASICITY
OF AMINES

Like ammonia, amines form basic (alkaline) solutions in water. The equilibrium that produces hydroxide ions is illustrated for a primary amine.

$$R-\overset{..}{N}H_2 \;+\; H-OH \;\rightleftharpoons\; R-\overset{+}{N}H_3 \;+\; OH^- \tag{12.7}$$

primary amine   an   hydroxide
                              alkylammonium   ion
                                   ion

The equilibrium constant for this reaction is called the *basicity constant*, K_b:

$$K_b = \frac{[R\overset{+}{N}H_3][OH^-]}{[RNH_2]}; \qquad pK_b = -\log K_b \tag{12.8}$$

(Compare these equations with eqs. 10.1 and 10.2 for carboxylic acids). Table 12.1 lists the K_b and pK_b values for several common amines.

Problem 12.6   **Write equations analogous to eq. 12.7 for a secondary and a tertiary amine. Write expressions for K_b in each case.**

Note in Table 12.1 that alkylamines are somewhat more basic than ammonia. For example, methylamine is 22 times more basic than ammonia ($K_b = 44 \times 10^{-5}$ and 2×10^{-5}, respectively). The compounds differ only by a methyl group versus a hydrogen atom. Because methyl is an electron-donating group compared to hydrogen, it stabilizes the positive charge on the alkylammonium ion (eq. 12.7) and shifts the equilibrium to the right. In general, electron-donating groups increase the basicity of amines, and electron-withdrawing groups decrease their basicity.

Aromatic amines are much weaker bases than aliphatic amines or ammonia. For example, aniline is less basic than cyclohexylamine by about a millionfold.

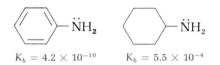

$K_b = 4.2 \times 10^{-10}$ $K_b = 5.5 \times 10^{-4}$

The reason for this huge difference is the resonance delocalization of the unshared electron pair that is possible in aniline but not in cyclohexylamine.

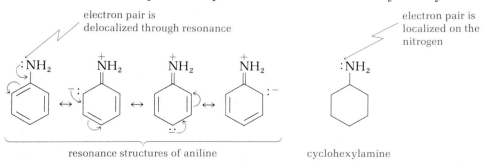

resonance structures of aniline   cyclohexylamine

Resonance stabilizes the unprotonated form of aniline (compared to cyclohexylamine) and shifts the equilibrium in eq. 12.7 to the left. Another way to view the situation is that the unshared electron pair in aniline is delocalized and therefore less available for donation to a proton than is the electron pair in cyclohexylamine.

Finally, a brief comparison of amines with amides is worthwhile. Whereas amines are basic, amides are very much less so. Their water solutions are essentially neutral. The reason for the difference is the exceedingly significant resonance in amides, which was mentioned in Sec. 10.19.

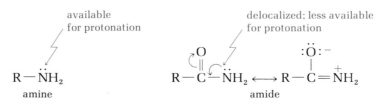

Problem 12.7 **Place the following compounds in order of increasing basicity: acetanilide (see eq. 12.11 for the structure), aniline, cyclohexylamine, N-ethylaniline.**

**12.7
REACTION OF
AMINES WITH
STRONG ACIDS;
AMINE SALTS**

Being bases, amines react with strong acids to form **alkylammonium salts.** An example of the reaction for a primary amine and HCl is

$$R-\ddot{N}H_2 \;+\; HCl \;\rightarrow\; R\overset{+}{N}H_3 \;\; Cl^- \tag{12.9}$$

primary amine   an alkylammonium
chloride

Example 12.3 **Complete the following acid–base reactions and name the products.**
a. $CH_3CH_2NH_2$ + HBr   b. $(CH_3)_3N$ + HCl

c. 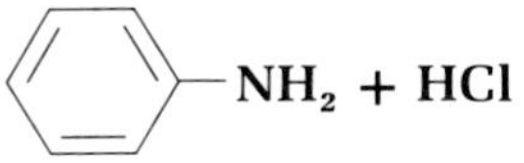 ⬡—NH₂ + HCl

Solution a. $CH_3CH_2\ddot{N}H_2 + HBr \rightarrow CH_3CH_2\overset{+}{N}{-}H + Br^-$

with H above and H below the N.

ethylammonium bromide

b. $CH_3-\ddot{N}-CH_3 + HCl \rightarrow CH_3-\overset{+}{N}{-}CH_3 + Cl^-$

with CH_3 below the N on the left, and H above / CH_3 below the N on the right.

trimethylammonium chloride

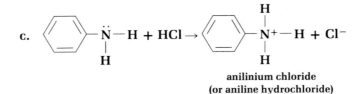

**anilinium chloride
(or aniline hydrochloride)**

Problem 12.8　Complete the following equation and name the product.

$$(CH_3CH_2)_2NH + H\!-\!OSO_3H \rightarrow$$

This reaction is particularly useful for separating or extracting amines from neutral or acidic water-insoluble substances. Consider, for example, a mixture of p-toluidine and p-nitrotoluene that might arise from a preparation of the amine (eq. 12.4) that for some reason did not go to completion. The amine can be separated from the nitro compound by the following scheme:

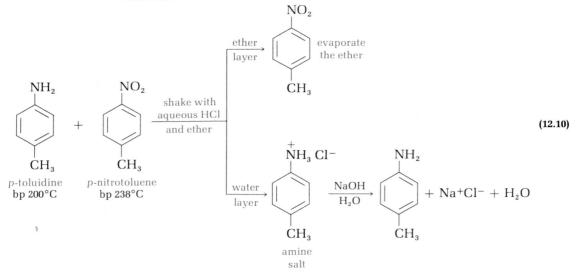

(12.10)

The mixture, neither component of which is water-soluble, is dissolved in an inert, low-boiling solvent such as ether and is shaken with aqueous hydrochloric acid. The amine reacts to form a salt, which, being ionic, dissolves in the water layer. The nitro compound does not react with HCl and remains in the ether layer. The two layers are then separated. The nitro compound can be recovered simply by evaporating the ether. The amine can be obtained from its salt by making the aqueous layer alkaline with a strong base such as NaOH.

A similar procedure is used to extract naturally occurring amines with biological activity, such as quinine, strychnine and morphine, from the plants that produce them.

Problem 12.9　Morphine is a tertiary amine whose structure is

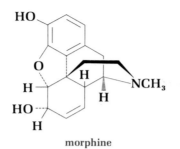

morphine

Its solubility in water is only 0.2 g/L. However, morphine hydrochloride has a water solubility of 57 g/L. Write equations to show how morphine could be extracted and isolated from opium.

12.8
REACTIONS OF
AMINES WITH
ACID DERIVATIVES
TO MAKE AMIDES

We have already seen that esters (eq. 10.19), acid chlorides (eq. 10.28), and anhydrides (eq. 10.29) react with ammonia to give simple amides. Similar **acylations** occur with primary and secondary amines. Examples are

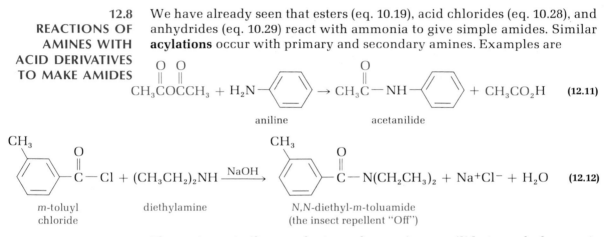

aniline   acetanilide

m-toluyl
chloride

diethylamine

N,N-diethyl-m-toluamide
(the insect repellent "Off")

The antipyretic (fever-reducing substance) **acetanilide** is made from aniline and acetic anhydride (eq. 12.11). The insect repellent "Off" is the amide shown in eq. 12.12. The mechanism of all such reactions involves, as the first step, nucleophilic attack by the amine on the carbonyl carbon atom of the acid derivative.

Problem 12.10 Write out the steps in the mechanisms of eqs. 12.11 and 12.12. (Use the answer to Problem 10.17 as a guide if necessary.) What is the purpose of the NaOH in eq. 12.12?

Problem 12.11 Complete the following equations.

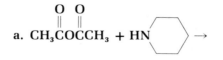

b. $CH_3CH_2CH_2CCl + 2\ CH_3CH_2CH_2NH_2 \rightarrow$

12.9
QUATERNARY
AMMONIUM
COMPOUNDS

Tertiary amines react with primary or secondary alkyl halides by an S_N2 mechanism to give **quaternary ammonium salts** (eq. 6.10). For example,

$$(CH_3)_3N \ + \ CH_3(CH_2)_{14}CH_2Cl \ \rightarrow \ (CH_3)_3\overset{+}{N}CH_2(CH_2)_{14}CH_3 \quad Cl^- \ \text{(12.13)}$$

trimethylamine   cetyl chloride   cetyltrimethylammonium chloride

The product is an ammonium salt, but the four hydrogens of the ammonium ion are replaced by alkyl groups. Compounds such as the product of eq. 12.13, which have a long carbon chain with a polar group at the end, are detergents. But they differ from the ordinary soaps and syndets that we have discussed, because the polar part of the molecule is positive instead of negative. They are used, for example, as fabric softeners and bacteriocides.

Quaternary ammonium compounds are important in certain biological processes. One of the most common of these compounds is **choline,** which, as we have already seen (Sec. 11.14), is present in phospholipids.

$$CH_3-\overset{\overset{\displaystyle CH_3}{|}}{\underset{\underset{\displaystyle CH_3}{|}}{\overset{+}{N}}}-CH_2CH_2OH \quad OH^- \qquad CH_3-\overset{\overset{\displaystyle CH_3}{|}}{\underset{\underset{\displaystyle CH_3}{|}}{\overset{+}{N}}}-CH_2CH_2-O-\overset{\overset{\displaystyle O}{\|}}{C}-CH_3 \quad OH^-$$

choline   acetylcholine

Choline is not only involved in various metabolic processes, but it is also the precursor of **acetylcholine,** which is essential to the transmission of nerve impulses.

12.10
REACTION OF
AMINES WITH
NITROUS ACID

Each class of amine reacts differently with nitrous acid. As we will see, the reaction is particularly useful synthetically with primary aromatic amines. **Nitrous acid** is prepared as needed by treating aqueous sodium nitrite with a strong acid at ice temperatures. At that temperature nitrous acid solutions are reasonably stable.

$$Na^+NO_2^- + H^+Cl^- \xrightarrow{0-5°C} H-\overset{..}{O}-\overset{..}{N}=\overset{..}{O}: + Na^+Cl^- \qquad \text{(12.14)}$$

sodium nitrite   nitrous acid

The reactive species in the reactions of nitrous acid is the NO^+ or **nitrosonium ion,** formed as follows (compare with eq. 4.19):

$$H\overset{..}{O}-\overset{..}{N}=\overset{..}{O}: + H^+ \rightleftharpoons H\overset{\oplus}{\overset{..}{O}}\overset{..}{N}=\overset{..}{O}: \rightleftharpoons H_2O + :\overset{\oplus}{N}=\overset{..}{O}: \qquad \text{(12.15)}$$

$$\underset{\displaystyle H}{}$$

nitrosonium ion

The reaction of *tertiary* amines with nitrous acid is important only if the amine is aromatic. In this case, electrophilic aromatic substitution occurs to give an **aromatic nitroso compound.**

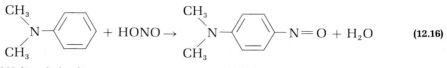

N,N-dimethylaniline
(a tertiary amine)

p-nitroso-N,N-dimethylaniline

With *secondary* amines, nitrosation occurs on the nitrogen, to give a **nitrosamine.**

$$\begin{matrix} R \\ \diagdown \ddot{N}H \\ \diagup \\ R \end{matrix} + HONO \rightarrow \begin{matrix} R \\ \diagdown N-N=O \\ \diagup \\ R' \end{matrix} + H_2O \qquad (12.17)$$

secondary amine   a nitrosamine

Nitrosamines are known to be dangerous carcinogens (cancer-producing compounds). For example, if inhaled or ingested by a rat, dimethylnitrosamine quickly causes cancer of the liver. Nitrosamines can be formed in our environment through the reaction of amines with oxides of nitrogen or with sodium nitrite that has been used as a meat preservative. The possible link between nitrosamines and human cancer is being carefully investigated.

Example 12.4   **Write a mechanism for eq. 12.17.**

Solution

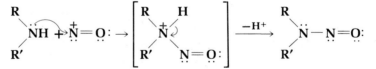

The amine acts as a nucleophile toward the nitrosonium ion. Simple loss of a proton completes the reaction.

Example 12.5   **Why cannot tertiary amines react analogously?**

Solution   **They have no hydrogen on the nitrogen atom to lose in the second step.**

Problem 12.12   **Write a mechanism for eq. 12.16.**

Primary amines react with nitrous acid to form **diazonium ions.**

$$RNH_2 + HONO + H^+ \xrightarrow{0°C} \left[R-\overset{+}{N}\equiv N: \right] + 2\,H_2O \qquad (12.18)$$

a diazonium ion

$$\text{products} \leftarrow R^+ \xleftarrow{-N_2} $$

a carbocation

*Alkyl*diazonium ions are exceedingly unstable, even at 0°C. They lose nitrogen to give a carbocation, which can either react with any nucleophile present or lose a proton to give an alkene. For example, isopropyl-amine reacts with nitrous acid as follows:

$$CH_3CHCH_3 + HONO \xrightarrow{0°C} CH_3CHCH_3 + CH_2{=}CHCH_3 + N_2 + H_2O \qquad \text{(12.19)}$$
$$\underset{NH_2}{|} \qquad\qquad\qquad \underset{OH}{|}$$

isopropylamine   isopropyl alcohol   propene

The products are derived from the isopropyl cation, which either reacts with water to form the alcohol or loses a proton to form the alkene. To summarize, the reaction of a primary aliphatic amine with nitrous acid produces a carbocation on the carbon atom to which the amino group was attached. The carbocation then gives products typical of carbocation reactions.

Example 12.6   **Write a mechanism that explains how the diazonium ion is formed in eq. 12.18.**

Solution   **The first steps are exactly the same as for a secondary amine, and they lead to a primary nitrosamine. The nitrosamine then rearranges as follows:**

$$\underset{H}{\overset{R}{\underset{\diagdown}{}}}\ddot{N}{-}N{=}\ddot{O}\colon \rightleftharpoons R{-}\ddot{N}{=}\ddot{N}{-}OH \xrightarrow{H^+} R{-}\ddot{N}{=}N{-}\underset{H}{\overset{+}{\underset{|}{\ddot{O}}}}{-}H \xrightarrow{-H_2O} R{-}\overset{+}{\ddot{N}}{\equiv}N\colon \qquad \text{(12.20)}$$

a primary nitrosamine

The first step is similar to keto–enol tautomerism (Sec. 9.14).

Problem 12.13   **What products are expected from the reaction of cyclopentylamine with nitrous acid?**

The diazonium ions formed from primary *aromatic* amines are more stable, and we will discuss their chemistry next.

12.11
AROMATIC
DIAZONIUM
COMPOUNDS

Primary aromatic amines react with nitrous acid at 0°C to give solutions of **aryldiazonium ions**. The process is called **diazotization:**

$$\langle\!\!\!\bigcirc\!\!\!\rangle{-}NH_2 + HONO + H^+Cl^- \xrightarrow[\substack{aqueous \\ solution}]{0{-}5°C}$$

aniline

$$\langle\!\!\!\bigcirc\!\!\!\rangle{-}N_2{}^+Cl^- + 2\,H_2O \qquad \text{(12.21)}$$

benzenediazonium
chloride

Solutions of aryldiazonium ions are moderately stable and can be kept at 0°C for several hours. However, they lose nitrogen on warming, providing a useful synthesis of phenols.

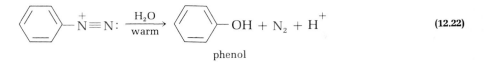

$$\text{(12.22)}$$

The nitrogen in diazonium ions can be replaced by a variety of other nucleophiles:

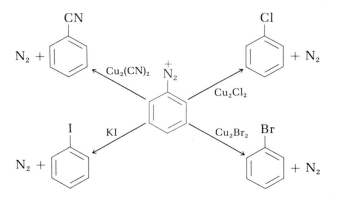

$$\text{(12.23)}$$

These reactions allow us to introduce substituents such as I and CN, a result not easily accomplished by direct electrophilic aromatic substitution. They also allow us to introduce a substituent in a position not easily accessible by electrophilic substitution, as shown in the following example.

Example 12.7   **Show how m-dibromobenzene could be prepared from nitrobenzene.**

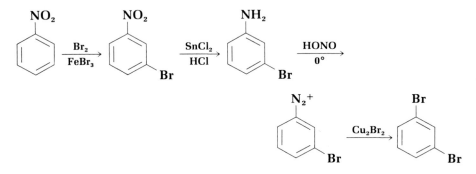

Note that the first step takes advantage of the m-directing effect of the nitro group. If we were to simply brominate bromobenzene, we would get a mixture of o- and p-dibromobenzene, not the *meta* isomer.

Aryldiazonium ions, being positively charged, are electrophiles. They react with strongly activated aromatic rings (those present in phenols or aromatic amines) to give **azo compounds.** For example,

(12.24)

benzenediazonium ion   phenol   p-hydroxyazobenzene yellow leaflets, mp 155–157°C

Note that the nitrogen atoms are retained in the product. This electrophilic substitution reaction is sometimes called **azo coupling,** because in the product two aromatic rings are coupled by the azo, or $-N=N-$, group. *Para* coupling is preferred but, if the *para* position is blocked by another substituent, *ortho* coupling can occur. All azo compounds are colored, and many are used commercially as dyes for cloth and in color photography.

Example 12.8   **Write the resonance contributors to the benzenediazonium ion that show how the nitrogen furthest from the benzene ring can become an electrophile.**

has only six electrons;
can react as an electrophile
toward an aromatic ring

Problem 12.14   **Equation 12.24 is an electrophilic aromatic substitution (Sec. 4.9). Write out the steps in the reaction mechanism.**

**12.12
DIAMINES AND
POLYAMIDES;
NYLON**

Two **diamines** (compounds with two amino groups) that occur in nature have rather vivid names—**putrescine** and **cadaverine.** They are formed when animal flesh decays and have odors much as one would expect from their names.

$$H_2NCH_2CH_2CH_2CH_2NH_2 \qquad H_2NCH_2CH_2CH_2CH_2CH_2NH_2$$

putrescine
(1,4-diaminobutane)   cadaverine
(1,5-diaminopentane)

The next higher homolog, **hexamethylenediamine,** is the most important commercial diamine, produced in annual quantities that exceed a million tons. It is one of the two raw materials needed to manufacture **nylon.**

Nylon is a **polyamide** made from hexamethylenediamine and adipic acid. When mixed, these reagents form a polysalt that, on heating, loses water (eq. 10.30) to form a polyamide.

$$H_2N(CH_2)_6NH_2 \quad + \quad HOC(CH_2)_4COH \xrightarrow[-nH_2O]{200-300°C} \left[NH(CH_2)_6NHC(CH_2)_4C \right]_n \quad \text{(12.25)}$$

hexamethylenediamine   adipic acid   nylon
(1,6-diaminohexane)   ($n \cong 50-100$)

This polymer was first made by W. H. Carothers in 1933 and was commercially introduced by the DuPont Company just a few years later. This particular nylon is known as **nylon-6,6** because each monomer (diamine and diacid) contains six carbon atoms. Other nylons are known, but this one is the most important.

The second most important nylon is **nylon-6,** made from **caprolactam:**

$$\xrightarrow{250-270°C} \left[NHCH_2CH_2CH_2CH_2CH_2C \right]_n \quad \text{(12.26)}$$

caprolactam   nylon-6

Lactams are cyclic amides (compare with lactones, Sec. 11.6). On heating, the seven-membered ring opens, and the amino group of one molecule reacts with the carboxyl group of the next. Its amino group in turn reacts with another carboxyl group, and so on, to produce the polyamide.

FIGURE 12.2

Nylon strings are widely used by classical guitarists. Unlike traditional gut strings, nylon strings are unaffected by changes in temperature and humidity and thus do not require constant retuning. (Courtesy of DuPont Magazine.)

Nylons are extremely versatile polymers that can be processed to give materials as delicate as sheer fabrics, as long-wearing as carpets, or as tough as molded automobile parts (see Figure 12.2).

a word about ——————————————————————————————

25. Aramids, the Latest in Polyamides

Aromatic polyamides (called aramids) are now being produced at a rapidly growing rate because of their

special properties of heat resistance and low flammability. The best-known is Kevlar (see Figure 12.3). Because of the aromatic rings, this type of polyamide has a much stiffer structure than the nylons. It is being used in place of steel to make tire cord for radial tires, in bullet-resistant vests, and in many other ways. It was used for the propeller of the *Gossamer Albatross*, the first human-powered craft to fly across the English Channel (June 12, 1979). Kevlar was selected because of its low-stretch, high-strength, and lightweight characteristics.

Nomex has a similar structure to Kevlar but uses *meta*- instead of *para*-oriented monomers. It is used in flame-resistant clothing because its fibers char rather than melt when exposed to flame. It has wide applications, ranging from fire-fighters' coats to racing-car drivers' uniforms. It has also been used in flame-resistant building materials. The combination of strength and light weight has made both Nomex and Kevlar popular in boat construction.

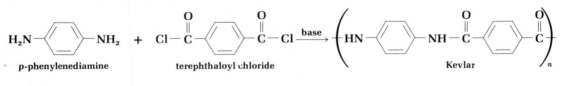

p-phenylenediamine   terephthaloyl chloride   Kevlar

FIGURE 12.3

12.13 HETEROCYCLIC AMINES IN NATURE	Nitrogen atoms often form part of a heterocyclic ring in natural products. Commonly these natural products have important physiological effects on humans or play a key role in some normal biological process. The subject is very broad. It will be our goal here to describe a few of the more prevalent ring systems and to illustrate each with well-known examples from nature.

12.13a Five-Membered Rings with One Nitrogen The three main parent compounds of this type are

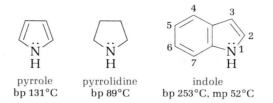

pyrrole	pyrrolidine	indole
bp 131°C	bp 89°C	bp 253°C, mp 52°C

The aromaticity of pyrrole has already been discussed (Sec. 4.16). The difference in basicities of pyrrole and pyrrolidine is truly striking. Pyrrolidine is an ordinary secondary amine, with $K_b = 1.3 \times 10^{-3}$. Pyrrole is hardly basic at all: $K_b = 4 \times 10^{-19}$. The reason is that the unshared electron pair on the nitrogen in pyrrole is delocalized as part of the 6 π aromatic system, whereas in pyrrolidine the electron pair is localized on the nitrogen and available for protonation.

Problem 12.15 **Do you expect indole to be strongly basic? Explain.**

Pyrrole rings form the building blocks of several biologically important pigments. The **porphyrins** contain four pyrrole rings linked by one-carbon bridges. The molecules are flat and have an 18 π-electron conjugated system shown in color in the parent molecule, **porphin.** Porphyrins are exceptionally stable and highly colored. They form complexes with metallic ions. In these complexes the two N—H hydrogens are absent and each of the four nitrogens donates an electron pair to the metal, which sits in the middle of the structure.

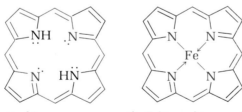

porphin
red crystals
darkens but does not
melt at 360°C

the Fe^{2+} porphin complex
brown cubic crystals

Problem 12.16 **Does porphin qualify as an aromatic compound, according to the Hückel rule? (See Sec. 4.15.)**

Porphin itself does not occur in nature, but several metalloporphyrins play key roles in life processes. The best-known of these are **heme,** an iron complex present in the red blood pigment hemoglobin, and **chlorophyll,** the green plant pigment essential to photosynthesis.

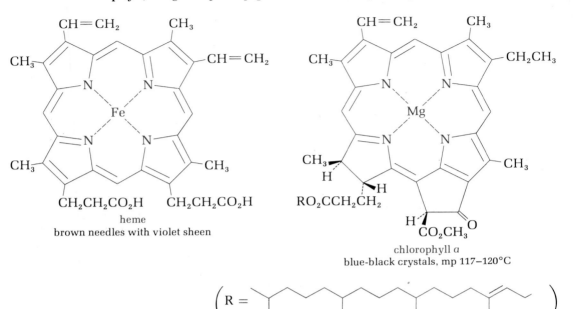

heme
brown needles with violet sheen

chlorophyll *a*
blue-black crystals, mp 117–120°C

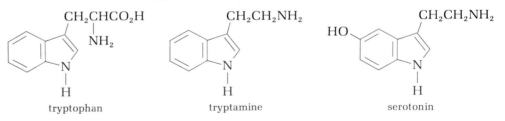

The indole ring system is present in many natural products, most of which are biosynthesized from the amino acid **tryptophan,** one of the building blocks of proteins. Indole itself and its 3-methyl derivative skatole are formed during protein decay. Both contribute to the odor of feces.

tryptophan

tryptamine

serotonin

Decarboxylation of tryptophan gives **tryptamine,** and many compounds that contain the tryptamine skeleton have a profound effect on the brain and nervous system. A simple example is **serotonin** (5-hydroxytryptamine), a neurotransmitter and vasoconstrictor active in the central nervous system. The tryptamine skeleton is disguised but present (shown in color) in more complex molecules such as **reserpine** and **lysergic acid.** Reserpine, present in Indian snake root (*Rauwolfia serpintina*), which grows wild on the foothills of the Himalayas, has been used medically for centuries. It is now used to calm schizophrenics and increase their accessibility to psychiatric treatment. Lysergic acid is present in the fungus ergot, which grows on rye and other grains. Conversion of the carboxyl group to its diethylamide gives the extremely potent hallucinogen LSD.

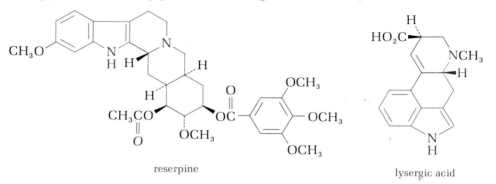

reserpine

lysergic acid

12.13b Six-Membered Rings with One Nitrogen The four main parent compounds of this type are

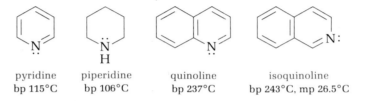

pyridine	piperidine	quinoline	isoquinoline
bp 115°C	bp 106°C	bp 237°C	bp 243°C, mp 26.5°C

The aromaticity of pyridine has already been discussed (Sec. 4.16). Pyridine is basic (eq. 4.30) but appreciably less basic than typical aliphatic amines. Its K_b is only 2.3×10^{-9}, compared with 1.6×10^{-3} for piperidine. The reason for the lower basicity of pyridine is the hybridization of the nitrogen. Because the nitrogen is sp^2-hybridized (compared with sp^3 for piperidine and other aliphatic amines), the unshared electron pair is held closer to the nitrogen nucleus and is less available to a proton.

Examples of natural products with these ring systems are

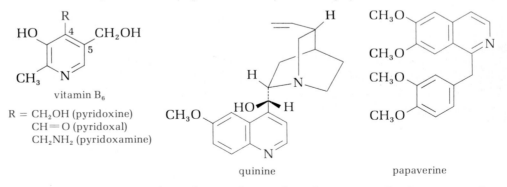

vitamin B$_6$

R = CH$_2$OH (pyridoxine)
CH=O (pyridoxal)
CH$_2$NH$_2$ (pyridoxamine)

quinine

papaverine

Vitamin B$_6$ is a relatively simple pyridine derivative. The R group at C-4 may be an alcohol, an aldehyde, or an amine. The vitamin functions as a coenzyme in the interconversion of keto and amino acids. **Quinine,** which occurs in cinchona bark and is used to treat malaria, contains the quinoline ring system. **Papaverine,** which is present in opium and is used as an antispasmodic drug, contains the isoquinoline ring system. **Nicotine** (eq. 12.3) is an example of a natural product containing both a pyridine and a pyrrolidine ring.

12.13c Rings with More Than One Nitrogen Three important heterocyclic ring systems with more than one nitrogen atom are

imidazole
bp 263°C, mp 91°C

pyrimidine
bp 124°C, mp 22°C

purine
mp 217°C

Imidazole is a little like pyrrole in that the unshared electron pair on the N—H nitrogen is delocalized and part of the aromatic 6 π-electron system. But the unshared electron pair on the other nitrogen is available for protonation. The K_b of imidazole is 1.2×10^{-7}, so it is about 100 times more basic than pyridine. The positive charge in protonated imidazole can be delocalized over both nitrogens through resonance:

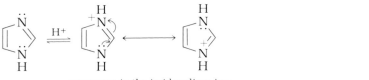

(12.27)

resonance in the imidazolium ion

The imidazole skeleton is present in the amino acid **histidine,** where it plays an important role in the reactions of many enzymes. Decarboxylation of histidine gives **histamine** (compare with tryptophan and tryptamine), a toxic substance present in combination with proteins in body tissues. It is released as a consequence of allergic hypersensitivity or inflammation (for example, in hay fever sufferers). Many **antihistamines,** compounds that counteract the effects of histamine, have been developed. One of the better known of these is **benadryl** (diphenylhydramine).

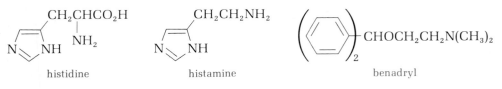

histidine

histamine

benadryl

The pyrimidine and purine ring systems are present in the DNA and RNA bases. Although we will discuss their chemistry in greater detail in

Chapter 16, we can give a few examples here. The well-known **barbiturates,** whose uses range from mild sedatives to hypnotics to anesthetics, are pyrimidine derivatives. They are made from substituted malonic esters and urea.

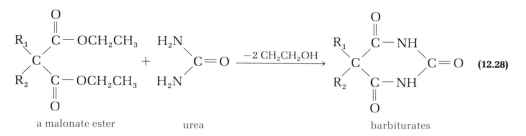

a malonate ester   urea   barbiturates   (12.28)

Examples include phenobarbital (R_1 = ethyl, R_2 = phenyl), nembutal (R_1 = ethyl, R_2 = $CH_3CH_2CH_2CHCH_3$) and seconal (R_1 = allyl, R_2 = $CH_3CH_2CH_2CHCH_3$). The thiobarbiturate sodium pentothal is used to induce general anesthesia by intravenous injection. A pyrimidine with an entirely different type of activity is **thiamin** (or vitamin B_1), a coenzyme required for certain metabolic processes and essential to life.

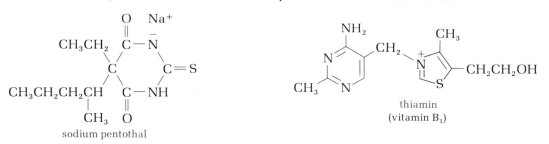

sodium pentothal

thiamin
(vitamin B_1)

Examples of well-known purines besides those we will encounter in Chapter 16 are **uric acid** (the chief product of nitrogen metabolism in birds and reptiles and a major component of guano), **caffeine** (present in coffee, tea, and cola beverages) and **theobromine** (found in cocoa).

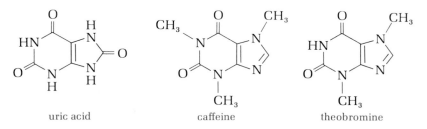

uric acid   caffeine   theobromine

Note that purines contain the imidazole and pyrimidine ring systems "fused" to one another.

a word about ———————————————

26. Morphine and Other Nitrogen-Containing Drugs

Morphine (named after Morpheus, the Greek god of dreams) is the major alkaloid present in opium. (An alkaloid is any basic, nitrogen-containing plant product, often with a complex structure and significant pharmacological properties. Quinine, papaverine, and caffeine are other examples of alkaloids already mentioned in this chapter.) Opium is the dried sap of the unripe seed capsule of the poppy *Papaver somniferum*, and its medical properties have been known since ancient times. However, morphine was not isolated in pure form until 1805, its correct structure was not established until 1925, and it was not synthesized in the laboratory until 1952.

Pain is a terrible problem in medicine, and relief of pain has long been a medical goal. Morphine is an analgesic, a substance that relieves pain without causing unconsciousness. It was used for the large-scale relief of pain from battle wounds during the American Civil War (largely as a consequence of the discovery, at about that time, of the hypodermic syringe). But morphine has serious side effects, particularly addiction. It also can cause nausea, a decrease in blood pressure, and a depressed breathing rate that can be fatal to the very young or the severely debilitated.

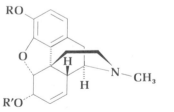

morphine (R=R′=H)
heroin (R=R′=—CCH₃
                  ‖
                  O
codeine (R=CH₃, R′=H)

The first attempts to find a better morphine, a substance with morphine's benefits but without its side effects, involved minor modification of its structure. Acetylation with acetic anhydride gave its di-

acetyl derivative heroin, which is a good analgesic with less of a respiratory depressant effect than morphine. But heroin is severely addictive and has become a narcotic problem. Partial methylation of morphine gave codeine, which is useful as an antitussive (anticough) agent. Unfortunately, it is less than one-tenth as effective as morphine as an analgesic.

Many compounds similar to various parts of the morphine structure have been synthesized and tested for their analgesic properties. Some of these are shown here. Their structural similarity to morphine is evident in the black parts of the structures.

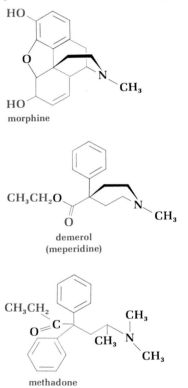

morphine

demerol
(meperidine)

methadone

Demerol is an effective analgesic with a relatively simple structure compared to that of morphine. Methadone was synthesized and used as an analgesic by the Germans during World War II, when natural sources of morphine were in short supply. Later it was used in substitution therapy for heroin addiction, but it too is addictive. The search for a perfect analgesic still goes on.

The pain associated with surgery or injury can sometimes be treated with local or regional anesthetics, many of which are nitrogen-containing drugs. Cocaine, an alkaloid found in the plant *Erythroxylum coca*, was one of the first anesthetics of this type. It constricts blood vessels, thus producing a bloodless surgical area. But cocaine is addictive and has other undesirable properties. Its medical use (of course it is used illegally for nonmedical purposes) has been supplanted largely by procain hydrochloride (Novocain).

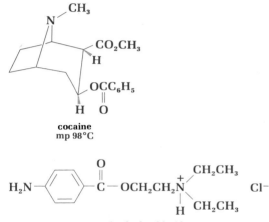

cocaine
mp 98°C

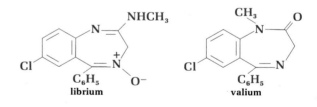

procaine hydrochloride
mp 153–156°C

Procaine is less toxic, easier to synthesize and sterilize, and has a desirably shorter period of action than cocaine. It is usually injected into a nerve to anesthetize a small region of the body. It acts by inhibiting nerve impulse transmission by acetylcholine (Sec. 12.9). Procaine hydrochloride is a widely used drug, sold to dentists and to doctors of human and veterinary medicine under at least 27 different trade names.

Lidocaine hydrochloride, which is somewhat similar to procaine in structure, is used not only as a local anesthetic but to treat abnormal heart rhythms by intravenous injection. Benzocaine (ethyl *p*-aminobenzoate), which has a very simple structure, is used as a mild topical anesthetic in ointments for burns and open wounds. Note that both procaine and benzocaine are esters of the same acid, *p*-aminobenzoic acid.

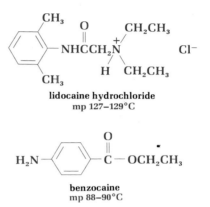

lidocaine hydrochloride
mp 127–129°C

benzocaine
mp 88–90°C

There are, of course, many types of pain. Sometimes, in this hectic, tension-filled world of ours, a mild tranquilizer can be medically useful. Two of the most commonly prescribed contain seven-membered heterocyclic rings and are the well-known twins of modern psychiatry, librium and valium.

librium   **valium**

ADDITIONAL PROBLEMS

12.17. Give an example of each of the following.

a. a primary amine   **b.** a cyclic secondary amine
c. a tertiary aromatic amine   **d.** a quaternary ammonium salt
e. an aryldiazonium salt   **f.** a heterocyclic amine
g. an azo compound   **h.** a nitrosamine
i. a primary amide   **j.** a lactam

12.18. Write structural formulas for each of the following compounds.
a. *m*-chloroaniline
b. *sec*-butylamine
c. 2-aminohexane
d. dimethyl-*n*-propylamine
e. benzylamine
f. 1,2-diaminopropane
g. *N,N*-dimethylaminocyclohexane
h. tetraethylammonium bromide
i. triphenylamine
j. *o*-toluidine

12.19. Write a correct name for each of the following compounds.

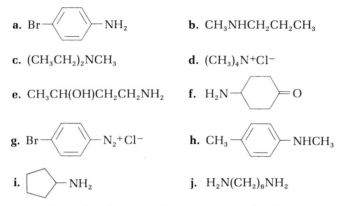

a. Br— ⟨ ⟩ —NH₂
b. CH₃NHCH₂CH₂CH₃

c. (CH₃CH₂)₂NCH₃
d. (CH₃)₄N⁺Cl⁻

e. CH₃CH(OH)CH₂CH₂NH₂
f. H₂N— ⟨ ⟩ =O

g. Br— ⟨ ⟩ —N₂⁺Cl⁻
h. CH₃— ⟨ ⟩ —NHCH₃

i. ⟨ ⟩ —NH₂
j. H₂N(CH₂)₆NH₂

12.20. Draw the structures for, name, and classify as primary, secondary, or tertiary the eight isomeric amines with the molecular formula $C_4H_{11}N$.

12.21. Explain why dimethylamine has a higher boiling point than trimethylamine (Table 12.1), even though the latter has an appreciably higher formula weight.

12.22. The formula weights of *n*-propylamine and ethylenediamine are nearly identical, yet they boil at temperatures over 60°C apart (Table 12.1). Explain.

12.23. Place the following substances, which have nearly identical formula weights, in order of increasing boiling point: 1-aminobutane, 1-butanol, methyl *n*-propyl ether, pentane.

12.24. Give equations for the preparation of the following amines from the indicated precursor.
a. *N,N*-diethylaniline from aniline
b. *m*-chloroaniline from benzene
c. *p*-chloroaniline from benzene
d. 1-aminopentane from 1-bromobutane

12.25. Complete the following equations.

a. ⟨ ⟩ —NH₂ + CH₂=CHCH₂Br $\xrightarrow{\text{heat}}$

b. CH₃ÖCCl + H₂NCH₂CH(CH₃)₂ → **A** $\xrightarrow{\text{LiAlH}_4}$ **B**

c. CH₃ÖC— ⟨ ⟩ $\xrightarrow[\text{H}^+]{\text{HONO}_2}$ **C** $\xrightarrow[\text{excess}]{\text{LiAlH}_4}$ **D**

d.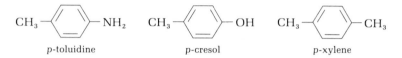

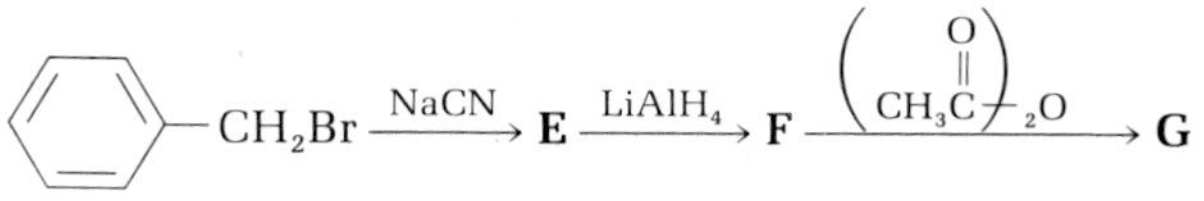

12.26. Write an equation for the reaction of nicotine (eq. 12.3) with *one* equivalent of HCl.

12.27. Tell which is the stronger base and why.
a. aniline or *p*-nitroaniline   **b.** aniline or diphenylamine

12.28. Write out a scheme similar to eq. 12.10 to show how you could separate a mixture of *p*-toluidine, *p*-cresol, and *p*-xylene.

CH_3—⟨⟩—NH_2 CH_3—⟨⟩—OH CH_3—⟨⟩—CH_3

p-toluidine   *p*-cresol   *p*-xylene

12.29. Explain why amines are more basic than amides.

12.30. Write an equation for the reaction of:
a. aniline with hydrochloric acid
b. triethylamine with sulfuric acid
c. diethylammonium chloride with sodium hydroxide
d. *N,N*-dimethylaniline with methyl iodide
e. cyclohexylamine with acetic anhydride

12.31. Write out the steps in the mechanism for the following reaction:

$$CH_3CH_2NH_2 + CH_3\overset{O}{\overset{\|}{C}}O\overset{O}{\overset{\|}{C}}CH_3 \rightarrow CH_3CH_2NH\overset{O}{\overset{\|}{C}}CH_3 + CH_3COOH.$$

Explain why only one of the hydrogens of the amine is replaced by an acetyl group, even if a large excess of acetic anhydride is used.

12.32. Choline (Sec. 12.9) can be prepared by the reaction of trimethylamine with ethylene oxide. Write an equation for the reaction, and show its mechanism.

12.33. Isopropylamine, methylethylamine, and trimethylamine are isomers. Show by means of equations how they can be distinguished by their reactions with nitrous acid.

12.34. Explain why aromatic diazonium ions are more stable toward loss of nitrogen than are aliphatic diazonium ions.

12.35. Write an equation for the reaction of CH_3—⟨⟩—$N_2^+HSO_4^-$ with:

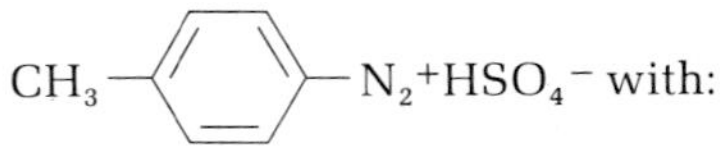

a. cuprous cyanide   **b.** water, heat
c. cuprous chloride   **d.** potassium iodide
e. *p*-cresol and OH⁻   **f.** *N,N*-dimethylaniline and base

12.36. Show how diazonium ions could be used to synthesize:
a. *p*-bromobenzoic acid from *p*-bromoaniline
b. *m*-iodobromobenzene from benzene

12.37. Methyl orange is used as an indicator in acid-base titrations. (It is yellow-orange above pH 4.5 and red below pH 3.) Show how it can be synthesized from p-aminobenzenesulfonic acid (sulfanilic acid) and N,N-dimethylaniline.

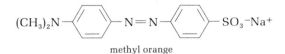

methyl orange

12.38. Congo red is used as a direct dye for cotton. Write equations to show how it could be synthesized from benzidine and 1-aminonaphthalene-4-sulfonic acid.

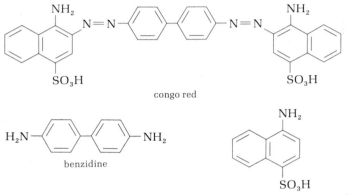

congo red

benzidine

1-aminonaphthalene-4-sulfonic acid

12.39. One commercial method for making hexamethylenediamine (for nylon-6,6 production) starts with the 1,4-addition of chlorine to 1,3-butadiene (see Sec. 3.16). Suggest a possibility for the remaining steps.

12.40. One commercial method for making hexamethylenediamine (for nylon-6,6 manufacture) starts by making the ammonium salt, then the diamide of adipic acid. Suggest a possibility for the remaining steps.

12.41. Using the structures of the parent compounds as given in Sec. 12.13 as a guide, write the formulas for:
a. 2,5-dimethylpyrrole
b. 2-propylpiperidine (Also called *coniine*; it is the poison in hemlock, taken by Socrates when he committed suicide.)
c. 5-hydroxyindole
d. N-methylpyrrolidine
e. 3-pyridinecarboxylic acid (Also called nicotinic acid or niacin, the antipellagra vitamin.)
f. 2-hydroxyquinoline
g. 2,4,6-trimethylpyridine
h. 2-methylimidazole
i. skatole (3-methylindole)
j. LSD (the diethylamide of lysergic acid)

12.42. Write an equation for each of the following reactions.
a. pyrrolidine + HCl
b. quinoline + CH_3I
c. piperidine + acetic anhydride
d. piperidine + nitrous acid
e. pyrimidine + HBr
f. tryptamine + HCl

12.43. Write a mechanism for the preparation of barbiturates from urea and a malonic ester (eq. 12.28).

12.44. The general formula for the penicillins, which are among the most important of the antibiotic drugs, is

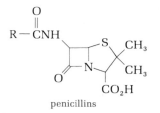

penicillins

In what types of functional groups are the nitrogens present?

thirteen

Spectroscopy and Structure

13.1
INTRODUCTION

In the early years of organic chemistry it was often quite a formidable task to determine the structure of a new compound. Since the 1940s, however, the increasing availability of various types of spectroscopy have greatly simplified this problem. Automated instruments have been developed that permit us to determine and record various spectroscopic properties of a molecule, often with little more effort than pushing a button. And these spectra, if properly interpreted, yield a great deal of structural information.

Spectroscopic methods have many advantages. Usually only a very small sample of material is required, and it can often be recovered if necessary. The methods are rapid, usually requiring only a few minutes. And frequently we can obtain more detailed structural information from spectra than from ordinary laboratory methods.

After a brief theoretical review, we will describe four of the most common spectroscopic techniques for structure determination.

13.2
THEORY

The equation $E = h\nu$ describes the relationship between the energy of light E (or any other radiation) and its **frequency** ν (nu, pronounced "new"). The equation says that there is a direct relationship between the frequency of light and its energy: the higher the frequency, the higher the energy. The proportionality constant between energy and frequency is known as **Planck's constant, h.** Because the frequency of light and its wavelength are *inversely* proportional, the equation can also be written $E = hc/\lambda$, where λ is the **wavelength** of light and c is the speed of light. In this form the equation tells us that the shorter the wavelength of light, the higher its energy.

Molecules can exist at various energy levels. For example, the bonds in a given molecule may stretch or bend, one part of the molecule may rotate with respect to another part (as in the conformations of ethane), and so on. These various molecular motions are quantized—that is, particular bonds may stretch or bend only at certain frequencies (or, because frequency and energy are proportional, only at certain energy levels).

The idea behind most forms of spectroscopy is very simple. A molecule at some energy level, say E_1, is exposed to radiation. The radiation passes through the molecule to a detector. As long as the molecule does not absorb the radiation, the amount of radiation detected will be equal to the amount of radiation emitted by the source. At a frequency that corresponds to some molecular energy transition, say from E_1 to E_2, the radia-

FIGURE 13.1

Radiation passes through the sample unchanged, except when its frequency corresponds to the energy difference between two energy states of the molecule.

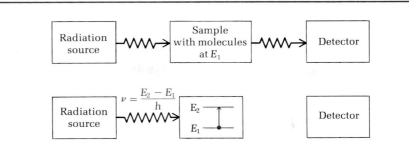

		Region of the spectrum			
Type of spectroscopy	**Radiation source**	**Frequency (hertz)**	**Wavelength (meters)**	**Energy (kcal/mol)**	**Type of transition**
infrared	infrared light	$0.2–1.2 \times 10^{14}$	$15.0–2.5 \times 10^{-6}$	2–12	molecule vibrations
visible-ultraviolet (electronic)	visible or ultraviolet light	$0.375–1.5 \times 10^{15}$	$8–2 \times 10^{-7}$	37–150	electronic states
nuclear magnetic resonance	radio waves	60×10^{6} (depends on magnet strength of the instrument)	5	6×10^{-6}	nuclear spin

TABLE 13.1   Types of spectroscopy and the electromagnetic spectrum

tion will be absorbed by the molecule and will not appear at the detector. These ideas are expressed schematically in Figure 13.1.

Of course, some transitions require more energy than others, so we must use radiation of the appropriate frequency to determine them. In this chapter, we will discuss three types of spectroscopy that depend on such transitions. Table 13.1 summarizes the regions of the electromagnetic spectrum in which these transitions can be observed.

Example 13.1   Verify that light of wavelength 15×10^{-6} meters corresponds to light of frequency 0.2×10^{14} hertz. (See the data for infrared light in Table 13.1.)

Solution   The relationship between frequency and wavelength is the inverse proportionality, $\nu = c/\lambda$, where c is the speed of light (3×10^8 meters per second). Hence

$$\nu = \frac{3 \times 10^8 \text{ m/s}}{15 \times 10^{-6} \text{ m}} = 0.2 \times 10^{14} \text{ cycles per second (or hertz)}$$

**13.3
INFRARED
SPECTROSCOPY**

As you can see from Table 13.1, infrared light has a wavelength in the region of 2.5–15.0 $\times$ 10^{-6} meters. A convenient unit for expressing wavelengths in this region of the spectrum is the **micron,** abbreviated μ. One micron equals 10^{-6} m, so the usual range of infrared spectroscopy is 2.5–15.0 μ. More commonly, however, we use a frequency unit, **wavenumber** (symbolized $\tilde{\nu}$), to describe infrared spectra. Wavenumber is defined as the *number of waves per centimeter* and is the reciprocal of the wavelength expressed in centimeters.

$$\tilde{\nu} = \frac{1}{\lambda \, (\text{cm})}$$

Example 13.2   **What frequency, in wavenumbers, corresponds to infrared radiation of wavelength 2.5 μ?**

Solution   **2.5 μ = 2.5 $\times$ 10^{-6} m = 2.5 $\times$ 10^{-4} cm. So**

$$\tilde{\nu} = \frac{1}{2.5 \times 10^{-4} \text{ cm}} = 4000 \text{ cm}^{-1} \quad (\text{cm}^{-1} \text{ is read "reciprocal centimeters")}$$

Problem 13.1   **Express the other limit of the infrared region, 15 μ, in wavenumbers.**

The infrared frequency range corresponds to energies of about 2–12 kcal/mol. This amount of energy is sufficient to excite covalent bonds from one vibrational state to another. The two basic types of bond vibrations are *bond stretching* and *bond bending,* exemplified for a CH_2 group as follows:

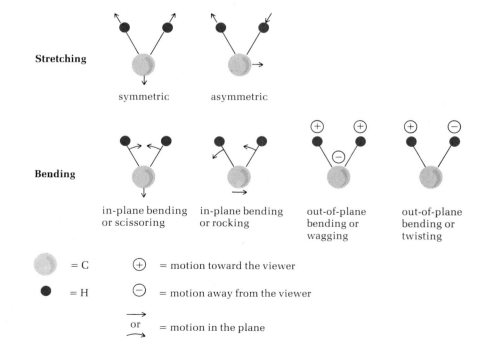

Stretching

symmetric   asymmetric

Bending

in-plane bending   in-plane bending   out-of-plane   out-of-plane
or scissoring   or rocking   bending or   bending or
                                         wagging   twisting

= C $\oplus$ = motion toward the viewer

= H $\ominus$ = motion away from the viewer

or   = motion in the plane

As we will see, different stretching and bending vibrations require different amounts of energies, and therefore correspond to different infrared frequencies. Furthermore, bonds between different kinds of atoms will vibrate with different frequencies. Consequently, infrared spectroscopy is particularly useful for determining what types of bonds are present in a molecule.

Example 13.3   **What is the relationship between the energies required for two different bond stretchings, one of which occurs at 3000 cm⁻¹ and the other at 1500 cm⁻¹?**

Solution   **Energy and frequency are directly proportional ($E = h\nu$). The bond stretching which occurs at 3000 cm⁻¹ therefore requires twice the energy of the bond stretching at 1500 cm⁻¹.**

Problem 13.2   **The ⌇C—H bond stretching frequency occurs at about 2850–3000 cm⁻¹.**

The ≡C—H bond stretching frequency occurs at about 3200–3350 cm⁻¹. Which carbon-hydrogen bond is harder to stretch (that is, which requires more energy)?

The infrared spectrum of a compound can easily be obtained in a few minutes. A small sample of the compound is placed in an instrument with

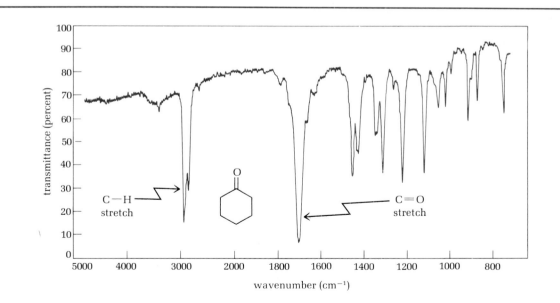

FIGURE 13.2   The infrared spectrum of cyclohexanone, a typical ketone. The band near 3000 cm⁻¹ is due to C—H stretching. The band near 1700 cm⁻¹ is due to the C=O stretch.

Bond type	Group	Class of compound	Frequency range (cm⁻¹)
Single bonds to hydrogen	C—H	alkane	2850–3000
	=C—H	alkene and aromatic compound	3030–3140
	O—H	alcohols and phenols	3500–3700 (free) 3200–3500 (hydrogen-bonded)
		carboxylic acids	2500–3000
	N—H	amines	3200–3600
	S—H	thiols	2550–2600
Double bonds	C=C	alkenes	1600–1680
	C=N	imines, oximes, etc.	1500–1650
	C=O	aldehydes, ketones, esters, acids, etc.	1650–1780
Triple bonds	C≡C	alkynes	2100–2260
	C≡N	nitriles	2200–2400

TABLE 13.2

Infrared stretching frequencies of some functional groups

an infrared radiation source. The spectrometer automatically scans the amount of radiation that passes through the sample over a given frequency range and records on a chart the percent of radiation that is transmitted. Radiation that is absorbed by the molecule appears as a band in the spectrum. Figure 13.2 shows a typical infrared spectrum, with the frequency expressed in wavenumbers.

The stretching frequencies of certain bonds usually lie within a particular range regardless of the rest of the molecular structure. For example, almost all carbon–oxygen double bonds (C=O) stretch in the same frequency range, 1650–1780 cm⁻¹, regardless of whether the compound is an aldehyde, ketone, acid, ester or whatever. When the C=O group is present in a molecule, its infrared spectrum will show a strong band somewhere in that region (the exact location, of course, depends on the particular molecular structure). Table 13.2 shows some of the more common group stretching frequencies.

The stretching frequency of a particular bond depends on several factors. It depends on the masses of the atom. A bond between a heavy atom and a light atom always vibrates at a higher frequency than a bond of the same type between two heavy atoms with comparable masses ($\tilde{\nu}_{C-H}$ = about 3000 cm⁻¹, $\tilde{\nu}_{C-C}$ = about 1200 cm⁻¹). The bond energy and multiplicity also affects the frequency. For example, double bonds vibrate at higher frequencies than single bonds between the same kinds of atoms.

Problem 13.3 **Using infrared spectroscopy, how could you quickly distinguish the structural isomers benzyl alcohol and methyl phenyl ether from one another?**

FIGURE 13.3

The infrared spectrum of cyclopentanone.

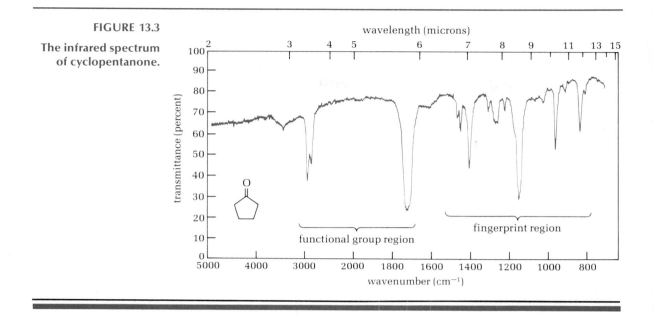

Look again at Figure 13.2. We clearly see the C—H stretch in the 3000 cm⁻¹ region and an intense band near 1700 cm⁻¹ for the C=O bond. But what does the rather complex pattern of bands at about 700–1500 cm⁻¹ mean? This low-frequency region of the spectrum is called the **fingerprint region.** The bands in this region result from combined bending and stretching motions of the atoms and are *unique* for each particular compound. For example, we expect many other ketones to have infrared spectra similar to that of cyclohexanone in the **functional group region** of the spectrum, between 1600 and 4000 cm⁻¹. They will show bands for C—H stretch and C=O stretch. But their spectra will almost always differ in the fingerprint regions, where combined atomic motions are displayed. Compare the infrared spectrum of cyclohexanone (Figure 13.2) with that of cyclopentanone shown in Figure 13.3. The two spectra are nearly identical in the functional group region but are vastly different and easily distinguishable in the fingerprint region.

To summarize, then, infrared spectra can be used to tell what types of bonds may be present in a molecule (by using the functional group region) and to tell whether two substances are identical or different (by using the fingerprint region).

**13.4
VISIBLE AND
ULTRAVIOLET
SPECTROSCOPY**

The visible region of the spectrum (visible to the human eye, that is) corresponds to light with wavelengths of 400–800 × 10⁻⁹ m. Light in the ultraviolet region has a somewhat shorter wavelength, about 200–400 × 10⁻⁹ m (by contrast, the infrared region that we just discussed begins at appreciably longer wavelengths, about 2500 × 10⁻⁹ m). Usually these wave-

TABLE 13.3	Visible (vis)	400–800 nm (or mμ)	4000–8000 Å
Units for visible-ultraviolet spectra	Ultraviolet (uv)	200–400 nm (or mμ)	2000–4000 Å

lengths are expressed in **nanometers** (1 nm = 10^{-9} m). Older methods that are sometimes still used report visible (vis) and ultraviolet (uv) spectra in millimicrons (mμ), which are identical to nm, or in **angstrom units** (10 Å = 1 nm). Table 13.3 summarizes these units.

The energies involved in ultraviolet radiation correspond to about 75–150 kcal/mol, and the energies in the visible region are 37–75 kcal/mol. These energies are much larger than those involved in infrared spectroscopy (2–12 kcal/mol).

Problem 13.4 **Which uv wavelength, 200 nm or 400 nm, corresponds to 75 kcal/mol? Which corresponds to 150 kcal/mol?**

Transitions in the vis–uv portion of the spectrum are **electronic transitions.** They correspond to having an electron jump from a filled molecular orbital to a higher-energy, vacant molecular orbital. Visible and uv light contain enough energy to bring about such transitions. [*Note:* These energies may be sufficient to break bonds. We saw in Chapter 2, for example, that ultraviolet light can catalyze the halogenation of alkanes by breaking a Cl—Cl bond or a Br—Br bond (eq. 2.16).]

Figure 13.4 shows a typical ultraviolet absorption spectrum. Unlike infrared spectra, vis-uv spectra are quite broad and generally show a small number of peaks. The peaks are reported as the wavelengths where

FIGURE 13.4

The absorption spectrum of 4-methyl-3-penten-2-one.

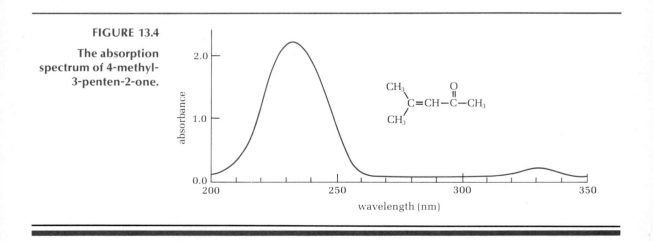

maxima occur. The conjugated, unsaturated ketone whose spectrum is shown in Figure 13.4 has an intense maximum at $\lambda = 232$ nm and a much weaker maximum at $\lambda = 330$ nm.

The intensity of an absorption band can be expressed quantitatively. Band intensity depends on the particular molecular structure and also on the number of absorbing molecules in the light path. Absorbance, which is the log of the ratio of light intensities entering and leaving the sample, is given by the equation

$$A = \epsilon cl \qquad \text{(Beer's Law)} \tag{13.1}$$

where ϵ is the **molar absorptivity** (sometimes called the **extinction coefficient**), c is the concentration of the solution in moles per liter, and l is the length in centimeters of the sample through which the light passes. The value of ϵ for any peak in the spectrum of a compound is a constant characteristic of that particular molecular structure. For example, the values of ϵ for the peaks in the spectrum of the unsaturated ketone shown in Figure 13.4 are $\lambda_{max} = 232$ nm ($\epsilon = 12,600$) and $\lambda_{max} = 330$ nm ($\epsilon = 78$).

Example 13.4 **What is the effect of doubling the concentration of a particular absorbing sample on A? on ϵ?**

Solution **The observed absorbance A will be doubled, since A is directly proportional to c. The value of ϵ, however, is a function of molecular structure and is a constant, independent of the concentration.**

Problem 13.5 **A particular solution of $(CH_3)_2C{=}CH{-}\overset{\displaystyle O}{\overset{\displaystyle \|}{C}}{-}CH_3$ (Fig. 13.4), placed in a 1-cm. absorption cell, shows a peak at $\lambda = 232$ nm with an observed absorbance $A = 2$. Calculate the concentration of the solution, using the value of ϵ given in the text.**

Vis-uv spectra are most commonly used to detect conjugation. In general, molecules with no double bonds and molecules with only one double bond do not absorb light in the region 200–800 nm. Conjugated systems do absorb here, and the greater the conjugation, the longer the wavelength of maximum absorption. For example,

$$CH_2{=}CH{-}CH{=}CH_2 \qquad\qquad CH_2{=}CH{-}CH{=}CH{-}CH{=}CH_2$$
$$\lambda_{max} = 220 \text{ nm} \qquad\qquad\qquad \lambda_{max} = 257 \text{ nm}$$
$$(\epsilon = 20,900) \qquad\qquad\qquad\qquad (\epsilon = 35,000)$$

$$CH_2{=}CH{-}CH{=}CH{-}CH{=}CH{-}CH{=}CH_2$$
$$\lambda_{max} = 287 \text{ nm}$$
$$(\epsilon = 52,000)$$

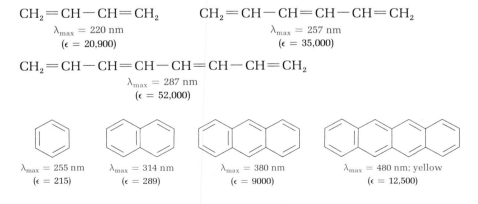

$\lambda_{max} = 255$ nm	$\lambda_{max} = 314$ nm	$\lambda_{max} = 380$ nm	$\lambda_{max} = 480$ nm; yellow
$(\epsilon = 215)$	$(\epsilon = 289)$	$(\epsilon = 9000)$	$(\epsilon = 12,500)$

Problem 13.6   **Which of the following aromatic compounds do you expect to absorb at the longer wavelength?**

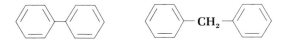

13.5 NUCLEAR MAGNETIC RESONANCE SPECTROSCOPY (NMR)

The kind of spectroscopy that has had the greatest impact by far on the determination of organic structures is **NMR spectroscopy.** Commercial instruments became available in the late 1950s, and since then NMR spectroscopy has become an indispensable tool for the organic chemist. Let us first look briefly at the theory and then see what practical information we can glean from an NMR spectrum.

13.5a Theory Certain nuclei behave as though they were spinning. Because the nuclei are charged, and a spinning charge creates a magnetic field, these nuclei behave in effect like tiny magnets. The most important nucleus that behaves like this, and the only one we will discuss in detail is ¹H (ordinary hydrogen). ¹²C and ¹⁶O, also present in many organic compounds, do not possess a spin and do not give NMR spectra. Certain of their isotopes do, however (¹³C, ¹⁷O), as do other nuclei present in some organic compounds (¹⁵N, ³¹P, ¹⁹F). Each of these types of NMR spectroscopy can be useful in its own way. ¹³C NMR is especially useful to organic chemists, and we will describe it briefly.

When nuclei with spin are placed between the poles of a very powerful magnet, they can align themselves *with* or *against* the field of the magnet. Nuclei aligned with the field have a slightly lower energy than those aligned against the field (Figure 13.5). By applying energy in the radio-frequency range (Table 13.1), it is possible to excite the nuclei and promote them from the lower to the higher energy spin state (we sometimes say that the spin "flips").

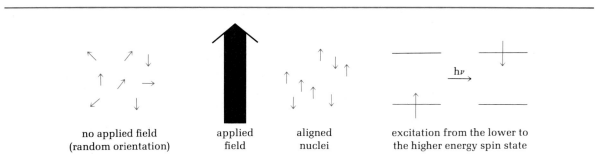

| no applied field (random orientation) | applied field | aligned nuclei | excitation from the lower to the higher energy spin state |

FIGURE 13.5   Orientation of nuclei in an applied field, and excitation of nuclei from the lower to the higher energy spin state.

The energy gap between the two spin states depends on the strength of the applied field; the stronger the field, the larger the energy gap. The spectra shown in this book were taken on an instrument with a magnet strength of 14,092 gauss (by comparison, the earth's magnetic field is only about 0.5 gauss). At that field strength, the energy gap corresponds to a frequency of 60×10^6 Hz (or 60 MHz, megahertz). This frequency corresponds to a wavelength of about 5 m, in the radio-frequency (rf) range. The energy gap between the spin states on such an instrument is only 0.000006 kcal/mol. Nevertheless, with modern technology we can detect such small changes. If the magnetic field strength were doubled, the energy gap between spin states would also be doubled, thus giving an instrument with higher resolution. For this reason there is a continuing effort to build NMR instruments with more powerful magnets.

13.5b Measuring an NMR Spectrum A proton NMR spectrum is usually obtained in the following way. A sample of the compound being studied is dissolved in some inert solvent that does not contain protons. Examples of such solvents are CCl_4 and solvents in which the hydrogens have been replaced by deuterium, such as $CDCl_3$ (deuterochloroform) and CD_3COCD_3 (hexadeuteroacetone). A small amount of a reference compound is also added (we will say more about this in the next section). The solution, contained in a thin glass tube, is placed in the center of a radio-frequency (rf) coil, between the pole faces of a powerful magnet. The nuclei align themselves with or against the field. Continuously increasing amounts of energy can then be applied to the nuclei by the rf coil. When this energy corresponds exactly to the energy gap between the lower and higher energy spin states, it will be absorbed by the nuclei. At this point, the nuclei are said to be in resonance with the applied frequency—hence the term **nuclear magnetic resonance.** A plot of the energy absorbed by the sample against the applied frequency of the rf coil gives an NMR spectrum.

In practice, it is usually easier to apply a *constant* rf frequency and vary slightly the strength of the applied magnetic field. One then measures exactly the strength of the magnetic field that corresponds to the applied rf frequency. The spectra given in this book were obtained in this way at an rf frequency of 60 MHz. The **applied magnetic field** increases as we go from left to right in the spectra.

13.5c Chemical Shifts and Peak Areas From what we have said so far, you might think that all the protons in a molecule would "flip" their spins at the same resonance frequency. If this were true, we would obtain very little structural information from an NMR spectrum because many different proton-containing compounds would have the same spectra. Fortunately, this is not the case. The radiation frequency that a proton absorbs depends on its immediate environment in the molecule. This is because the electrons that surround the various proton nuclei shield them slightly from the applied magnetic field. The amount of shielding depends on the density of the various electrons (all of them: the nonbonding, the π, and the σ electrons). At a fixed rf frequency, the more shielded a particular

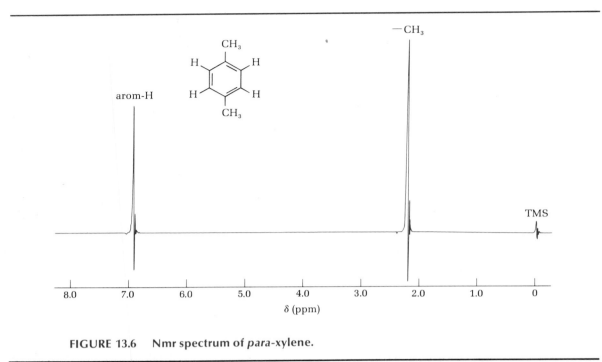

FIGURE 13.6   Nmr spectrum of *para*-xylene.

proton, the higher the applied magnetic field required to achieve resonance.

Now let us examine some spectra. Figure 13.6 shows the proton NMR spectrum of *para*-xylene. Note that the spectrum is very simple and consists of only two peaks. The positions of the peaks are measured in δ (delta) units from a reference compound, which is **tetramethylsilane (TMS)**, $(CH_3)_4Si$. The reasons for selecting TMS as a reference are that (1) all twelve of its protons are chemically equivalent, so it shows only one sharp NMR signal which serves as a reference point; (2) its protons appear at higher field than those of most protons in organic compounds, thus making it easy to identify the TMS peak; and (3) TMS is inert, so it does not react with most organic compounds, and it is low-boiling and can be removed easily at the end of a measurement.

Most organic compounds have peaks downfield from TMS and are given positive δ values. A **δ value** of 1.00 means that a peak appears 1 part per million (ppm) downfield from the TMS peak. If the spectrum is measured at 60 MHz (60×10^6 Hz), then 1 ppm is 60 Hz (one millionth of 60 MHz) downfield from TMS. (If the spectrum is run at 100 MHz, a δ value of 1 ppm is 100 Hz downfield from TMS, and so on.) The **chemical shift** of a particular kind of proton is its δ value with respect to TMS. It is called a *chemical* shift because it depends on the chemical environment of the protons.

In the spectrum of *para*-xylene, we see a peak at δ 2.20 and another peak at δ 6.95. It seems reasonable that these peaks would be caused by the two different "kinds" of protons in the molecule, the methyl protons and the aromatic ring protons. How can we tell which is which?

One way is to integrate the area under each peak. The **peak area** is *directly proportional to the number of protons responsible for the particular peak.* Thus, we find that the relative areas of the peaks at δ 2.20 and δ 6.95 in the *para*-xylene spectrum are 3:2 (or 6:4). These areas allow us to assign the peak at δ 2.20 to the six methyl protons and the peak at δ 6.95 to the four aromatic ring protons.

Example 13.5   How many peaks do you expect to see in the NMR spectra of the following compounds? If you expect more than one, what will their relative areas be?

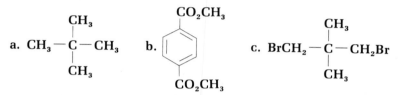

Solution   a. All twelve protons are equivalent and appear as a single peak.
b. All four aromatic protons are equivalent and the six methyl protons on the ester functions are equivalent. There will be two peaks in the spectrum, with an area ratio of 4:6 (or 2:3).
c. There are two kinds of protons, $CH_3 - C$ and $CH_2 - Br$. There will be two peaks with the area ratio 6:4 (or 3:2).

Problem 13.7   Which of the following compounds show only a single peak in their NMR spectra?

a. CH_3OCH_3 b. $CH_3CH_2OCH_2CH_3$ c. ⬠

Problem 13.8   Each of the following compounds shows more than one peak in its NMR spectrum. What will their relative areas be?

a. CH_3OH b. $CH_3\overset{\overset{\displaystyle O}{\|}}{C}OCH_3$ c. $CH_3CH_2\overset{\overset{\displaystyle O}{\|}}{C}CH_2CH_3$

A more general way to assign peaks is to compare the chemical shifts with those of similar protons in a reference compound. For example, benzene itself has six equivalent protons and shows a single peak in its NMR spectrum, at δ 7.24. Other aromatic compounds also show a peak in this region. We can conclude that most protons on an aromatic ring will have chemical shifts of about δ 7.

Investigators have determined the chemical shifts of protons in various chemical environments by measuring the NMR spectra of a large number of compounds with known, relatively simple structures. Table 13.4 gives the expected chemical shifts for several common types of protons.

TABLE 13.4	Type of proton	δ (ppm)	Type of proton	δ (ppm)
Typical proton chemical shifts (relative to tetramethylsilane)	$C-CH_3$	0.85–0.95	$-CH_2-F$	4.3–4.4
	$C-CH_2-C$	1.20–1.35	$-CH_2-Br$	3.2–3.3
	$C-\overset{\mid}{\underset{\mid}{C}}H-C$	1.40–1.65	$CH_2=C$	4.6–5.0
	$CH_3-C=C$	1.6–1.9	$-CH=C$	5.2–5.7
	CH_3-Ar	2.2–2.5	$Ar-H$	6.6–8.0
	$CH_3-\overset{\mid}{\underset{\mid}{C}}=O$	2.1–2.6	$-C\equiv C-H$	2.4–2.7
	$CH_3-N\diagdown$	2.1–3.0	$-\overset{O}{\overset{\|}{C}}-H$	9.5–9.7
	CH_3-O-	3.5–3.8	$-\overset{O}{\overset{\|}{C}}-OH$	10–13
	$-CH_2-Cl$	3.4–3.5	$R-OH$	0.5–5.5
	$-CHCl_2$	5.8–5.9	$Ar-OH$	4–8

Example 13.6   Using Table 13.4, describe the expected NMR spectrum of:

a. $CH_3\overset{O}{\overset{\|}{C}}-OCH_3$ b. $CHCl_2-\overset{CH_3}{\underset{CH_3}{\overset{\mid}{\underset{\mid}{C}}}}-CH_2Cl$

Solution   a. The spectrum will consist of two peaks, equal in area, at about δ 2.3 (for the $CH_3\overset{O}{\overset{\|}{C}}-$ protons) and δ 3.6 (for the $-OCH_3$ protons).

b. The spectrum will consist of three peaks, with relative areas 6:2:1 at δ 0.9 (the two methyls), δ 3.5 (the $-CH_2-Cl$ protons), and δ 5.8 (the $-CHCl_2$ proton).

Problem 13.9   Describe the expected NMR spectrum of:

a. $CH_3\overset{O}{\overset{\|}{C}}OH$ b. $(CH_3)_2C=CH_2$

Problem 13.10   An ester is suspected of being either $(CH_3)_3C\overset{O}{\overset{\|}{C}}OCH_3$ or $CH_3\overset{O}{\overset{\|}{C}}-OC(CH_3)_3$. Its NMR spectrum consists of two peaks at δ 0.9 and δ 3.6 (relative areas 3:1). Which compound is it? Describe the expected spectrum if it had been the other ester.

Many compounds show more complex spectra than just single peaks (called **singlets**) for each type of proton. Let us examine some spectra of this type to see what additional structural information they convey.

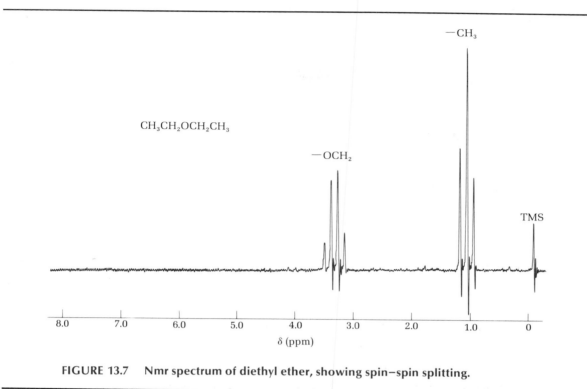

FIGURE 13.7   Nmr spectrum of diethyl ether, showing spin–spin splitting.

13.5d Spin–Spin Splitting Figure 13.7 shows the NMR spectrum of diethyl ether. From the information given in Table 13.4, we might have expected the NMR spectrum of diethyl ether, $CH_3CH_2OCH_2CH_3$, to consist of two lines: one in the region of δ 0.9 for the six equivalent CH_3 protons and one at about δ 3.5 for the four equivalent CH_2 protons adjacent to an oxygen atom, with relative areas 6:4. Indeed, in Figure 13.7 we see absorptions in each of these regions, with the expected total area ratios. But we do not see singlets! Instead the methyl signal is split into three peaks, a **triplet** with relative areas 1:2:1, and the CH_2 (methylene) signal is split into four peaks, a **quartet** with relative areas 1:3:3:1. These **spin–spin splittings** as they are called, tell us quite a bit about molecular structure. The splittings arise in the following way.

We know that each proton in the molecule acts as a tiny magnet. When we run an NMR spectrum, each proton "feels" not only the very large applied magnetic field but also a tiny field due to its neighboring protons. At the time we are sweeping through one signal, the protons on neighboring carbons can be in either the lower or the higher energy spin state, with nearly equal probabilities (nearly equal because, as we have said, the energy difference between the two states is exceedingly small). So the

magnetic field of the protons whose peak we are sweeping through is per-
turbed slightly by the tiny fields of its neighboring protons.

We can predict the splitting pattern, then, by the so-called **n + 1 rule:** if
a proton or a set of equivalent protons has n proton neighbors with a substantially different chemical shift, its NMR signal will be split into $n + 1$
peaks. In diethyl ether, each CH_3 proton has *two* proton neighbors (on
the CH_2 group). Therefore the CH_3 signal is split into $2 + 1 = 3$ peaks.
At the same time, each CH_2 proton has *three* proton neighbors (on the CH_3
group). The CH_2 signal is therefore split into $3 + 1 = 4$ peaks. Let us see
why this rule works and why the split peaks have the area ratios that
they do.

If proton H_a has *one* nonequivalent proton neighbor H_b, then at the time
we pass through the H_a signal, H_b can be in either the lower or the higher
energy spin state. Because there are two nearly equal possibilities, the H_a
signal will be split into two *equal* peaks: a **doublet.**

If proton H_a has *two* nonequivalent proton neighbors H_b, then, at the time
we pass through the H_a signal, there are three possibilities for the two
H_b protons:

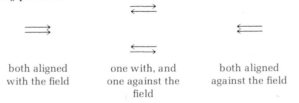

both aligned	one with, and	both aligned
with the field	one against the	against the field
	field	

Both can be in the lower energy state, both can be in the higher energy
state, or one can be in each state, the latter arrangement being possible
in two ways. Hence the H_a signal will be a triplet with relative areas
1:2:1.

Example 13.7 **Explain why the signal of H_a with *three* nonequivalent neighboring protons H_b is a *quartet* with relative areas 1:3:3:1.**

Solution **The possibilities for spin states of the three H_b protons are**

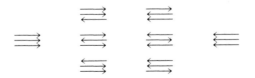

Hence the H_a signal will appear as four peaks, area ratio 1:3:3:1.

Problem 13.11 **Predict the NMR spectrum of CH_3CHCl_2. Give the approximate chemical
shifts and the splitting patterns of the various protons.**

Protons that split one another's signals are said to be **coupled.** The extent of the coupling, or the number of hertz that the signals are split, is

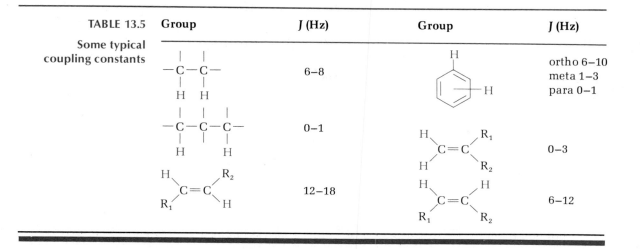

TABLE 13.5	Group	J (Hz)	Group	J (Hz)
Some typical coupling constants	$-\overset{\mid}{\underset{\mid H}{C}}-\overset{\mid}{\underset{\mid H}{C}}-$	6–8	(benzene ring with two H)	ortho 6–10 meta 1–3 para 0–1
	$-\overset{\mid}{\underset{\mid H}{C}}-\overset{\mid}{\underset{\mid}{C}}-\overset{\mid}{\underset{\mid H}{C}}-$	0–1	$\underset{H}{\overset{H}{\diagdown}}C=C\underset{R_2}{\overset{R_1}{\diagup}}$	0–3
	$\underset{R_1}{\overset{H}{\diagdown}}C=C\underset{H}{\overset{R_2}{\diagup}}$	12–18	$\underset{R_1}{\overset{H}{\diagdown}}C=C\underset{R_2}{\overset{H}{\diagup}}$	6–12

called the **coupling constant** (abbreviated J). A few typical coupling constants are shown in Table 13.5

Note that spin–spin splitting falls off rapidly with distance. Whereas protons on adjacent carbons show appreciable splitting (J = 6–8 Hz), protons farther apart hardly "feel" each other's presence (J = 0–1 Hz). As seen in Table 13.5, coupling constants can sometimes be used to distinguish between *cis-trans* isomers or between positions of substituents on a benzene ring.

Chemically equivalent protons do not split each other. For example, $BrCH_2CH_2Br$ shows only a sharp singlet in its NMR spectrum for all four protons. Even though they are on adjacent carbons, the protons do not split each other because they have identical chemical shifts. For the same reason, equivalent protons attached to a single carbon atom normally do not split each other.

Problem 13.12   **Describe the NMR spectra of**
a. $BrCH_2CH_2Cl$ b. $ClCH_2CH_2Cl$

NMR spectra may sometimes be quite complex. This complexity can arise when adjacent protons have nearly the same but not identical chemical shifts. An example is phenol (see Figure 13.8). We can easily distinguish the aromatic protons (δ 6.6–7.4) from the hydroxyl proton (δ 5.85), but the splitting pattern of the complex **multiplet** seen for the aromatic protons cannot be analyzed with the simple $n + 1$ rule. Such spectra can, however, be thoroughly analyzed by specially designed computer programs.

In summary, then, proton NMR spectroscopy can give us the following kinds of structural information:

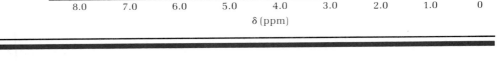

1. The number of signals and their chemical shifts can be used to identify the kinds of chemically different protons in the molecule.
2. The peak areas tell us how many protons of each kind are present.
3. The spin–spin splitting pattern gives us information about the number of nearest proton neighbors a particular kind of proton may have.

13.5e ^{13}C NMR Spectroscopy Just as proton NMR spectroscopy gives us information about the arrangement of protons in a molecule, ^{13}C NMR spectroscopy gives us information about the carbon skeleton. The ordinary isotope of carbon, ^{12}C, does not have a nuclear spin, but carbon-13 does. ^{13}C constitutes only 1.1% of naturally occurring carbon atoms. Also, the energy gap between the higher and lower spin states of ^{13}C is very small. For these two reasons, ^{13}C NMR spectrometers must be exceedingly sensitive, much more so than the standard proton spectrometer. Such instruments have been designed, and their use has become fairly routine in recent years.

^{13}C spectra differ from proton spectra in several ways. ^{13}C chemical shifts occur over a wider range than for protons. They are measured against the same reference compound, TMS, whose methyl carbons are all equivalent and give a sharp signal. Chemical shifts for ^{13}C are reported in δ units, but the usual range is about 0–200 ppm downfield from TMS (instead of the smaller range of 1–10 ppm observed for protons). This wide range of chemical shifts tends to simplify ^{13}C spectra relative to ^{1}H spectra.

Because of the low natural abundance of ^{13}C, the areas under the peaks do not give an accurate measure of the numbers of carbon atoms respon-

FIGURE 13.9

The ^{13}C NMR spectrum of 2-butanol without (bottom) and with (top) ^{13}C — ^{1}H coupling. δ-values are shown in the lower spectrum. Reprinted with permission from University Science Books.

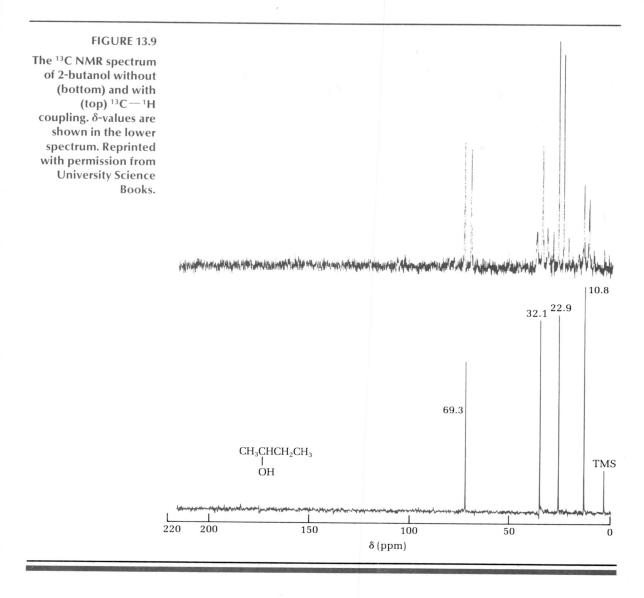

FIGURE 13.9

The ^{13}C NMR spectrum of 2-butanol without (bottom) and with (top) ^{13}C — ^{1}H coupling. δ-values are shown in the lower spectrum. Reprinted with permission from University Science Books.

sible for a particular peak. Therefore relative areas are normally not very useful.

Another consequence of the low natural abundance of ^{13}C is that the chance of finding two adjacent ^{13}C atoms in the same molecule is very small. Hence ^{13}C — ^{13}C spin–spin splitting is ordinarily not seen. This feature simplifies ^{13}C spectra. However, ^{13}C — ^{1}H spin–spin splitting can occur. The spectra can be run in such a way as to show this splitting or not, as desired. Figure 13.9 shows the ^{13}C spectrum of 2-butanol measured with and without ^{13}C — ^{1}H splitting. The spectrum without proton splitting shows four sharp singlets, one for each type of carbon atom. The hydroxyl-bearing carbon occurs at lowest field (δ 69.3) and the two methyl carbons are well separated (δ 10.8 and 22.9). In the spectrum with $^{13}C - {}^1H$ splitting, the $n + 1$ rule applies. Both methyl signals are quartets (three hydrogens, therefore $n + 1 = 4$), the CH_2 carbon is a triplet, and the CH carbon is a doublet.

Example 13.8   **Describe the ^{13}C spectrum of CH_3CH_2OH.**

Solution   **The spectrum without ^1H coupling will consist of two lines (in fact, they come at δ 18.2 and 57.8). With ^1H splitting, the signal at δ 18.2 is a quartet, and the one at δ 57.8 is a triplet.**

Problem 13.13   **Describe the main features of the ^{13}C spectrum of $CH_3CH_2CH_2OH$.**

The combination of ^1H and ^{13}C NMR spectroscopy provides a very powerful tool for determining organic structure.

13.6
MASS
SPECTROMETRY

Mass spectra, unlike the other types of spectra discussed in this chapter, do not depend on excitation of a molecule from one energy state to another. Yet they are easily obtained and routinely used for determining structure.

A mass spectrometer is an instrument that converts molecules to ions, sorts them according to their mass-to-charge ratio (m/e), and determines the relative amounts of each ion present. A small sample of the substance whose spectrum is to be determined is introduced into a high-vacuum chamber, where it is vaporized and bombarded with high-energy electrons. These bombarding electrons eject an electron from the molecule M, giving a positive **molecular ion** M^+ (sometimes called the **parent ion**).

$$M + e^- \rightarrow M^+ + 2\, e^- \tag{13.2}$$

    molecular
      ion

Methanol, for example, forms a molecular ion in the following way:

$$e^- + CH_3\ddot{O}H \rightarrow \left[CH_3\dot{O}H \right]^+ + 2\, e^- \tag{13.3}$$

        methanol molecular
         ion ($m/e = 32$)

The beam of these parent ions then passes between the poles of a powerful magnet, which deflects the beam. The extent of the deflection depends on the mass of the ion. Since M^+ has a mass that is essentially identical to the mass of the molecule M (the mass of the ejected electron is trivial compared to the mass of the rest of the molecule), mass spectrometers can be used to determine molecular weights. High-resolution mass spectrometers can measure the mass of a parent ion very precisely, to four decimal places. This accuracy permits one to deduce molecular formulas. For example, carbon monoxide, nitrogen, and ethylene all have an approxi-

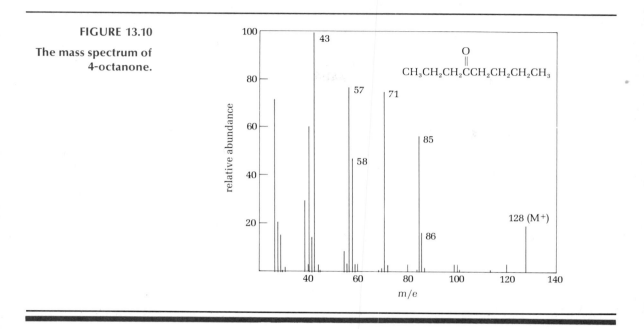

FIGURE 13.10

The mass spectrum of 4-octanone.

mate molecular weight of 28. But the precise molecular weights are slightly different and can be distinguished by a high resolution instrument.

CO	N_2	$CH_2=CH_2$
$^{12}C = 12.0000$	$^{14}N = 14.0031$	$2 \times {}^{12}C = 24.0000$
$^{16}O = \underline{15.9949}$	$^{14}N = \underline{14.0031}$	$4 \times {}^{1}H = \underline{4.0312}$
$M^+ = 27.9949$	$M^+ = 28.0062$	$M^+ = 28.0312$

If the bombarding electrons have enough energy, they produce not only parent ions but also fragment ions called **daughter ions.** That is, the original ion breaks into smaller fragments, some of which are ionized and get sorted on an m/e basis by the spectrometer. A prominent peak in the mass spectrum of methanol, for example, is the $M^+ -1$ peak at $m/e = 31$. This peak arises through loss of a hydrogen atom from the molecular ion:

$$H-\overset{\displaystyle H}{\underset{\displaystyle H}{\overset{+}{C}}}-\overset{..}{\underset{..}{O}}-H \longrightarrow H-\overset{\displaystyle H}{\underset{\displaystyle H}{C}}=\overset{+}{\underset{..}{O}}-H + H\cdot \qquad \textbf{(13.4)}$$

$$m/e = 32 \qquad\qquad m/e = 31$$

You will recognize the daughter ion as protonated formaldehyde, a stabilized carbocation (compare with eq. 9.4).

A mass spectrum consists, therefore, of a series of signals of varying intensities at different m/e ratios. In practice, most of the ions are singly charged ($e = 1$), so that we can readily obtain their masses, m. Figure 13.10 shows a mass spectrum as it might be printed from the computer output of a mass spectrometer. It is the mass spectrum of a typical ketone, 4-octanone. Note the peak at $m/e = 128$. It is the highest mass peak in the spectrum, and it corresponds to the molecular weight of the ketone. In addition, we see certain prominent daughter ion peaks. For example, the peaks at $m/e = 85$ and $m/e = 71$ correspond in mass to $C_4H_9CO^+$ and $C_3H_7CO^+$, respectively. This suggests that one easy fragmentation path for the parent ion is to break the carbon–carbon bond adjacent to the carbonyl group. Ion fragmentation paths depend on ion structure, and the interpretation of mass spectral fragmentation patterns can give significant information about molecular structure.

Example 13.9   **The most intense peak (called the *base peak*) in Figure 13.10 occurs at $m/e = 43$. Suggest how it might arise.**

Solution   **This peak corresponds to the m/e for $C_3H_7^+$, suggesting that the daughter ion $C_3H_7CO^+$ loses carbon monoxide to give $C_3H_7^+$. This explanation becomes more plausible when we consider that the spectrum also contains an intense peak at $m/e = 57$, corresponding to the analogous process for $C_4H_9CO^+$. We can summarize these conclusions in the following "family tree" of ions:**

$$
\begin{array}{c}
& \xrightarrow{-C_3H_7\cdot} C_4H_9CO^+ \xrightarrow{-CO} C_4H_9^+ \\
C_8H_{16}O^+ & \qquad\qquad (85) \qquad\qquad\quad (57) \\
M^+ (128) & \xrightarrow{-C_4H_9\cdot} C_3H_7CO^+ \xrightarrow{-CO} C_3H_7^+ \\
& \qquad\qquad (71) \qquad\qquad\quad (43)
\end{array}
$$

The four types of spectroscopy that we have described in this chapter are all routinely used in research laboratories. With modern instrumentation, each type of spectrum can be obtained in a few minutes to an hour, including time to prepare the sample. Interpretation of the spectra may take a little longer, but investigators with experience can often deduce the structure of even complex molecules from their spectra alone in a relatively short time.

ADDITIONAL PROBLEMS

13.14. What wavelengths, expressed in meters, are involved in infrared spectroscopy? in visible spectroscopy? in ultraviolet spectroscopy? What are the corresponding frequencies, expressed as wavenumbers (waves/cm)? Which of these three types of spectroscopy corresponds to the lowest-energy transitions? to the highest-energy transitions?

13.15. The C—H stretching frequency occurs at about 3000 cm^{-1}. Do you expect the C—D band (replace H by its heavy isotope, deuterium) to appear at a higher or lower frequency, or do you expect the absorption to remain the same? Explain.

13.16. A compound C_3H_6O has no bands in the infrared region around 3500 or 1720 cm^{-1}. What structures can be eliminated by these data? Suggest a possible structure and tell how you could determine whether it is correct.

13.17. A very dilute solution of ethanol in carbon tetrachloride shows a sharp infrared band at 3580 cm^{-1}. As the solution is made more concentrated, a new, rather broad band appears at 3250–3350 cm^{-1}. Eventually the sharp band disappears and is replaced entirely by the broad band. Explain. (*Hint:* Review Sec. 7.5).

13.18. In the following pairs of compounds, list at least one major peak in the infrared spectrum of one member of the pair that would enable you to distinguish it from the other member.

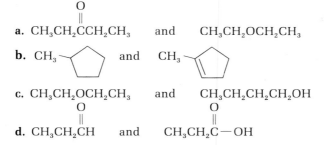

a. $CH_3CH_2CCH_2CH_3$ and $CH_3CH_2OCH_2CH_3$

b. and

c. $CH_3CH_2OCH_2CH_3$ and $CH_3CH_2CH_2CH_2OH$

d. CH_3CH_2CH and $CH_3CH_2C—OH$

13.19. A compound $C_5H_{10}O$ has an intense infrared band at 1725 cm^{-1}. Its NMR spectrum consists of a quartet of δ 2.7 and a triplet at δ 0.9 with relative areas 2:3. What is its structure?

13.20. What features would be similar in the infrared spectra of the following compounds, and how would their infrared spectra differ?

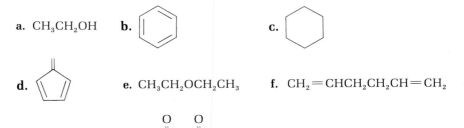

$CH_3CCH_2CH_2CH_2OH$ and $CH_3CH_2CCH_2CH_2OH$?

13.21. Which of the following compounds are *not* likely to absorb ultraviolet radiation in the range of 200–400 nm?

a. CH_3CH_2OH **b.**   **c.**

d.   **e.** $CH_3CH_2OCH_2CH_3$ **f.** $CH_2=CHCH_2CH_2CH=CH_2$

13.22. The compound $CH_3CCH_2CCH_3$ exists in equilibrium with a tautomer. Its solutions show a strong uv absorption with a λ_{max} at 272 nm. What is the structure of the tautomer? How could infrared spectroscopy be used to confirm your suggestion?

13.23. The unsaturated aldehydes $CH_3(CH=CH)_nCH=O$ have uv absorption spectra that depend on the value of n, the λ_{max} being 220, 270, 312, and 343 nm as n changes from 1 to 4. Explain.

13.24. A sample of cyclohexane is suspected of being contaminated with benzene, from which it had been prepared by hydrogenation. At $\lambda = 255$ nm, benzene has a molar absorptivity $\epsilon = 215$, whereas cyclohexane does not absorb at that wavelength ($\epsilon = 0$). A uv spectrum of the contaminated cyclohexane (obtained in a 1.0 cm cell) shows an absorbance $A = 0.43$ at 255 nm. Calculate the concentration of benzene in the cyclohexane.

13.25. Draw the structure of a compound with each of the following molecular formulas that will show only one peak in its proton NMR spectrum.
a. C_6H_{12} **b.** $C_3H_6Cl_2$ **c.** C_4H_6
d. $C_{12}H_{18}$ **e.** C_2H_6O **f.** C_5H_{12}

13.26. The proton NMR spectrum of a compound C_4H_9Br consists of a single sharp peak. What is its structure? The spectrum of an isomer of this compound consists of a doublet at δ 3.2, a complex pattern at δ 1.9, and a doublet at δ 0.9, with relative areas 2:1:6. What is its structure?

13.27. How could you distinguish between the following pairs of isomers by proton NMR spectroscopy?
a. CH_3CCl_3 and $CH_2ClCHCl_2$
b. $CH_3CH_2CH_2OH$ and $(CH_3)_2CHOH$

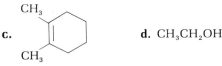

c. $CH_3CH_2\overset{\overset{\displaystyle O}{\|}}{C}-OCH_3$ and $CH_3\overset{\overset{\displaystyle O}{\|}}{C}-OCH_2CH_3$

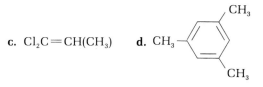

d. ⬡$-CH_2-CH=O$ and   ⬡$-\overset{\overset{\displaystyle O}{\|}}{C}-CH_3$

13.28. How many chemically different types of protons are present in:
a. $(CH_3)_2CHCH_2CH_3$ **b.** $(CH_3)_2NCH_2CH_3$

c. [cyclohexene with CH₃ groups]   **d.** CH_3CH_2OH

13.29. The proton NMR spectrum of cyclohexane, measured at room temperature, consists of a single peak. At very low temperatures, however, its NMR spectrum consists of two sets of peaks that are equal in area. Suggest an explanation.

13.30. The NMR spectrum of methyl p-toluate consists of a singlet at δ 2.35, a singlet at δ 3.82, and two doublets, at δ 7.15 and δ 7.87, relative areas 3:3:2:2. Draw the structure and determine which protons are responsible for each peak, as far as you can tell (use Table 13.4).

13.31. Using the information in Table 13.4, sketch the anticipated proton NMR spectrum of each of the following compounds. Be sure to show all splitting patterns.
a. CH_3CHO **b.** $(CH_3)_2CHOCH(CH_3)_2$

c. $Cl_2C=CH(CH_3)$ **d.** [benzene ring with three CH₃ groups]

13.32. A compound $C_5H_{10}O_3$ has a strong infrared band at 1745 cm^{-1}. Its proton NMR spectrum consists of a quartet at δ 4.15 and a triplet at δ 1.20, relative areas 2:3. What is the correct structure?

13.33 The proton NMR spectrum of a compound $C_3H_3Cl_5$ consists of a triplet at δ 4.5 and a doublet at δ 6.0, J = 7 Hz, with relative areas 1:2. What is its structure?

13.34. Sketch the expected proton NMR spectrum of the insect repellant "Off" (N,N-diethyl-m-toluamide).

13.35. A compound is known to be the methyl ester of a toluic acid, but the orientation of the two substituents (CH_3 and $-CO_2CH_3$) on the aromatic ring is not known. The ^{13}C NMR spectrum shows 7 peaks. Which isomer is it? What would the proton NMR spectrum look like?

13.36. A hydrocarbon shows a parent ion peak in its mass spectrum at $m/e = 102$. Its proton NMR spectrum shows peaks at δ 2.7 and δ 7.4, relative areas 1:5. What is the correct structure?

13.37. Write a formula for the molecular ion of ethanol.

13.38. The mass spectrum of 1-pentanol shows an intense daughter ion peak at $m/e = 31$. Explain how this peak might arise.

13.39. An alcohol $C_5H_{12}O$ shows, in its mass spectrum, a daughter ion peak at $m/e = 59$. An isomeric alcohol shows no daughter ion at $m/e = 59$, but does have a peak at $m/e = 45$. Suggest possible structures for each isomer. How could you confirm these structures by proton NMR spectroscopy? by ^{13}C NMR spectroscopy?

fourteen
Carbohydrates

**14.1
INTRODUCTION**

Carbohydrates are natural products that perform many vital functions in both plants and animals. Through photosynthesis, plants convert carbon dioxide to carbohydrates. The most common of these are cellulose, starch, and the various sugars. Cellulose is the main structural component of plants, used to construct rigid cell walls, fibers, and woody tissue. Starch is the chief form for storing carbohydrates for later use as a food or energy source. Some plants (cane and sugar beets) produce large amounts of sucrose, the main sugar that is harvested commercially.

In higher animals, the sugar glucose is an essential component of blood. Two sugars, D-ribose and 2-deoxyribose, are essential in genetic material. Other carbohydrates are important components of coenzymes, antibiotics, cartilage, the shells of crustaceans, and bacterial cell walls.

In this chapter we will describe the structures and a few reactions of some of the more important carbohydrates.

**14.2
DEFINITIONS AND
CLASSIFICATION**

The term **carbohydrate** arose because the formulas of many compounds of this type can be expressed as hydrates of carbon, $C_n(H_2O)_m$. Glucose, for example, has the molecular formula $C_6H_{12}O_6$, which could also be expressed as $C_6(H_2O)_6$. Although this type of formula is almost useless in studying the chemistry of carbohydrates, the old name persists.

We now define carbohydrates as *polyhydroxyaldehydes, polyhydroxyketones,* or substances that give such compounds on hydrolysis. For the most part, then, the chemistry of carbohydrates is the combined chemistry of two functional groups, the hydroxyl group and the carbonyl group.

Carbohydrates are usually classified according to their structure as **monosaccharides, oligosaccharides,** and **polysaccharides.** The term *saccharide* comes from the Latin (*saccharum,* sugar) and refers to the sweet taste of many simple carbohydrates. The three main classes of carbohydrates are related to each other through hydrolysis:

$$\text{Polysaccharide} \xrightarrow[\text{H}^+]{\text{H}_2\text{O}} \text{oligosaccharides} \xrightarrow[\text{H}^+]{\text{H}_2\text{O}} \text{monosaccharides} \tag{14.1}$$

A typical specific example is the hydrolysis of starch to maltose and ultimately to glucose:

$$[\underset{\substack{\text{starch}\\ \text{(a polysaccharide)}}}{C_{12}H_{20}O_{10}}]_n \xrightarrow{n\text{H}_2\text{O}} n\underset{\substack{\text{maltose}\\ \text{(a disaccharide)}}}{C_{12}H_{22}O_{11}} \xrightarrow{n\text{H}_2\text{O}} 2\ n\underset{\substack{\text{glucose}\\ \text{(a monosaccharide)}}}{C_6H_{12}O_6} \tag{14.2}$$

Monosaccharides, or *simple sugars,* as they are sometimes called, are carbohydrates that cannot be hydrolyzed to simpler compounds. Polysaccharides contain many monosaccharide units—sometimes hundreds or even thousands. Usually, but not always, the units are identical. Two of the most important polysaccharides, starch and cellulose, for example, contain linked glucose units. Oligosaccharides (from the Greek *oligos,* few) contain at least 2 and generally no more than 8 or 10 linked monosaccharide units. They may be called **disaccharides, trisaccharides,** and so on, depending on the number of linked monosaccharide units. The linked units may be the same or different. Maltose, for example, is a disaccharide made of 2 glucose units, but sucrose (ordinary table sugar) is a disaccharide made of 2 different monosaccharide units, glucose and fructose.

In the next section we will describe the structures of monosaccharides. Later in the chapter we will see how these units are linked together to form oligosaccharides and polysaccharides.

14.3
MONOSACCHARIDES

Monosaccharides are classified according to the number of carbon atoms present (**triose, tetrose, pentose, hexose,** and so on) and according to whether the carbonyl group is present as an aldehyde (**aldose**) or as a ketone (**ketose**).

There are only two trioses, **glyceraldehyde** and **dihydroxyacetone.** Each triose has 2 hydroxyl groups, on separate carbon atoms, and 1 carbonyl group.

$$
\begin{array}{ll}
^1CH{=}O & CH_2OH \\
| & | \\
^2CHOH & C{=}O \\
| & | \\
^3CH_2OH & CH_2OH \\
\text{glyceraldehyde} & \text{dihydroxyacetone} \\
\text{an aldose} & \text{a ketose}
\end{array}
$$

Glyceraldehyde is the simplest aldose, and dihydroxyacetone is the simplest ketose.

Example 14.1 **Why do we need *at least* 3 carbons for a monosaccharide?**

Solution **According to the definition (Sec. 14.2), monosaccharides are polyhydroxyaldehydes or polyhydroxyketones. *Poly-* means at least 2. Because the hydroxyl groups cannot be attached to the same carbon atom (review Sec. 9.8), we need at least 3 carbon atoms: 2 for the hydroxyl groups and 1 for the carbonyl group.**

The remaining aldoses or ketoses can be derived from glyceraldehyde or dihydroxyacetone by adding carbon atoms, each with a hydroxyl group.

$$
\begin{array}{cccccc}
^1CH{=}O & ^1CH{=}O & ^1CH{=}O & ^1CH_2OH & ^1CH_2OH & ^1CH_2OH \\
^2CHOH & ^2CHOH & ^2CHOH & ^2C{=}O & ^2C{=}O & ^2C{=}O \\
^3CHOH & ^3CHOH & ^3CHOH & ^3CHOH & ^3CHOH & ^3CHOH \\
^4CH_2OH & ^4CHOH & ^4CHOH & ^4CH_2OH & ^4CHOH & ^4CHOH \\
 & ^5CH_2OH & ^5CHOH & & ^5CH_2OH & ^5CHOH \\
 & & ^6CH_2OH & & & ^6CH_2OH \\
\end{array}
$$

| tetrose | pentose | hexose | | tetrose | pentose | hexose |

aldoses   ketoses

In aldoses, the chain is numbered consecutively from the aldehyde carbon. In most ketoses the carbonyl group is located at carbon-2.

14.4
CHIRALITY IN MONOSACCHARIDES

You will notice that glyceraldehyde has one chiral carbon atom (C-2), and hence can exist in two enantiomeric forms:

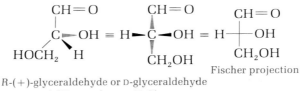

R-(+)-glyceraldehyde or D-glyceraldehyde
$[\alpha]_D^{25}$ +8.7 (c = 2, H_2O)

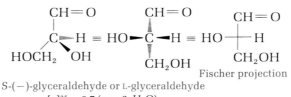

S-(−)-glyceraldehyde or L-glyceraldehyde
$[\alpha]_D^{25}$ −8.7 (c = 2, H_2O)

The dextrorotatory form has the R absolute configuration.

Before the R/S convention (Sec. 5.9) was developed, an older system for designating chirality was devised by carbohydrate chemists, and it is still used. Emil Fischer, a nineteenth-century German chemist who did much to elucidate the structures of carbohydrates, proposed what are now called **Fischer projection formulas.** The carbon chain is written vertically, with the most oxidized carbon at the top (with glyceraldehyde, this is the CH=O carbon) and the most reduced carbon at the bottom (with glyceraldehyde, this is the CH_2OH carbon). *Horizontal* lines in the formula show groups that project *above* the plane of the paper *toward* the viewer; *vertical* lines show groups that project *below* the plane of the paper *away* from the viewer. Note that in Fischer projection formulas, the symbol for the chiral carbon is omitted. It is this omission that lets you know that a Fischer projection formula, and not an ordinary structural formula, is meant.

The configuration of glyceraldehyde is then designated as D if the hydroxyl group on the chiral carbon is on the right in the Fischer projection formula, and L if the hydroxyl group is on the left. The system has been extended to other monosaccharides in the following way. If the chiral carbon *farthest* from the aldehyde or ketone group has the same configuration as D-glyceraldehyde (hydroxyl on the right), the compound is a D-monosaccharide. If the configuration at the remote carbon has the same configuration as L-glyceraldehyde (hydroxyl on the left), the compound is an L-monosaccharide.

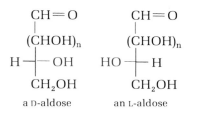

a D-aldose   an L-aldose

Table 14.1 shows the Fischer projection formulas for all of the D-aldoses through the hexoses. Starting with D-glyceraldehyde, one CHOH group at a time is inserted in the chain. This carbon, which adds a new chiral center to the structure, is shown in black. In each case, the new chiral center can have the hydroxyl group at the right or at the left in the Fischer projection formula (R or S absolute configuration).

Example 14.2 Using Table 14.1 as a guide, write the Fischer projection formula for L-erythrose.

Solution L-Erythrose is the enantiomer of D-erythrose. Consequently, its Fischer projection formula is

$$
\begin{array}{c}
\text{CH}=\text{O} \\
\text{HO}-\!\!\!-\text{H} \\
\text{HO}-\!\!\!-\text{H} \\
\text{CH}_2\text{OH}
\end{array}
$$

Example 14.3 Convert the Fischer projection formula for D-erythrose to a three-dimensional structural formula.

Solution Recalling that in Fischer projection formulas horizontal groups extend toward the viewer and vertical groups extend away from the viewer, we can write

$$
\begin{array}{c}
\text{CH}=\text{O} \\
\text{H}-\!\!\!-\text{OH} \\
\text{H}-\!\!\!-\text{OH} \\
\text{CH}_2\text{OH}
\end{array}
\quad=\quad
\begin{array}{c}
\text{CH}=\text{O} \\
\text{H}-\text{C}-\text{OH} \\
\text{H}-\text{C}-\text{OH} \\
\text{CH}_2\text{OH}
\end{array}
$$

D-erythrose

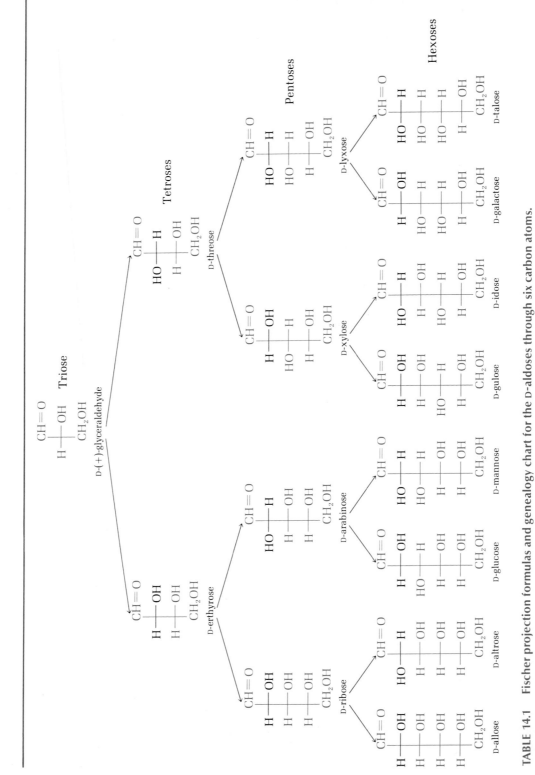

TABLE 14.1   Fischer projection formulas and genealogy chart for the D-aldoses through six carbon atoms.

The three-dimensional structure can be rewritten as

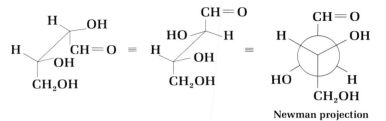

Newman projection

Molecular models may help you follow these interconversions.

Problem 14.1   Using Table 14.1 as a guide, write the Fischer projection formula for
a. L-threose   b. L-glucose

Problem 14.2   Convert the Fischer projection formula for D-threose to three-dimensional representations.

Problem 14.3   How many D-aldoheptoses are possible?

The term **epimers** is used to describe a pair of stereoisomers that differ in configuration at only one chiral center. For example, D-glucose and D-mannose are epimers (at C-2), and D-glucose and D-galactose are epimers (at C-4). Each pair has the same configurations at all chiral centers except one.

Problem 14.4   What pairs of D-pentoses are epimeric at C-3?

Finally, note that the designations D and L refer only to the configuration of the highest-numbered chiral carbon. They are not used for the other chiral centers.

Problem 14.5   Why is D-talose called a D-monosaccharide when 3 of the 4 hydroxyl groups at the chiral carbons are on the left?

The symbols D and L are not used to designate the sign of optical rotation. For example, D-glyceraldehyde and D-glucose are dextrorotatory (+), but D-erythrose and D-ribose are levorotatory (−).

**14.5
THE CYCLIC
HEMIACETAL
STRUCTURES OF
MONOSACCHARIDES**
Although the structures described so far for monosaccharides are consistent with much of their known chemistry, they are oversimplified. It is time to examine the true structures of these compounds. We learned earlier (Sec. 9.7) that alcohols react with aldehydes to form hemiacetals (eq. 9.5). Monosaccharides contain both of the functional groups needed for this reaction. It is not surprising, then, that the two groups should interact. Indeed, studies on simple hydroxyaldehydes in which a hydroxyl

group is on the fourth or fifth carbon in the chain show that the compounds exist preferably in the cyclic, hemiacetal form:

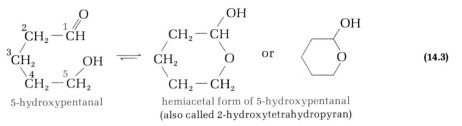

5-hydroxypentanal   hemiacetal form of 5-hydroxypentanal
(also called 2-hydroxytetrahydropyran)

$$(14.3)$$

Problem 14.6   Write a reasonable mechanism for eq. 14.3.

The same is true for monosaccharides. The following equation shows how the chain in D-glucose can be arranged so that the hydroxyl group on C-5 of the chain comes within reacting distance of the aldehyde carbon (C-1).

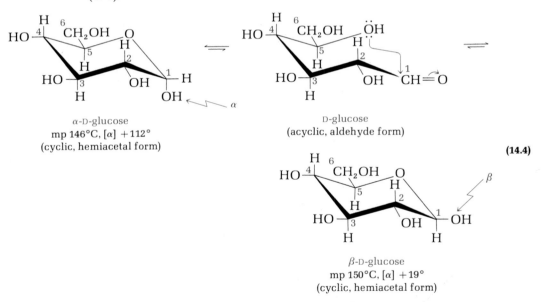

α-D-glucose
mp 146°C, [α] +112°
(cyclic, hemiacetal form)

D-glucose
(acyclic, aldehyde form)

$$(14.4)$$

β-D-glucose
mp 150°C, [α] +19°
(cyclic, hemiacetal form)

Reaction of the two groups gives the cyclic, six-membered ring hemiacetal forms of the monosaccharide. Note that C-1 in the cyclic structures is a hemiacetal carbon. It is connected to an OH group and an OR group.

In the acyclic, aldehyde form of glucose, C-1 is achiral, but this carbon is chiral in the cyclic structure. Consequently, two hemiacetal structures are possible, depending on the configuration at the new chiral center. The hemiacetal carbon, the carbon that forms the new chiral center, is called the **anomeric carbon.** Two monosaccharides that differ only at the anomeric center are called **anomers** (a special kind of epimers). Anomers are called α or β, depending on the position of the hydroxyl group. For monosaccharides in the D-series, the OH group is "down" in the α isomer and "up" in the β isomer (eq. 14.4). Note that in the cyclic forms of D-glucose the hydroxyl group at the anomeric carbon is axial in the α-anomer and equatorial in the β-anomer.

14.6
MUTAROTATION

How do we know that monosaccharides exist mainly as cyclic hemiacetals? There is direct physical evidence. For example, if D-glucose is crystallized from methanol, the pure α form is obtained, and x-ray analysis of the crystals shows that it has the structure depicted at the left of eq. 14.4. On the other hand, crystallization of glucose from acetic acid gives a different type of crystal, the β form. The α and β forms of D-glucose are diastereomers. They differ in configuration *only* at C-1. Being diastereomers, they have different physical properties, as shown under their structures in eq. 14.4.

The α and β forms of D-glucose interconvert in aqueous solution. For example, if crystalline α-D-glucose is dissolved in water, the specific rotation drops gradually from an initial value of $+112°$ to an equilibrium value of $+52°$. Starting with the pure crystalline β form results in a gradual change in specific rotation from an initial $+19°$ to the same equilibrium value of $+52°$. These changes in optical rotation are called **mutarotation.** The phenomenon is explained by the equilibria shown in eq. 14.4. Starting with either pure hemiacetal form, the ring can open to the acyclic aldehyde, which can then re-cyclize to give either the α or β form. Eventually an equilibrium mixture is obtained.

At equilibrium, an aqueous solution of D-glucose contains 36.4% of the α form and 63.6% of the β form. There is only about 0.003% of the open-chain aldehyde form present. Note that, in the chair conformation (eq. 14.4), in the predominant isomer, β-D-glucose, all of the substituents occupy equatorial positions.

Mutarotation occurs with all monosaccharides that exist in α and β forms.

Example 14.4

Show that the percentage of α- and β-D-glucose at equilibrium can be calculated from the specific rotation of the pure α and β forms and from the specific rotation of the solution at equilibrium.

Solution

The equilibrium rotation is $+52°$ and the rotations of pure α and β forms are $+112°$ and $+19°$, respectively. Assuming that no other forms are present at equilibrium, we can express these values graphically as follows:

$+112°$ $+52°$ $+19°$

├──────────────┼──────────────┤

100% α equilibrium   100% β

The percentage of the β form at equilibrium is then

$$\frac{112 - 52}{112 - 19} \times 100 = \frac{60}{93} \times 100 = 64.5\%$$

Problem 14.7   **α-D-Galactose and β-D-galactose have specific rotations of +151° and −53°, respectively. Each pure anomer mutarotates to an equilibrium rotation of +84°. Assuming that no other forms are present, calculate the percentage of each anomer present at equilibrium.**

14.7
CONVENTIONS
FOR WRITING
CYCLIC MONO-
SACCHARIDE
STRUCTURES

The cyclic structures of monosaccharides are commonly represented by three types of formulas. The conformational formulas used in eq. 14.4 give the closest representation of the true molecular geometry and shape, but they are a little difficult to draw.

Fischer projection formulas can be adapted to show cyclic structures as follows:

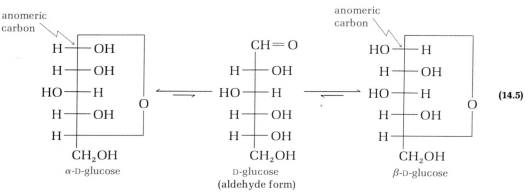

$$(14.5)$$

This equation represents the same structures as those shown in eq. 14.4. In the center we have the open-chain aldehyde form of D-glucose, as shown in Table 14.1. At the right and left we show that the hydroxyl at C-5 reacts with the carbonyl group at C-1, which becomes the hemiacetal or anomeric carbon. C-1 has the OH at the right in the α form and at the left in the β form, for D-monosaccharides. Although the Fischer convention is fairly easy to use for cyclic structures, it contains long, distorted, and awkwardly bent bonds connecting C-1 and C-5 through the ring oxygen atom.

Haworth formulas* use planar hexagons to represent the cyclic structures:

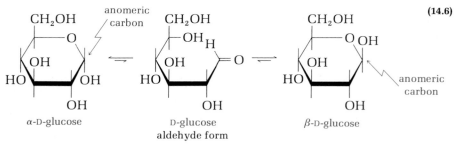

$$(14.6)$$

*First introduced in the 1920s by W. N. Haworth, British carbohydrate chemist at the University of Birmingham, who won the Nobel Prize in 1937.

The undesignated bonds in Haworth formulas are to hydrogen atoms, and they are sometimes omitted. The Haworth formulas clearly show the configuration at each chiral center, but they do not show conformations. They are somewhat of a compromise between the Fischer and the conformational formulas.

Example 14.5   Draw the Fischer and Haworth projection formulas for the six-membered cyclic structure of β-D-mannose.

Solution   Note in Table 14.1 that D-mannose differs from D-glucose *only* in the configuration at C-2. Consequently, we can use the formulas for β-D-glucose as a guide and change only the configuration at C-2.

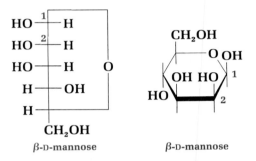

β-D-mannose   β-D-mannose

Problem 14.8   **Draw the Fischer and Haworth projection formulas for the six-membered cyclic structure of α-D-galactose.**

14.8 PYRANOSE AND FURANOSE STRUCTURES

The six-membered cyclic form of most monosaccharides is the preferred structure. These structures are called **pyranose** forms after the six-membered oxygen heterocycle known as **pyran.**

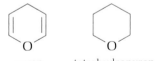

pyran   tetrahydropyran

Thus the formula at the extreme left of eq. 14.6 is more completely named **α-D-glucopyranose.**

In some cases, however, the cyclic hemiacetal may have a five-membered ring, in which case it is called a **furanose,** after the parent five-membered oxygen heterocycle **furan.**

furan   tetrahydrofuran

The ketose **D-fructose,** for example, exists in solution mainly in the furanose forms. Equation 14.7 shows these furanose forms, using the Haworth convention.

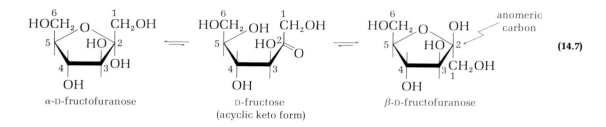

$$\text{(14.7)}$$

Example 14.6 Draw the Fischer projection formulas for the open-chain and furanose forms of D-fructose.

Solution Compare the structure with that of D-glucose, noting that the configurations are identical at C-3, C-4, and C-5. C-1, C-2, and C-6 in fructose are achiral in the acyclic form, but C-2 becomes chiral in the cyclic forms.

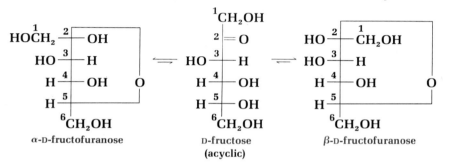

14.9
REACTIONS OF
MONOSACCHARIDES

14.9a Oxidation to Aldonic and Aldaric (Saccharic) Acids Although aldoses exist primarily in cyclic, hemiacetal forms, these cyclic structures are in equilibrium with a small but finite amount of the open-chain aldehyde form. It is not surprising, then, that these aldehyde groups can be easily oxidized to acids (Sec. 9.13). For example, D-glucose is easily oxidized to D-gluconic acid.

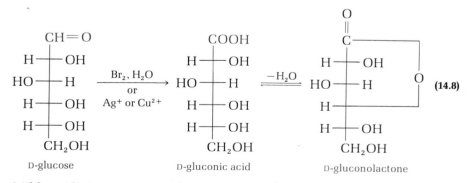

$$\text{(14.8)}$$

Mild oxidizing agents such as aqueous bromine are sufficient. Being γ-hydroxyacids, these **aldonic acids** are usually in equilibrium with their cyclic esters, or lactones (compare eq. 14.8 and eq. 11.11).

The oxidation of aldoses to aldonic acids is so easy that such mild oxidizing agents as Ag^{1+} and Cu^{2+} will work. Therefore aldoses (such as glucose) give positive tests with Tollens', Fehling's, and Benedict's reagents (Sec. 9.13).

Problem 14.9 Write an equation for the reaction of D-mannose (Table 14.1) with Fehling's reagent (Cu^{2+}) to give D-mannonic acid (use eq. 9.34 as a guide).

Stronger oxidizing agents, such as aqueous nitric acid, convert the aldehyde group *and* the primary alcohol group to carboxylic acids called **aldaric acids.** For example, D-glucose gives D-glucaric acid.

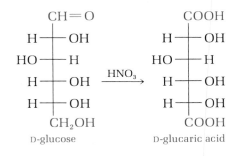

(14.9)

D-glucose   D-glucaric acid

Problem 14.10 D-Glucaric acid is in equilibrium (through loss of water) with two different lactones. What are their structures?

14.9b Reduction to Alditols The aldehyde group of aldoses and the keto group of ketoses can be reduced by various reagents (review Sec. 9.12). The products are **polyols,** called in general **alditols.** For example, catalytic hydrogenation or reduction with sodium borohydride ($NaBH_4$) converts D-glucose to D-glucitol.

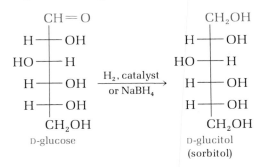

(14.10)

D-glucose   D-glucitol (sorbitol)

D-glucitol (an older name is sorbitol) is used as a sweetening agent in foods for diabetics.

Problem 14.11 D-Mannitol occurs naturally in olives, onions, and mushrooms. It can be made by $NaBH_4$ reduction of D-mannose (Table 14.1). Draw its structure.

14.9c Esterification Monosaccharides contain hydroxyl groups. It is not surprising, then, that they undergo some reactions typical of alcohols. For example, they can be converted to esters by reaction with acid derivatives. The conversion of β-D-glucose to its pentaacetate with acetic anhydride is typical.

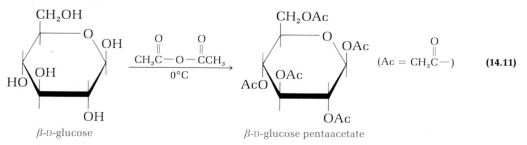

$$(Ac = CH_3\overset{O}{\overset{\|}{C}}-) \qquad \textbf{(14.11)}$$

Note that the hemiacetal hydroxyl at C-1 and all the other hydroxyls are esterified. This same type of reaction is used to convert cellulose to cellulose acetate for the manufacture of rayon (Sec. 14.12b).

14.10 FORMATION OF GLYCOSIDES (ACETALS) FROM MONO-SACCHARIDES

Because monosaccharides exist mainly in cyclic, hemiacetal structures, they react with one equivalent of an alcohol to form acetals (review Sec. 9.7). An example is the reaction of β-D-glucose with methanol.

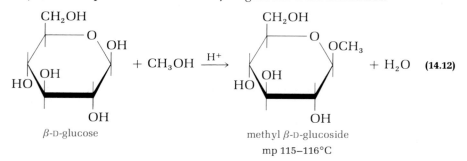

Note that only the anomeric OH is replaced by an OR group. Such acetals are called **glycosides,** and the bond from the anomeric carbon to the OR group is called the **glycosidic bond.** Particular glycosides are named from the corresponding monosaccharide by changing the -e ending to -ide. Thus glucose (eq. 14.12) gives glucosides, mannose gives mannosides, and so on.

Example 14.7   Write the Haworth projection formula for ethyl α-D-mannoside.

Solution

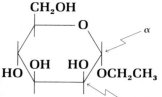

mannose differs from glucose in the configuration at C-2

Problem 14.12 **Write an equation for the acid-catalyzed reaction of β-D-galactose (Table 14.1) with methanol.**

Naturally occurring alcohols or phenols often occur in cells combined as a glycoside with some sugar, most commonly glucose. The reason is that the many hydroxyl groups of the sugar portion of the glycoside help make soluble certain compounds that would otherwise be incompatible with cellular protoplasm. An example is the bitter-tasting glucoside **sali-cin,** which occurs in willow bark, and whose fever-reducing power was known to the ancients.

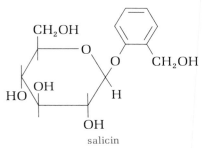

salicin
(the β-D-glucoside of salicyl alcohol)

Example 14.8 **What products would be expected from the hydrolysis of salicin?**

Solution **The acetal bond at C-1 is hydrolyzed (reverse of eq. 9.10). The products are**

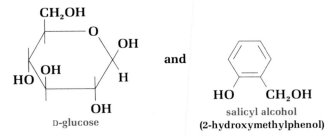

D-glucose

and

salicyl alcohol
(2-hydroxymethylphenol)

Note the close structural relationship between salicyl alcohol and aspirin (sec. 11.7).

We shall now see that the glycosidic bond is the key to understanding the structure of oligosaccharides and polysaccharides.

14.11 DISACCHARIDES The most common oligosaccharides are disaccharides. In a disaccharide, two monosaccharides are linked together by a glycosidic bond between the anomeric carbon of one monosaccharide unit and a hydroxyl group on the other unit. In this section we will describe the structure and properties of four important disaccharides.

14.11a Maltose **Maltose** is a disaccharide obtained by the partial hydrolysis of starch. Further hydrolysis of maltose gives only D-glucose. Hence

maltose must consist of two linked glucose units. The anomeric carbon
of one unit is linked to the C-4 hydroxyl group of the other unit. The configuration at the anomeric carbon of the first unit is α. In the crystalline
form the second unit has the β configuration. Both units are in the pyranose
form.

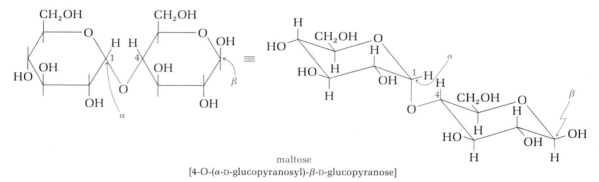

maltose
[4-O-(α-D-glucopyranosyl)-β-D-glucopyranose]

The systematic name for maltose describes the structure fully, including
the name of each unit (D-glucose), the ring sizes (pyranose), the configuration at each anomeric carbon (α or β), and the location of the hydroxyl
group involved in the glycosidic link (4-O).

Note that the anomeric carbon of the second glucose unit in maltose
is a hemiacetal. Naturally this hemiacetal function will be in equilibrium
with the open-chain aldehyde form. Consequently maltose gives a positive
Tollens' test and other reactions similar to those of the anomeric carbon
in glucose.

Problem 14.13   **When crystalline maltose is dissolved in water, the initial specific rotation
changes and gradually reaches an equilibrium value. Explain.**

14.11b Cellobiose **Cellobiose** is a disaccharide obtained via the partial hy-
drolysis of cellulose. Further hydrolysis of cellobiose gives only D-glucose.
Hence cellobiose must be an isomer of maltose. In fact, it differs from
maltose only in having the β configuration at C-1 of the first glucose unit.
Otherwise all other structural features are identical, including a link from
C-1 of the first unit to the hydroxyl group at C-4 in the second unit.

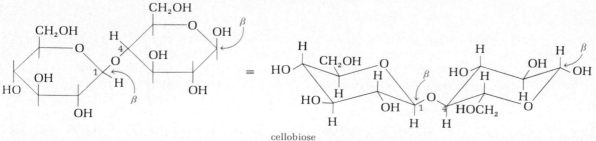

cellobiose
[4-O(β-D-glucopyranosyl)-β-D-glucopyranose]

Note that in the conformational structure of cellobiose we have drawn one ring-oxygen to the 'rear' and one to the 'front' of the molecule. This is the way the rings exist in the cellulose chain (Sec. 14.12b).

Problem 14.14 Draw a conformational structure of cellobiose in which the right hand ring is rotated 180° about the C-4-to-O bond, to show that this structure is in fact equivalent to the Haworth structure at the left, and that the configuration of the hydroxyl group at C-1 of the second glucose unit is β.

14.11c Lactose **Lactose** is the major sugar found in human and cow's milk (4–8% lactose). Hydrolysis of lactose gives equimolar amounts of D-galactose and D-glucose. The anomeric carbon of the galactose unit has the β configuration at C-1 and is linked to the hydroxyl group at C-4 of the glucose unit. The crystalline α anomer (at the glucose unit), made commercially from cheese whey, is shown.

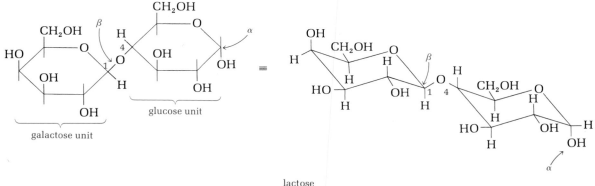

lactose
[4-O-(β-D-galactopyranosyl)-α-D-glucopyranose]

Problem 14.15 **Will lactose give a positive Fehling's test? Will it mutarotate?**

Some human infants inherit a disease called *galactosemia*. They lack an enzyme that isomerizes galactose to glucose and hence cannot digest milk. If milk is excluded from the infant's diet, the disease symptoms caused by accumulation of galactose in the tissues can be avoided.

14.11d Sucrose Perhaps the most important of all disaccharides is **sucrose**, ordinary table sugar. Sucrose occurs in all photosynthetic plants, where it functions as an energy source. It is obtained commercially from sugar cane and from sugar beets, where it constitutes 14–20% of the plant juices.

Hydrolysis of sucrose gives equimolar amounts of D-glucose and the keto sugar D-fructose. Sucrose differs from the disaccharides we have discussed before in that the anomeric carbons of *both* units are involved in the glycosidic link. That is, C-1 of the glucose unit is linked, via oxygen, to C-2 of the fructose unit. A further difference is that the fructose unit is present in the furanose form.

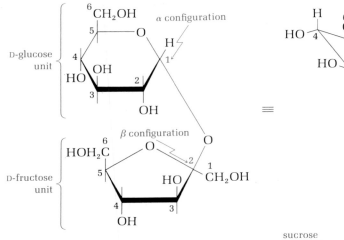

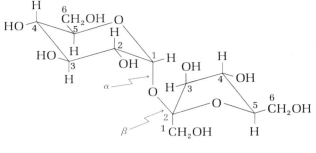

sucrose
α-D-glucopyranosyl-β-D-fructofuranoside
(or β-D-fructofuranosyl-α-D-glucopyranoside)

Both anomeric carbons are linked, so no hemiacetal group remains in either monosaccharide unit. Therefore neither unit of sucrose is in equilibrium with an acyclic form. Sucrose cannot mutarotate. And, because there is no free or potentially free aldehyde group, sucrose cannot reduce Tollens', Fehling's, or Benedict's reagent. Sucrose is therefore referred to as a *nonreducing sugar*, in contrast with the other disaccharides and monosaccharides we have discussed, which are reducing sugars.

Sucrose has an optical rotation $[\alpha] = +66°$. When sucrose is hydrolyzed to an equimolar mixture of D-glucose and D-fructose, the optical rotation changes sign and becomes $[\alpha] = -20°$. This is because the equilibrium mixture of D-glucose anomers (α and β) has a rotation of $+52°$, but the mixture of fructose anomers has a strong negative rotation, $[\alpha] = -92°$. In the early days of carbohydrate chemistry, glucose was called **dextrose** (because it was dextrorotatory), and fructose was called **levulose** (because it was levorotatory). Because hydrolysis of sucrose changes or inverts the sign of optical rotation, enzymes that bring about that hydrolysis are called **invertases,** and the resulting equimolar mixture of glucose and fructose is called *invert sugar*. A number of insects, including the honeybee, contain invertases. Honey, then, is largely a mixture of D-glucose, D-fructose, and sucrose. It also contains flavors from the particular flowers whose nectars are collected.

a word about ————————————————————————

27. Sweetness and Sweeteners

Sweetness is literally a matter of taste. Although individuals vary greatly in their sensory perceptions,

it is possible to make some quantitative comparisons. For example, we can take some standard sugar solution (say 10% sucrose in water) and compare its sweetness with that of solutions containing other sugars or sweetening agents. If only a 1% solution of compound X compares in sweetness with the 10% sucrose solution, we can say that compound X is 10 times sweeter than sucrose.

D-Fructose is the sweetest of the simple sugars — almost twice as sweet as sucrose. D-Glucose is almost as sweet as sucrose. On the other hand, many so-called sugars (such as lactose and galactose) have less than 1% of the sweetness of sucrose.

Many synthetic sweeteners are known, perhaps the most familiar being saccharin, which was discovered in 1879 in the laboratory of Professor Ira Remsen at the Johns Hopkins University. Although its structure has no relation whatever to that of saccharides, saccharin is perhaps 300 times sweeter than sucrose. For most tastes, 0.5 grain (0.03 g) of saccharin is equivalent in sweetness to a heaping teaspoon (10 g) of sucrose. Saccharin is made commercially from toluene (shown below in Figure 14.1). Saccharin is very sweet, yet it has virtually no calorific content. It is useful as a sugar substitute for diabetics and others who are forced to restrict their sugar intake and also for those who wish to control their weight yet have a desire for sweets. Unfortunately, experiments with mice have yielded some evidence that in very large doses it is a carcinogen.

Other common synthetic sweeteners are calcium cyclamate (sucaryl), Dulcin (sucrol), and aspartame. Cyclamates were discovered at the University of Illinois in 1937 and are about 30 times sweeter than cane sugar. They have the advantage of lacking the "aftertaste" that offends some saccharin users. Unfortunately, there is some evidence that cyclamates may produce cancers, and their use in the United

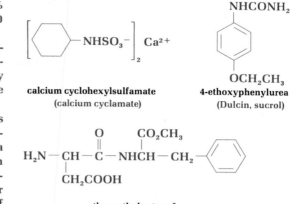

calcium cyclohexylsulfamate
(calcium cyclamate)

4-ethoxyphenylurea
(Dulcin, sucrol)

the methyl ester of
N-L-α-aspartyl-L-phenylalanine
(aspartame)

States has been banned since 1970. Dulcin is about 100 times sweeter than sucrose but it is too toxic for use in foods. In 1981, aspartame became the first new sweetener to be approved by the Food and Drug Administration in nearly 25 years. It is about 160 times sweeter than sucrose. Structurally, aspartame is the methyl ester of a dipeptide of two amino acids that occur naturally in proteins, aspartic acid, and phenylalanine (Table 15.1), so it is expected to be quite safe to use in reasonable amounts.

Note the considerable structural variation in sugars and artificial sweeteners — yet they all have the property of sweet taste. Existing theories about what chemical structures are necessary for sweetness are still incomplete and inconclusive.

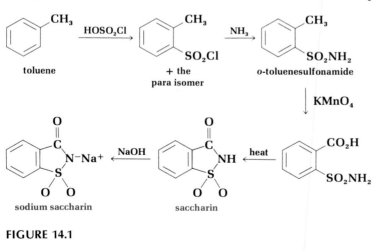

FIGURE 14.1

14.12
POLYSACCHARIDES
Polysaccharides vary in chain length and molecular weight. The mono-saccharide units may be linked in a linear manner, or the chains may be branched. Most polysaccharides give a single monosaccharide on complete hydrolysis, but there are exceptions to this rule. In this section we will describe a few of the more important polysaccharides.

14.12a Starch and Glycogen **Starch** is the energy-reserve carbohydrate of plants. It constitutes large percentages of cereals, potatoes, corn, and rice. It is the form in which glucose is stored by plants for later use.

Starch is made up of glucose units joined mainly by 1,4-α-glycosidic bonds, although the chains may contain a number of branches due to 1,6-α-glycosidic bonds. Partial hydrolysis of starch gives maltose, and complete hydrolysis gives only D-glucose.

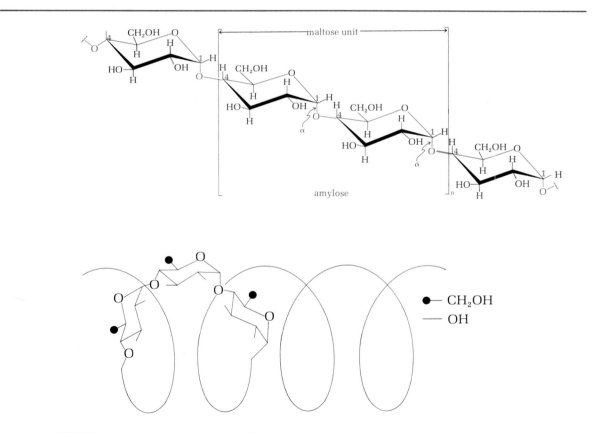

FIGURE 14.2   Structure of the amylose fraction of starch. (Adapted from R. J. Ferrier and P. M. Collins, *Monosaccharide Chemistry,* Penguin Books, Ltd., England.)

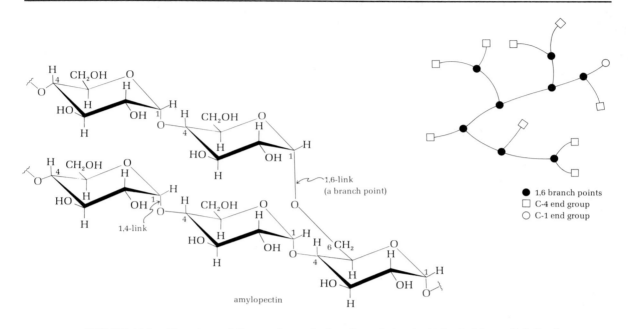

FIGURE 14.3   Structure of the amylopectin fraction of starch. (Adapted from R. J. Ferrier and P. M. Collins, *Monosaccharide Chemistry*, Penguin Books, Ltd., England.)

Starch can be separated by various solution and precipitation techniques into two fractions, amylose and amylopectin. In **amylose,** which constitutes about 20% of starch, the glucose units (50–300) are in a continuous chain, linked 1,4 (Figure 14.2). In solution, the chain adopts a helical shape because of the α configuration at each glucoside link. The tubular shape, with about 6 glucose units per turn of the helix, permits amylose to form complexes with various small molecules that fit within the coil. The deep blue color that starch gives with iodine is due to such a complex.

Amylopectin (Figure 14.3) is highly branched. Although each molecule may contain as many as 300–5000 glucose units, chains with consecutive 1,4-links average only 25–30 units in length. These chains are connected at branch points by 1,6-linkages. No doubt it is because of this highly branched structure that starch granules swell and eventually form colloidal solutions in water.

Glycogen is the reserve carbohydrate of animals. Like starch, it is made of 1,4- and 1,6-linked glucose units. Glycogen has a high molecular weight (perhaps 100,000 glucose units). Its structure is even more highly branched than that of amylopectin, with a branch every 8–12 glucose units. Glycogen is produced when glucose is absorbed from the intestines into the

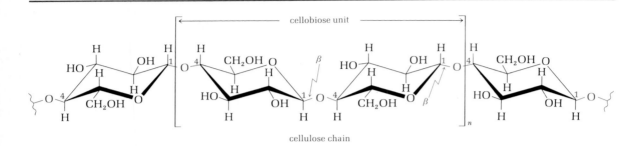

FIGURE 14.4   Partial structure of a cellulose molecule showing the β-linkage of glucose units.

blood and transported to the liver, muscles, and elsewhere, and then polymerized enzymatically. Glycogen helps maintain the glucose balance of the body by removing and storing excess glucose derived from ingested food and by supplying it to the blood when body cells need it for energy.

14.12b Cellulose **Cellulose** is an unbranched polymer of glucose joined by 1,4-β-glycosidic bonds. X-ray examination of cellulose shows that it consists of linear chains of cellobiose units, in which the ring oxygens alternate in "forward" and "backward" positions (Figure 14.4). These linear molecules, with an average of 5000 glucose units, aggregate to give fibrils bound together by hydrogen bonds between hydroxyls on adjacent chains. Cellulose fibers with considerable physical strength are then built up from these fibrils, wound spirally in opposite directions around a central axis. Wood, cotton, hemp, linen, straw, and corn cobs are mainly cellulose.

Although humans and other animals can digest starch and glycogen, they cannot digest cellulose. This is a truly striking example of the specificity of biochemical reactions. The only chemical difference between starch and cellulose is the stereochemistry of the glucosidic link—more precisely, the stereochemistry at C-1 of each glucose unit. Our digestive systems contain enzymes that catalyze the hydrolysis of α-glucosidic bonds but lack the enzymes (β-glucosidases) necessary to hydrolyze β glucosidic bonds. Many bacteria, however, do contain β-glucosidases and can hydrolyze cellulose. Termites have such bacteria in their intestines and thrive on wood (cellulose) as their main food. Ruminants (cud-chewing animals such as cows) can digest grasses and other forms of cellulose only because they harbor the necessary microorganisms in their rumen.

Cellulose is the raw material for several commercially important derivatives. Note from its structure (Figure 14.4) that each glucose unit in cellulose contains 3 hydroxyl groups. These hydroxyl groups can be modified by the usual reagents that react with alcohols. For example, reaction of cellulose with acetic anhydride gives **cellulose acetate:**

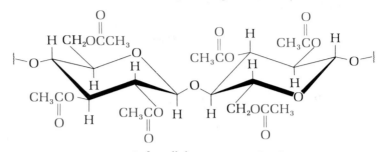

segment of a cellulose acetate molecule

Cellulose with about 97% of the hydroxyl groups acetylated is used to make acetate rayon. A viscous solution of cellulose acetate in acetone is forced through fine openings in a current of warm air that evaporates the acetone. The resulting long fibers can be spun and woven. Cellulose in which only 80–85% of the hydroxyls are acetylated is cast from dichloromethane solutions into films that are used in making motion pictures.

Cellulose nitrate is another useful cellulose derivative. Like glycerol (eq. 7.31), cellulose can be converted with nitric acid to a nitrate ester. The number of hydroxyl groups nitrated per glucose unit determines the properties of the product. Guncotton, a highly nitrated cellulose, is an efficient explosive used in smokeless powders.

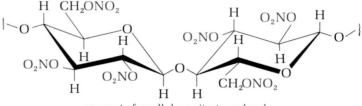

segment of a cellulose nitrate molecule

14.12c Other Polysaccharides Chitin is a nitrogen-containing polysaccharide that forms the shells of crustacea and the exoskeletons of insects. It is similar to cellulose, except that the hydroxyl group at C-2 of each glucose unit is replaced by the acetylamino group CH_3CONH-. **Pectins,** which are obtained from fruits and berries, are polysaccharides used in making jellies. They are linear polymers of D-galacturonic acid linked with 1,4-α-glycosidic bonds. D-Galacturonic acid has the same structure as D-galactose except that the C-6 primary hydroxyl group has been oxidized to a carboxyl group.

Numerous other polysaccharides are known, including gum arabic and other gums and mucilages; chondroitin sulfate in cartilage; heparin, which is found in the liver and heart and is a blood coagulant; the dextrans, used as blood plasma substitutes; and many others.

14.13 SACCHARIDES WITH MODIFIED STRUCTURES

Some saccharides have structures that differ somewhat from the usual polyhydroxyaldehyde or polyhydroxyketone pattern. In this final section we describe a few such modified saccharides that are important in nature.

14.13a Sugar Phosphates Phosphate esters of monosaccharides are found in all living cells, where they act as intermediates in carbohydrate metabolism. Some of the more common **sugar phosphates** are

D-glyceraldehyde-3-phosphate

dihydroxyacetone-phosphate

α-D-glucose-6-phosphate

α-D-fructose-6-phosphate

α-D-fructose-1,6-diphosphate

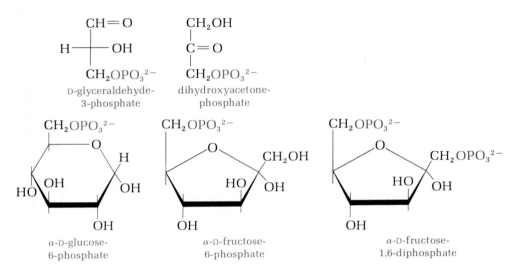

Sugar phosphates are also involved in DNA and RNA structure (Chapter 16).

14.13b Deoxy Sugars In **deoxy sugars,** one or more of the hydroxyl groups are replaced by a hydrogen atom. The most important example is **2-deoxyribose,** the sugar component of DNA. It lacks the hydroxyl group at C-2 and occurs in DNA in the furanose form:

β-D-deoxyribofuranose
(the sugar of DNA)

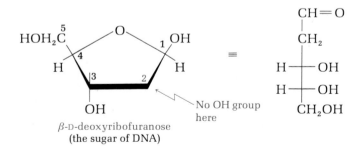

14.13c Amino Sugars In **amino sugars,** a hydroxyl group at some position is replaced by an amino group. In most natural amino sugars the $-NH_2$ group is acetylated. **D-Glucosamine** is one of the more abundant of these amino sugars.

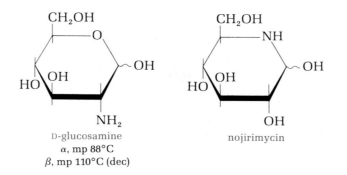

D-glucosamine
α, mp 88°C
β, mp 110°C (dec)

nojirimycin

In its N-acetyl form, D-glucosamine is the monosaccharide unit of chitin, which forms the shells of lobsters, crabs, and shrimp, and other shellfish. Amino sugars are also part of antibiotic structures such as streptomycin and erythromycin. At least one simple amino sugar, nojirimycin, is itself an antibiotic.

14.13d Ascorbic Acid (Vitamin C) **L-Ascorbic acid (vitamin C)** has a structure resembling that of a monosaccharide, but the structure has several unusual features. The compound has a five-membered unsaturated lactone ring with two hydroxyl groups on the double bond. This enediol structure is relatively uncommon.

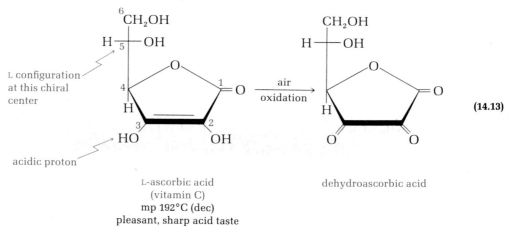

L-ascorbic acid
(vitamin C)
mp 192°C (dec)
pleasant, sharp acid taste

dehydroascorbic acid

(14.13)

As a consequence of this group, ascorbic acid is easily oxidized (eq. 14.13) to dehydroascorbic acid. Both forms are biologically effective as a vitamin.

There is no carboxyl group in ascorbic acid, but it is in fact an acid. The hydroxyl at C-3 is strongly acidic, because the anion resulting from proton loss is resonance-stabilized and similar to a carboxylate anion.

Humans, monkeys, guinea pigs, and a few other vertebrates lack an enzyme essential for the biosynthesis of ascorbic acid from D-glucose, though most other higher animals and plants possess the required enzyme. Hence ascorbic acid must be included in the diet of humans and other species unable to synthesize it. Ascorbic acid is abundant in citrus fruits and tomatoes. Lack of ascorbic acid in the diet causes scurvy, a terrible disease that results in weak blood vessels, hemorrhaging, loosening of teeth, lack of ability to heal wounds, and eventual death. Ascorbic acid is probably essential for collagen synthesis (collagen is the structural protein of skin, connective tissue, tendon, cartilage, and bone). In the eighteenth century, British sailors were required to eat fresh limes to prevent outbreaks of the dread scurvy; hence the nickname "limeys" for the British.

ADDITIONAL PROBLEMS

14.16. Define and give the structural formula for an example of each of the following.

a. aldohexose	**b.** ketopentose
c. monosaccharide	**d.** disaccharide
e. polysaccharide	**f.** furanose
g. pyranose	**h.** glycoside
i. anomeric carbon	**j.** reducing sugar

14.17. Explain, using formulas, the difference between a D-sugar and an L-sugar.

14.18. What is the absolute configuration (R or S) of the chiral centers at C-2 and C-3 of D-erythrose. (Use Table 14.1 as a guide for this and other problems if you have not yet memorized the structures of the monosaccharides).

14.19. What is the absolute chirality (R or S) at each chiral center in the acyclic form of D-glucose? at the new chiral center in β-D-glucose?

14.20. What term would you use to describe the stereochemical relationship between D-gulose and D-idose? (See Table 14.1.)

14.21. Construct a table analogous to Table 14.1 for the D-ketoses through the ketohexoses. Dihydroxyacetone should be at the head of the table, in place of glyceraldehyde.

14.22 Using Table 14.1 if necessary, write Fischer projection formulas and Haworth formulas for:

a. methyl α-D-glucopyranoside   **b.** α-D-gulopyranose

c. β-D-arabinofuranose   **d.** methyl α-L-glucopyranoside

14.23. Draw the Fischer projection formulas for:

a. L-mannose   **b.** L-(+)-fructose

14.24. At equilibrium in aqueous solution, D-ribose exists as a mixture containing 20% α-pyranose, 56% β-pyranose, 6% α-furanose, and 18% β-furanose forms. Draw Haworth projection formulas for each of these forms.

14.25. Write Fischer, Haworth, and conformational structures for β-D-allose.

14.26. The solubilities of α- and β-D-glucose in water at 25°C are 82 and 178 g/100 mL, respectively. Why are their solubilities not identical?

14.27. The specific rotations of pure α- and β-D-fructofuranose are +21° and −133°, respectively. Solutions of each isomer mutarotate to an equilibrium specific rotation of −92°. Assuming that no other forms are present, calculate the equilibrium concentrations of the two forms.

14.28. Starting with β-D-glucose and using acid (H⁺) as a catalyst, write out all the steps in the mechanism for the mutarotation process. Use Haworth projection formulas for the cyclic structures.

14.29. Oxidation of either D-erythrose or D-threose with nitric acid gives tartaric acid. In one case, the tartaric acid is optically active; in the other, it is optically inactive. How can these facts be used to assign stereo structures to erythrose and threose?

14.30. Write structures for:
a. D-galactonic acid   **b.** D-galactaric acid

14.31. Write structures for the reaction of D-mannose with:
a. bromine water   **b.** nitric acid
c. sodium borohydride   **d.** acetic anhydride

14.32. Reduction of D-fructose gives a mixture of D-glucitol and D-mannitol. What does this result prove about the configurations of D-fructose, D-mannose, and D-glucose?

14.33. Write equations that show all the steps in the mechanism for eq. 14.12.

14.34. Emil Fischer carried out the acid-catalyzed reaction of D-glucose ($C_6H_{12}O_6$) with methanol, expecting to get a product that would analyze for $C_8H_{18}O_7$. Instead he obtained two products, both of which analyzed for $C_7H_{14}O_6$. Explain the difference between his expectations and his results.

14.35. Although D-galactose contains five chiral centers, its oxidation with nitric acid gives an optically inactive dicarboxylic acid (called galactaric or mucic acid). What is the structure of this acid and why is it optically inactive?

14.36. Write equations that clearly show the mechanism for the acid-catalyzed hydrolysis of:
a. maltose to glucose
b. lactose to galactose and glucose
c. sucrose to fructose and glucose

14.37. Write equations for the reaction of maltose with:
a. methanol and H⁺   **b.** Tollens' reagent
c. bromine water   **d.** acetic anhydride

14.38. Trehalose is a disaccharide that is the main carbohydrate component in the blood of insects. Its structure is

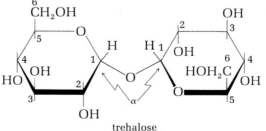

trehalose

a. What are its hydrolysis products?
b. Will trehalose give a positive or a negative test with Fehling's reagent? Explain.

14.39. Lactose exists in α and β forms, with specific rotations of $+92.6°$ and $+34°$, respectively.
a. Draw their structures.
b. Solutions of each isomer mutarotate to an equilibrium value of $+52°$. What is the concentration of each isomer at equilibrium?

14.40. Write a balanced equation for the reaction of $D(+)$-glucose (use either an acyclic or a cyclic structure, whichever seems most appropriate) with each of the following.
a. acetic anhydride (excess) **b.** bromine water
c. hydrogen, catalyst **d.** hydroxylamine to form an oxime
e. methanol, H^+ **f.** hydrogen cyanide, to form a cyanohydrin
g. Fehling's reagent

14.41. Explain why sucrose is a nonreducing sugar, whereas maltose is a reducing sugar.

14.42. Given the descriptions in Sec. 14.12c, write formulas for:
a. chitin **b.** pectin

14.43. L-Fucose is a component of bacterial cell walls. It is also called 6-deoxy-L-galactose. Write its Fischer projection formula.

14.44. Write the main contributors to the resonance hybrid anion formed when ascorbic acid acts as an acid (loss of the proton from the OH at C-3).

14.45. Hemicelluloses are noncellulose materials produced by plants and found in straw, wood, and other fibrous tissues. Xylans are the most abundant hemicelluloses. They consist of $1,4$-β-linked D-xylopyranose. Draw the structure for the repeating unit in xylans.

14.46. Inositols are hexahydroxycyclohexanes, with one hydroxyl group on each carbon atom of the ring. Although not strictly carbohydrates, they are obviously similar to pyranose sugars, and they do occur in nature. There are nine possible isomers. Using Haworth projections, draw all possibilities (all are known), and tell which are chiral.

fifteen

Amino Acids, Peptides, and Proteins

**15.1
INTRODUCTION**

The word *protein* comes from the Greek word *protos,* meaning first or of prime importance. Indeed, proteins are of prime importance to the structure, function, and reproduction of living matter.

Proteins are natural polymers composed of **amino acid** units joined one to another by amide (or peptide) bonds. In this chapter we will first discuss the structure and properties of amino acids. Then we will describe the properties of **peptides,** which have a few amino acids linked together, and proteins, which have many such units.

**15.2
NATURALLY
OCCURRING
AMINO ACIDS**

The amino acids obtained when a protein is hydrolyzed are **α-amino acids.** That is, the amino group is on the carbon atom adjacent or α to the carboxyl group:

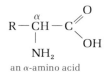

an α-amino acid

With the exception of glycine, where R=H, the α carbon is a chiral center. Hence all the amino acids derived from proteins (except glycine) are optically active. They have the L- configuration relative to glyceraldehyde, as shown in Figure 15.1. Note that the Fischer conventions used with carbohydrates are also applied to the amino acids. Table 15.1 lists the twenty α-amino acids commonly found in proteins.

The amino acids are known by their common names. Each also has a three-letter abbreviation used when writing the formulas of peptides or proteins. The amino acids are grouped in Table 15.1 to emphasize structural similarities. Of the 20 amino acids listed in the table and present in proteins, 12 can be synthesized from other foods. However, 8 amino acids (abbreviations shown in color) cannot be synthesized by adult humans and must be included (via proteins) in the diet. For this reason, they are

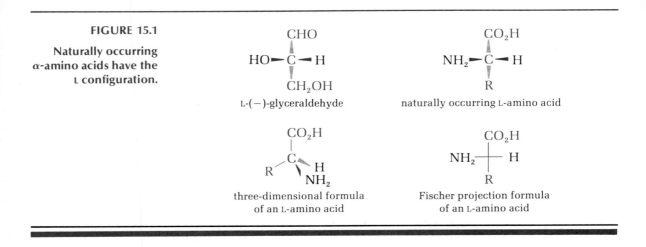

FIGURE 15.1

Naturally occurring
α-amino acids have the
L configuration.

L-(−)-glyceraldehyde

naturally occurring L-amino acid

three-dimensional formula
of an L-amino acid

Fischer projection formula
of an L-amino acid

referred to as **essential amino acids.** Of course, all 20 amino acids are necessary for growth, development, and maintenance of a healthy body.

**15.3
THE ACID–BASE
PROPERTIES OF
AMINO ACIDS**

So far we have studied the acidity of carboxylic acids (Chapter 10) and the basicity of amines (Chapter 12) separately. Both are *simultaneously* present in amino acids, and we might well ask whether they are mutually compatible. Although we have represented amino acids in Table 15.1 as having amino and carboxyl groups, we will now see that those structures are an oversimplification.

Amino acids with one amino group and one carboxyl group are best represented by a **dipolar ion structure.** *

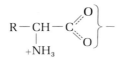

dipolar structure of an α-amino acid

That is, the amino group is present as an ammonium ion and the carboxyl group is present as a carboxylate ion. This dipolar ion structure is consistent with the rather high melting points of amino acids and with their relatively low solubilities in organic solvents. As we will see, the dipolar structure is also consistent with the electrical properties of amino acids at various pH values.

Amino acids are **amphoteric.** That is, they can behave as acids and do-

*Such structures are sometimes called zwitterions (from a German word that literally means "double ions").

TABLE 15.1   Names and formulas of the common amino acids

Name	Abbreviation	Formula	R
A. One amino group and one carboxyl group			

A. One amino group and one carboxyl group

1. glycine — Gly

$$H-CH-CO_2H$$
$$| $$
$$NH_2$$

2. alanine — Ala

$$CH_3-CH-CO_2H$$
$$|$$
$$NH_2$$

3. valine — Val

$$CH_3CH-CH-CO_2H$$
$$|\quad\quad|$$
$$CH_3\ \ NH_2$$

4. leucine — Leu

$$CH_3CHCH_2-CH-CO_2H$$
$$|\quad\quad\quad|$$
$$CH_3\quad\quad NH_2$$

5. isoleucine — Ile

$$CH_3CH_2CH-CH-CO_2H$$
$$|\quad\quad|$$
$$CH_3\ \ NH_2$$

R = H or alkyl

6. serine — Ser

$$CH_2-CH-CO_2H$$
$$|\quad\quad|$$
$$OH\quad NH_2$$

7. threonine — Thr

$$CH_3CH-CH-CO_2H$$
$$|\quad\quad|$$
$$OH\quad NH_2$$

R contains an alcohol function

8. cysteine — Cys

$$CH_2-CH-CO_2H$$
$$|\quad\quad|$$
$$SH\quad NH_2$$

9. methionine — Met

$$CH_3S-CH_2CH_2-CH-CO_2H$$
$$|$$
$$NH_2$$

Two sulfur-containing amino acids

10. proline — Pro

$$CH_2-CH-CO_2H$$
$$|\quad\quad|$$
$$CH_2\quad NH$$
$$\diagdown\quad\diagup$$
$$CH_2$$

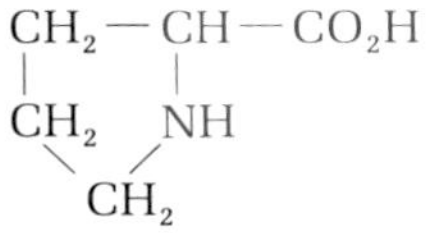

The amino group is secondary and in a ring

11. phenylalanine — Phe

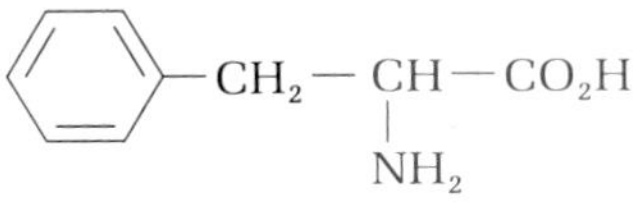

$$\text{C}_6\text{H}_5-CH_2-CH-CO_2H$$
$$|$$
$$NH_2$$

12. tyrosine — Tyr

$$HO-\text{C}_6\text{H}_4-CH_2-CH-CO_2H$$
$$|$$
$$NH_2$$

13. tryptophan — Trp

$$\text{(indole)}-CH_2-CH-CO_2H$$
$$|$$
$$NH_2$$

One hydrogen in alanine is replaced by an aromatic or heteroaromatic (indole) ring

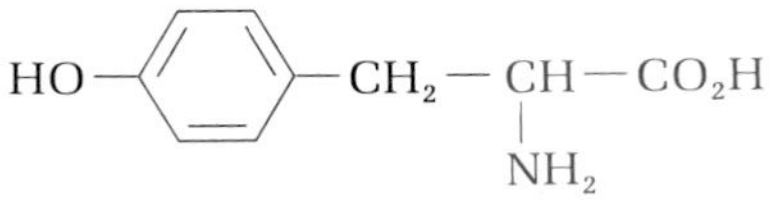

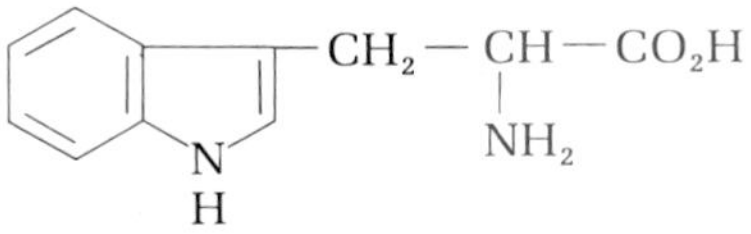

Name	Abbreviation	Formula	R

B. One amino group and two carboxyl groups

14. aspartic acid   Asp

$$HOOC-CH_2-\underset{\underset{NH_2}{|}}{CH}-CO_2H$$

15. glutamic acid   Glu

$$HOOC-CH_2CH_2-\underset{\underset{NH_2}{|}}{CH}-CO_2H$$

16. asparagine   Asn

$$H_2N-\overset{\overset{O}{\|}}{C}-CH_2-\underset{\underset{NH_2}{|}}{CH}-CO_2H$$

the amides of aspartic acid and glutamic acid

17. glutamine   Gln

$$H_2N-\overset{\overset{O}{\|}}{C}-CH_2CH_2-\underset{\underset{NH_2}{|}}{CH}-COOH$$

C. One carboxyl group and two basic groups

18. lysine   Lys

$$\underset{\underset{NH_2}{|}}{CH_2}CH_2CH_2CH_2-\underset{\underset{NH_2}{|}}{CH}-CO_2H$$

19. arginine   Arg

$$\underset{NH}{\overset{NH_2}{\diagdown}}C-NH-CH_2CH_2CH_2-\underset{\underset{NH_2}{|}}{CH}-CO_2H$$

20. histidine   His

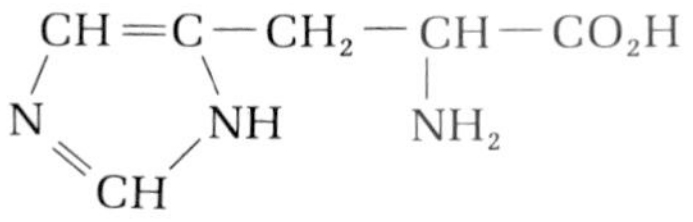

$$\underset{N\diagdown_{CH}\diagup NH}{CH=C}-CH_2-\underset{\underset{NH_2}{|}}{CH}-CO_2H$$

TABLE 15.1   Names and formulas of the common amino acids (*continued*)

nate a proton to a strong base, or they can behave as bases and accept a proton from a strong acid. These behaviors are expressed in the following equilibria for an amino acid with one amino and one carboxyl group:

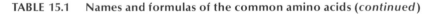

$$\underset{\underset{^+NH_3}{|}}{RCHCO_2H} \underset{H^+}{\overset{OH^-}{\rightleftharpoons}} \underset{\underset{^+NH_3}{|}}{RCHCO_2^-} \underset{H^+}{\overset{OH^-}{\rightleftharpoons}} \underset{\underset{NH_2}{|}}{RCHCO_2^-} \qquad (15.1)$$

amino acid   dipolar ion   amino acid
at low pH   form   at high pH
(acid)   (neutral)   (base)

FIGURE 15.2

Titration curve for alanine.

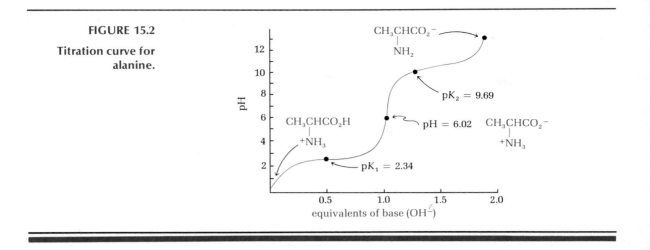

Figure 15.2 shows a titration curve for alanine, a typical amino acid of this kind. At low pH (acidic solution), the amino acid is in the form of a substituted ammonium ion. At high pH (basic solution), it is present as a substituted carboxylate ion. At an intermediate pH (for alanine, pH 6.02), the amino acid is present as the dipolar ion.

Example 15.1 **Starting with alanine as its hydrochloride, write equations for the reaction with 1 equivalent of sodium hydroxide and with 2 equivalents of sodium hydroxide.**

Solution $CH_3CHCO_2H + Na^+OH^- \rightarrow CH_3CHCO_2^- + Na^+Cl^- + H_2O$ (15.2)

$\quad\quad\quad$ | $\quad\quad\quad\quad\quad\quad\quad\quad\quad$ |

$\quad\quad$ $^+NH_3\ Cl^-$ $\quad\quad\quad\quad\quad\quad\quad$ $^+NH_3$

$\quad\quad$ ammonium salt $\quad\quad\quad\quad\quad\quad$ dipolar ion

$CH_3CHCO_2^- + Na^+OH^- \rightarrow CH_3CHCO_2^-Na^+ + H_2O$ (15.3)

$\quad\quad$ | $\quad\quad\quad\quad\quad\quad\quad\quad\quad\quad\quad$ |

$\quad\quad$ $^+NH_3$ $\quad\quad\quad\quad\quad\quad\quad\quad\quad$ NH_2

$\quad\quad$ dipolar ion $\quad\quad\quad\quad\quad\quad\quad$ carboxylate salt

One equivalent of base gives the dipolar ion, and a second equivalent of base gives the sodium carboxylate.

Problem 15.1 **Starting with the sodium carboxylate salt of alanine, write equations for the reaction with 1 equivalent and with 2 equivalents of hydrochloric acid.**

Note that the sign of the charge on an amino acid such as alanine changes as the pH changes. At low pH the sign is positive, at high pH it is negative, and near neutrality the ion is dipolar. If placed in an electric

FIGURE 15.3

The migration of an amino acid (such as alanine) in an electric field depends on pH.

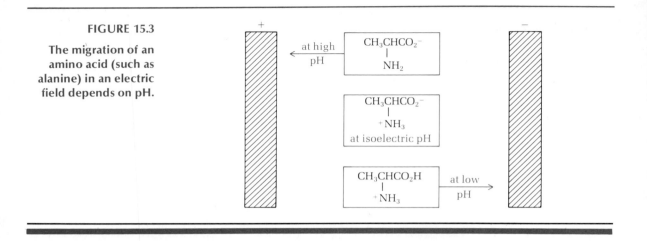

field, the amino acid will migrate toward the cathode (negative electrode) at low pH and toward the anode (positive electrode) at high pH (Fig. 15.3). At some intermediate pH, called the **isoelectric point,** the amino acid will be dipolar and have a net charge of zero. It will be unable to move toward either electrode. For alanine, the isoelectric point is pH 6.02.

Example 15.2   **Write structures for valine at**
a. the isoelectric point   **b. high pH**   **c. low pH**

Solution   **a. $(CH_3)CHCHCO_2{}^-$**   **b. $(CH_3)_2CHCHCO_2{}^-$**   **c. $(CH_3)_2CHCHCO_2H$**

                $^+NH_3$ **NH_2**   **$^+NH_3$**

            (dipolar)   **(negative)**   **(positive)**

Problem 15.2   **Write structures for the predominant form of the following amino acids at the indicated pH. If placed in an electric field, toward which electrode (+ or −) will the amino acid migrate?**
a. phenylalanine at its isoelectric point
b. methionine at low pH
c. serine at high pH

An important method, called **electrophoresis,** for separating amino acids (and proteins) is based on their differential rates and directions of migration in an electric field at a controlled pH.

In general, amino acids like alanine, with one amino group and one carboxyl group and no other acidic or basic groups in their structure, have two pK_a values, one near 2–3 and the other near 9–10, with isoelectric points near 6. But, as we will now see, the situation is more complex with amino acids containing two acidic or two basic groups.

**15.4
THE ACID–BASE
PROPERTIES OF
AMINO ACIDS
WITH MORE THAN
ONE ACIDIC OR
BASIC GROUP**

Aspartic and glutamic acids (numbers 14 and 15 in Table 15.1) have two carboxyl groups and one amino group. In strong acid (low pH) all three of these groups are in their acidic form. As the pH is raised and the solution becomes more basic, each group in succession gives up a proton. The equilibria are shown for aspartic acid, with the three pK_a values over the equilibrium arrows:

$$HO_2CCH_2CHCO_2H \overset{2.09}{\rightleftharpoons} HO_2CCH_2CHCO_2^- \overset{3.86}{\rightleftharpoons} {}^-O_2CCH_2CHCO_2^- \overset{9.82}{\rightleftharpoons} {}^-O_2CCH_2CHCO_2^-$$

with substituents $^+NH_3$, $^+NH_3$, $^+NH_3$, NH_2 respectively.   (15.4)

low pH ⟶ high pH

Example 15.3   Which is the most acidic group in aspartic acid?

Solution   As shown in eq. 15.4, the first proton to be removed from the most acidic form of aspartic acid (at the extreme left of the equation) is the proton on the carboxyl group nearest the $^+NH_3$ substituent. The $^+NH_3$ group is electron-withdrawing due to its positive charge. Electron-withdrawing substituents enhance acidity, and the effect falls off with distance (review Sec. 10.6). Also, the resulting dipolar ion has the opposite charges closer to each other than would be the case if the other carboxyl group had given up its proton.

Problem 15.3   Which is the least acidic group in aspartic acid? Why?

The first and third pK_a's of aspartic acid are not so different from those of alanine, but between them there is a second pK_a (3.86) for the remote carboxyl group. The pH at which the dipolar form predominates is therefore much lower than for the monoamino monocarboxylic acids. The isoelectric points for aspartic and glutamic acids come at about pH = 3.

Problem 15.4   Explain why the first and third pK_a's of aspartic acid are similar to the first and second pK_a's, respectively, of alanine. (*Hint:* compare the acidic groups).

The situation differs for amino acids with two basic groups and only one carboxyl group (numbers 18, 19, and 20 in Table 15.1). With lysine, for example, the equilibria are

$$CH_2(CH_2)_3CHCO_2H \overset{2.18}{\rightleftharpoons} CH_2(CH_2)_3CHCO_2^- \overset{8.95}{\rightleftharpoons}$$

with substituents $^+NH_3$, $^+NH_3$ and $^+NH_3$, $^+NH_3$

$$CH_2(CH_2)_3CHCO_2^- \overset{10.53}{\rightleftharpoons} CH_2(CH_2)_3CHCO_2^-$$

with substituents $^+NH_3$, NH_2 and NH_2, NH_2 (15.5)

low pH ⟶ high pH

The isoelectric point comes at a relatively high pH, 9.74.

Example 15.4   Compare the charge on the predominant species at pH = 6 for alanine, aspartic acid, and lysine.

Solution $CH_3CHCO_2^-$ $^-O_2CCH_2CHCO_2^-$ $CH_2(CH_2)_3CHCO_2^-$
          | | | |
        $^+NH_3$ $^+NH_3$ $^+NH_3$ $^+NH_3$

        alanine   aspartic acid   lysine
      (neutral)   (net negative charge)   (net positive charge)

The predominant form can be selected using the data in Fig. 15.2 and eq. 15.4 and 15.5.

Problem 15.5   Show how the method of electrophoresis could be used to separate a mixture of alanine, aspartic acid and lysine.

The second basic groups in arginine and histidine are not simple amino groups. They are a **guanidine** group and an **imidazole** ring, respectively. The most protonated forms of these two amino acids are

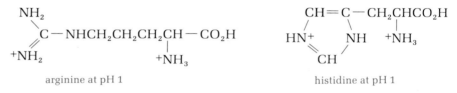

            arginine at pH 1   histidine at pH 1

Example 15.5   The least acidic group in arginine is the guanidinium group. Using resonance, explain how this group is stabilized.

Solution   The + charge can be delocalized over all three nitrogens, as follows:

$$: NH_2 \qquad\qquad ^+NH_2 \qquad\qquad NH_2$$
$$\ce{C-\overset{..}{N}HR} \leftrightarrow \ce{C-\overset{..}{N}HR} \leftrightarrow \ce{C=\overset{+}{N}HR}$$
$$^+NH_2 \qquad\qquad :NH_2 \qquad\qquad NH_2$$

The group is therefore much less acidic than the $^+NH_3$ group, where the charge is localized on one nitrogen atom.

Problem 15.6   How can the positive charge in the protonated imidazole ring of histidine be delocalized? Do you expect this group to be more or less acidic than the guanidinium ion in arginine? Explain.

Problem 15.7   Arginine shows three pK_a's, at 1.82 (the COOH group), at 8.99 (the $^+NH_3$ group), and at 13.20. Write equilibria (similar to eq. 15.5) for its dissociation. At approximately what pH will the isoelectric point come, and what will be the structure of the dipolar ion?

**15.5
NON-PROTEIN-
DERIVED AMINO
ACIDS**   Well over 150 amino acids besides those listed in Table 15.1 occur in nature. For example, **β-alanine** (a β- instead of an α-amino acid) forms part of the coenzyme-A structure (page 254). **γ-Aminobutyric acid** (GABA) is involved in the transmission of nerve impulses. **Homocysteine** and **ornithine** are two of many amino acid intermediates in metabolism.

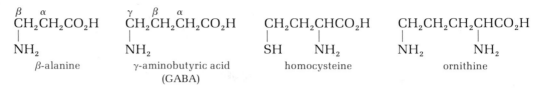

$$\overset{\beta}{C}H_2\overset{\alpha}{C}H_2CO_2H$$
$$|$$
$$NH_2$$

β-alanine

$$\overset{\gamma}{C}H_2\overset{\beta}{C}H_2\overset{\alpha}{C}H_2CO_2H$$
$$|$$
$$NH_2$$

γ-aminobutyric acid
(GABA)

$$CH_2CH_2CHCO_2H$$
$$| \quad\quad |$$
$$SH \quad NH_2$$

homocysteine

$$CH_2CH_2CH_2CHCO_2H$$
$$| \quad\quad\quad |$$
$$NH_2 \quad\quad NH_2$$

ornithine

A few amino acids with stereochemistry opposite to that usually found in proteins are also known. For example, D-serine has been found in the earthworm, D-alanine in octopus muscle and guinea pig blood, and D-ornithine in shark liver.

15.6 REACTIONS OF AMINO ACIDS

In addition to their acidic and basic properties, amino acids undergo many other reactions typical of carboxyl groups or amino groups. For example, the carboxyl group can be esterified.

$$R{-}CH{-}CO_2{}^- + R'OH + H^+ \xrightarrow{heat} R{-}CH{-}CO_2R' + H_2O \qquad (15.6)$$
$$\quad\;\; | \quad\quad\quad\quad\quad\quad\quad\quad\quad\quad\quad\quad | $$
$$\quad\;\; {}^+NH_3 \quad\quad\quad\quad\quad\quad\quad\quad\quad\quad {}^+NH_3$$

The amino group can be converted to an amide.

$$R{-}CH{-}CO_2{}^- + R'{-}\overset{O}{\overset{||}{C}}{-}Cl \xrightarrow{OH^-} R{-}CH{-}CO_2{}^- + H_2O + Cl^- \qquad (15.7)$$
$$\quad\;\; | \quad\quad\quad\quad\quad\quad\quad\quad\quad\quad\quad\quad | $$
$$\quad\;\; {}^+NH_3 \quad\quad\quad\quad\quad\quad\quad\quad\quad HNC{-}R'$$
$$\quad\quad\quad\quad\quad\quad\quad\quad\quad\quad\quad\quad\quad || $$
$$\quad\quad\quad\quad\quad\quad\quad\quad\quad\quad\quad\quad\quad O$$

an amide

These types of reactions are useful in temporarily modifying or protecting either of the two functional groups, especially during the controlled linking of amino acids to form peptides or proteins.

Problem 15.8 Using eqs. 15.6 and 15.7 as models, write equations for the following reactions.
a. phenylalanine + CH_3OH + HCl →
b. valine + benzoyl chloride + OH^- →
c. glycine + acetic anhydride $\xrightarrow{heat}$

15.7 THE NINHYDRIN REACTION

Ninhydrin is a useful reagent for detecting amino acids and determining the concentration of their solutions. It is the hydrate of a cyclic triketone and, when it reacts with amino acids, a violet dye is produced. The overall reaction, whose mechanism is complex, is

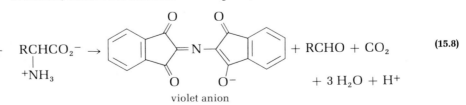

(15.8)

ninhydrin

violet anion

Note that the amino acid contributes only the nitrogen atom of the violet dye. The rest of the amino acid is converted to an aldehyde and carbon dioxide. The same violet dye is produced from *all* α-amino acids with a primary NH_2 group, and the intensity of its color is directly proportional to the concentration of the amino acid present. Only proline (Table 15.1), which has a secondary rather than a primary amino group, does not give the violet dye. It reacts differently to give a yellow dye, which can be used for its analysis.

Problem 15.9 Write an equation for the reaction of alanine with ninhydrin.

Problem 15.10 Show, by writing appropriate resonance structures, that the negative charge in the violet dye formed from ninhydrin and amino acids can be delocalized over all four oxygens.

15.8 Amino acids are linked together in peptides and proteins through the for-
PEPTIDES mation of an amide between the carboxyl group of one amino acid and the α-amino group of another amino acid. Emil Fischer, who first proposed the structure, called this amide bond a **peptide bond.** A molecule containing only *two* amino acids (aa's) joined in this way is called a **dipeptide:**

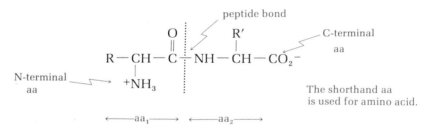

By convention, the peptide bond is written in the manner shown. The amino acid having a free $\overset{+}{N}H_3$ group is at the left, and the amino acid with a free CO_2^- group is at the right. These amino acids are called the **N-terminal amino acid** and the **C-terminal amino acid,** respectively.

Example 15.6 Write the possible structures of a dipeptide made by joining alanine and glycine by a peptide bond.

Solution There are two possibilities:

$$\overset{+}{H_3}N-CH_2-\overset{\overset{\displaystyle O}{\|}}{C}-NH-\underset{\underset{\displaystyle CH_3}{|}}{CH}-CO_2^- \qquad\qquad \overset{+}{H_3}N-\underset{\underset{\displaystyle CH_3}{|}}{CH}-\overset{\overset{\displaystyle O}{\|}}{C}-NHCH_2CO_2^-$$

glycylalanine   alanylglycine

In glycylalanine, glycine is the N-terminal amino acid and alanine is the C-terminal amino acid. In alanylglycine, these roles are reversed. The two dipeptides are structural isomers.

Problem 15.11   Write the structural formulas for:
a. valylalanine   b. alanylvaline

We often write the formulas for peptides in a kind of shorthand by simply linking the three-letter abbreviations for each amino acid, starting with the N-terminal end at the left. For example, glycylalanine would be Gly–Ala, whereas alanylglycine would be Ala–Gly.

Example 15.7   Consider the tripeptide abbreviated formula Gly–Ala–Ser. Which is the N-terminal amino acid and which is the C-terminal amino acid?

Solution   Such formulas always read from the N-terminal amino acid at the left to the C-terminal amino acid at the right. Glycine is the N-terminal amino acid, and serine is the C-terminal amino acid. Both the amino group *and* the carboxyl group of the middle amino acid, alanine, are tied up in peptide bonds.

Problem 15.12   Write out the complete structural formula for Gly–Ala–Ser.

Problem 15.13   Write out the *abbreviated* structural formulas for all possible tripeptide isomers of Gly–Ala–Ser.

The complexity that is possible in peptide and protein structures is truly astounding. For example, Problem 15.13 showed that there are 6 possible arrangements of 3 different amino acids in a tripeptide. For a tetrapeptide this number jumps to 24, and for an octapeptide (constructed from 8 different amino acids) there are 40,320 possible arrangements! This calculation neglects the complication that would arise if we included the chirality of the α or other carbon atoms.

Now we must introduce one small additional complication before we consider the structures of particular peptides and proteins.

15.9 THE DISULFIDE BOND   Aside from the peptide bond, the only other type of covalent bond between amino acids in peptides and proteins is the **disulfide bond.** It links two **cysteine** units. Recall (from Sec. 7.17) that thiols are easily oxidized to disulfides (eq. 7.41). If two cysteine units are close in space, they may become linked in a disulfide bond.

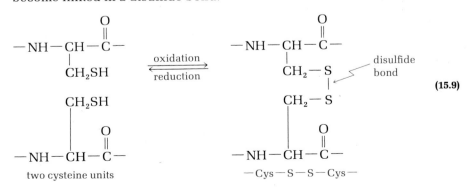

(15.9)

If the two cysteine units are in different parts of the *same* peptide chain, a disulfide bond between them will form a "loop" or large ring. If the two units are on different chains, the disulfide bond will link the two chains together. We will see examples of both arrangements. S—S bonds can always be easily broken by mild reducing agents (the reverse of eq. 15.9; see also A Word About Hair on page 189).

a word about

28. Some Naturally Occurring Peptides

Numerous peptides containing a relatively small number of linked amino acids per molecule have been isolated from living matter, and they often perform important roles in biology. We mention here only a few examples. Bradykinin is a nonapeptide present in blood plasma and involved in the regulation of blood pressure. Several peptides have been found in the brain, where they may be chemical transmitters of nerve impulses. One of these is the decapeptide substance P, which is thought to be a transmitter of pain impulses.

Arg—Pro—Pro—Gly—Phe—
                    Ser—Pro—Phe—Arg
                 bradykinin

Arg—Pro—Lys—Pro—Gln—
              Phe—Phe—Gly—Leu—Met
           substance P

Oxytocin (Figure 15.4) and vasopressin are two cyclic nonapeptide hormones produced by the pos-

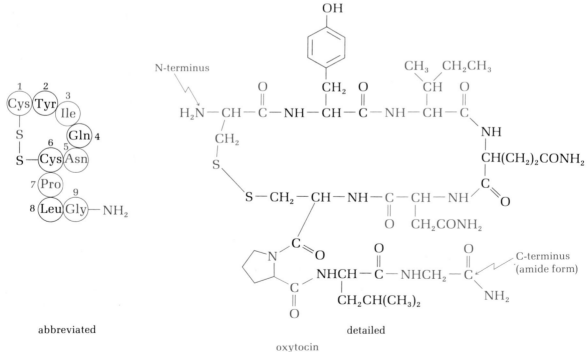

abbreviated   detailed

oxytocin
white powder, $[\alpha]_D^{22}$ −26.2°C (water)

FIGURE 15.4   The nonapeptide hormone oxytocin.

terior pituitary gland. Oxytocin regulates uterine contraction and lactation and may be administered when necessary to induce labor at childbirth. Note that its structure includes two cysteine units joined by a disulfide bond. Note also that the C-terminal amino acid, glycine, is present as the amide. This is common in peptide chains.

Vasopressin differs from oxytocin only in replacement of Ile by Phe and Leu by Arg. Vasopressin regulates the excretion of water by the kidneys and also affects blood pressure. The disease *diabetes insipidus*, in which too much urine is excreted, is a consequence of vasopressin deficiency and can be treated by administering this hormone.

In the remainder of this chapter, we will describe the principal features of peptide and protein structure. This description can be given at several levels of detail. We can say which amino acids are present, and how many of each per peptide or protein molecule. Or we can give their sequence in the chain. Or we can describe more gross aspects of their structure, such as molecular shape. Are the molecules helical, spherical, or sheet-like? Do the molecules aggregate?

One usually considers four levels of description, referred to as the primary, secondary, tertiary, and quaternary structure. We begin with the primary structure.

15.10
THE PRIMARY
STRUCTURE OF
PEPTIDES AND
PROTEINS

The first things we must know about a peptide or protein if we are to be able to write down its structure are (1) which amino acids are present and how many of each, and (2) what the sequence of the amino acids is in the chain. In this section, we will briefly describe ways of obtaining this kind of information, usually called the primary structure.

15.10a Amino Acid Analysis Complete hydrolysis of a peptide or protein converts it to a mixture of amino acids. This hydrolysis is typically accomplished by heating with 6N HCl at 110°C for 24 hours. Analysis of the resulting amino acid mixture requires a procedure for separating the amino acids from one another, a method for identifying each amino acid present, and a way of determining the amount of each amino acid present.

Nowadays an instrument called an **amino acid analyzer** performs these tasks automatically in the following way. The amino acid mixture from the complete hydrolysis of a few milligrams of the peptide or protein is placed at the top of a column packed with material that selectively absorbs amino acids. The packing is an insoluble resin that contains strongly acidic groups, which protonate the amino acids. Next a buffer solution of known pH is pumped through the column. The amino acids pass through at different rates, depending on their structure and basicity, and are thus separated.

The column effluent is met by a stream of ninhydrin reagent (Sec. 15.7). Therefore the effluent is alternately violet or colorless, depending on whether an amino acid is being eluted from the column. The intensity of the color is automatically recorded as a function of the volume of effluent. Calibration with known amino acid mixtures allows each amino acid to be identified by the appearance time of its peak. Furthermore, the intensity

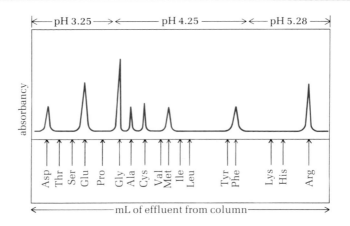

FIGURE 15.5   Different amino acids in a peptide hydrolyzate are separated on an ion-exchange resin. Buffers with different pH's (shown at top) are used to elute the amino acids from the column. Each amino acid is identified by comparing it with the standard elution profile that is shown by the arrows near the bottom on the figure and was determined using a known amino acid mixture. The amount of each amino acid is proportional to the area under each peak.

of each peak gives a quantitative measure of the amount of each amino acid present. Figure 15.5 shows a typical plot that might be obtained from an automatic amino acid analyzer.

15.10b Sequence Determination In 1953 Frederick Sanger (Cambridge, England) published his report on the amino acid sequence of insulin, a protein hormone with 51 amino acid units. His work was truly a landmark in chemistry, and in 1958 he received his first (of two) Nobel Prize. His idea was relatively simple.

Consider a polypeptide chain. The N-terminal amino acid differs from all other amino acids in the sequence by having a free amino group. If that amino group were to react with some reagent *prior* to hydrolysis, then after hydrolysis the N-terminal amino acid would be labeled and could be identified:

$$aa_1 - aa_2 - aa_3 - aa_4 - aa_5$$
$$\downarrow \text{ reagent X (reacts with amino groups)}$$
$$X - aa_1 - aa_2 - aa_3 - aa_4 - aa_5$$
$$\downarrow \text{ complete hydrolysis}$$
$$X - aa_1 + aa_2 + aa_3 + aa_4 + aa_5$$

Sanger's reagent was 2,4-dinitrofluorobenzene, which reacts with the NH_2 group of amino acids and peptides to give yellow 2,4-dinitrophenyl (DNP) derivatives.

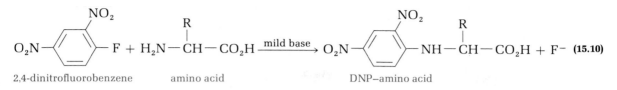

2,4-dinitrofluorobenzene   amino acid   DNP–amino acid

Of course Sanger's method identifies only one amino acid in the sequence, the N-terminal amino acid. Although Sanger was able to use this method effectively (by partially hydrolyzing insulin and identifying the N-terminal amino acids of the separated smaller peptides), a better method can be imagined. It would be nice to clip off just one amino acid at a time from the end of a peptide chain and identify it.

In 1950 such a method was devised by Pehr Edman (University of Lund, Sweden), and it is now widely used. Edman's reagent is phenyl isothiocyanate, $C_6H_5N=C=S$. The steps in selectively labeling and releasing the N-terminal amino acid are

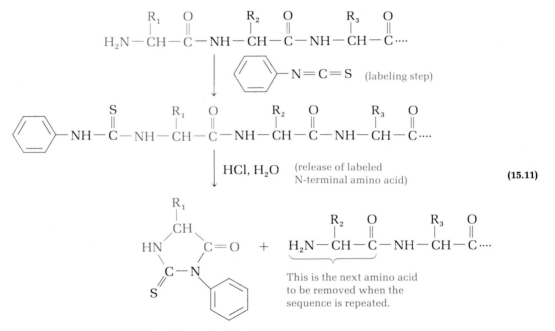

a phenylthiohydantoin

The N-terminal amino acid is removed in the form of a heterocyclic compound, a phenylthiohydantoin, which can be identified by comparison with reference compounds prepared from known amino acids. At present, automated amino acid "sequenators" can easily determine, in a day, the sequence of the first 20 or so amino acids from the N-terminal end.

15.10c Selective Cleavage of Peptide Bonds
Even the Edman method cannot be used indefinitely due to the accumulation of impurities. Consequently, if a protein contains several hundred amino acid units, it is best

TABLE 15.2	Reagent	Cleavage site
Specific cleavage of polypeptides	trypsin	Carboxyl side of Lys, Arg
	chymotrypsin	Carboxyl side of Phe, Tyr, Trp
	cyanogen bromide (CNBr)	Carboxyl side of Met
	carboxypeptidase	C-terminal amino acid

to partially hydrolyze the chain to smaller fragments that can be separated and sequenced. Certain reagents and enzymes are used to cleave proteins selectively. Some of these reagents are listed in Table 15.2.

Example 15.8   Consider the following peptide:

Ala—Gly—Tyr—Trp—Ser—Lys—Gly—Leu—Met—Gly.

Determine what fragments will be obtained when this peptide is hydrolyzed with:
a. trypsin
b. chymotrypsin
c. cyanogen bromide

Solution   a. The enzyme trypsin will split the peptide on the carboxyl side of lysine giving

Ala—Gly—Tyr—Trp—Ser—Lys and Gly—Leu—Met—Gly.

b. The enzyme chymotrypsin will split the peptide on the carboxyl sides of tyrosine and tryptophan, giving

Ala—Gly—Tyr and Trp and Ser—Lys—Gly—Leu—Met—Gly.

c. Cyanogen bromide will split the peptide on the carboxyl side of methionine, thus splitting off the C-terminal glycine and leaving the rest untouched. Carboxypeptidase would do the same thing, confirming that the C-terminal amino acid is glycine.

Problem 15.14   Determine what fragments would be obtained if bradykinin (see A Word About Peptides, page 380) were hydrolyzed enzymatically with:
a. trypsin
b. chymotrypsin

15.10d The Logic of Sequence Determination   Let us illustrate, with a specific example, the reasoning that is used to fully determine the sequence of amino acids in a particular peptide containing 30 amino acids units. First we hydrolyze the peptide completely, subject it to amino acid analysis, and find that it has the formula $Ala_2ArgAsnCys_2GlnGlu_2Gly_3His_2$-$Leu_4LysPhe_3ProSerThrTyr_2Val_3$. Using the Sanger method, we find that the N-terminal amino acid is Phe. Suppose that the chain is too long to degrade completely by the Edman method. So, we decide to simplify the problem by digesting the peptide with chymotrypsin. (We select chymotrypsin because we note that the intact peptide contains three Phe's and two Tyr's and will undoubtedly be cleaved by chymotrypsin.) When we do, we find that we get three fragment peptides. In addition, we get 2 equivalents of Phe and 1 of Tyr. We subject the three fragment peptides to Edman degradation and obtain their structures:

A. Leu—Val—Cys—Gly—Glu—Arg—Gly—Phe

B. Val—Asn—Gln—His—Leu—Cys—Gly—Ser—His—Leu—Val—Glu—Ala—Leu—Tyr

C.
$$\begin{matrix} 27 & 28 & 29 & 30 \\ \text{Thr} & \text{Pro} & \text{Lys} & \text{Ala} \end{matrix}$$
C. Thr—Pro—Lys—Ala

We still cannot write a unique structure, but we can now say that the C-terminal amino acid must be Ala and that the last four amino acids must be in the sequence shown for fragment C. We can say this because we know that Ala is not cleaved at its carboxyl end by chymotrypsin. (Note that the C-terminal amino acids in fragments A and B are Phe and Tyr, both cleaved at the carboxyl ends by chymotrypsin.) That the C-terminal amino acid is Ala can be confirmed using carboxypeptidase. We can number the amino acids in fragment C as 27–30 in the chain.

What to do next? Cyanogen bromide is no help, because the peptide does not contain Met. But the peptide does contain Lys and Arg, so we go back to the beginning and digest the intact peptide with trypsin, which cleaves peptides at these amino acid units. We obtain (not surprisingly) some Ala (the C-terminal amino acid) because it comes right after a Lys. We also obtain two peptides. One of them is relatively short, so we determine its sequence by the Edman method and find it to be

D.
$$\begin{matrix} 23 & 24 & 25 & 26 & 27 & 28 & 29 \\ \text{Gly} & \text{Phe} & \text{Phe} & \text{Tyr} & \text{Thr} & \text{Pro} & \text{Lys} \end{matrix}$$
D. Gly—Phe—Phe—Tyr—Thr—Pro—Lys

Because we recognize the last three amino acids in fragment D as 27, 28, and 29, we can number the rest of the chain, back to 23. We now note that amino acids 23 and 24 appear at the end of fragment A, so originally A must have been connected to C. The only place left for fragment B is in front of A. This leaves only one of the Phe's to account for, and it must occupy the N-terminal position (recall the Sanger result). We can now write out the complete structure.

Phe—Val—Asn—Gln—His—Leu—Cys—Gly—Ser—His—Leu—Val—Glu—Ala—Leu—Tyr

Leu—Val—Cys—Gly—Glu—Arg—Gly—Phe—Phe—Tyr—Thr—Pro—Lys—Ala

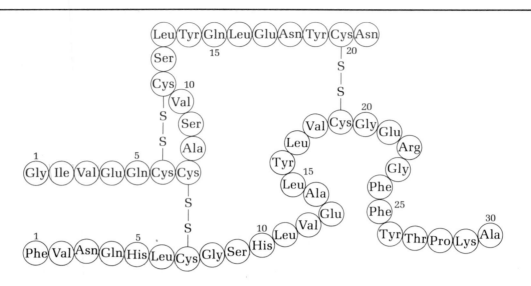

FIGURE 15.6   Primary structure of beef insulin. The A chain is in color and the B chain
whose structure determination was described in the text is shown in black.

The colored arrows show the cleavage points with chymotrypsin, and the
black arrows show the cleavage points with trypsin.

In fact, the peptide chosen for illustration is the B chain of the protein
hormone **insulin,** whose structure was first determined by Sanger and is
shown schematically in Figure 15.6. Insulin consists of an A chain with 21
amino acid units and a B chain with 30 amino acid units. The two chains
are joined by two disulfide bonds, and the A chain also contains a small
disulfide "loop."

Problem 15.15   **What peptides would be obtained if the A chain of insulin were digested
with chymotrypsin?**

**15.11
PEPTIDE SYNTHESIS**   Once we know the amino acid sequence in a peptide or protein, we are
in a position to synthesize the compound from its amino acid components.
Why would we want to do this? There are several reasons. We may wish
to verify a particular peptide structure by comparing the properties of the
synthetic and natural substances. We might wish to study the effect of sub-
stituting one amino acid for another on the biological properties of the
peptide. Or we might want to modify an enzyme by altering its amino
acid sequence in a specific way.

Many methods have been developed for linking amino acids in a controlled manner. They require careful strategy. Amino acids are bifunc-

tional. To link the carboxyl group of one amino acid to the amino group of a second amino acid, we must first prepare each compound by "protecting" the amino group of the first and the carboxyl group of the second:

$$\underset{aa_1}{H_2N-\overset{\overset{\displaystyle R_1}{|}}{CH}-CO_2H} \xrightarrow[\text{amino group}]{\text{protect the}} \boxed{P_1}-NH-\overset{\overset{\displaystyle R_1}{|}}{CH}-CO_2H$$

(15.12)

$$\underset{aa_2}{H_2N-\overset{\overset{\displaystyle R_2}{|}}{CH}-CO_2H} \xrightarrow[\text{carboxyl group}]{\text{protect the}} H_2N-\overset{\overset{\displaystyle R_2}{|}}{CH}-\overset{\overset{\displaystyle O}{\|}}{C}-\boxed{P_2}$$

In this way we can control the linking of the two amino acids so that the carboxyl group of aa_1 combines with the amino group of aa_2:

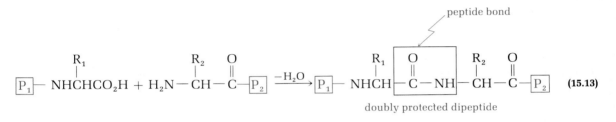

doubly protected dipeptide   (15.13)

Example 15.9

What would happen if we tried to combine aa_1 with aa_2 without using protecting groups?

Solution

Since each amino acid could react either as an amine or as a carboxylic acid, we could get not only $aa_1 - aa_2$ but also $aa_2 - aa_1$, $aa_1 - aa_1$, and $aa_2 - aa_2$. Furthermore, since the resulting dipeptides would still have a free amino and a free carboxyl group, we could also get trimers, tetramers, and so on.

After the peptide bond is formed, we must be able to remove the protecting groups under conditions that do not also hydrolyze the peptide bond, or all would be lost. If more amino acids are to be added to the chain, we must be able to selectively remove one of the two protecting groups from the doubly protected dipeptide. All of this can be quite tricky and tedious business. Yet these methods were used by Vincent du Vigneaud and his colleagues (Cornell University; Nobel Prize, 1955) to synthesize oxytocin and vasopressin, the first naturally occurring polypeptides to be synthesized.

In 1965 R. B. Merrifield (Rockefeller University) developed the **solid-phase technique** for peptide synthesis. This technique avoids many of the tedious aspects of previous methods and is now universally used. The principle of the technique is to assemble the peptide chain while one end of it is chemically attached to an insoluble inert solid. In this way excess reagents and by-products can be removed simply by washing and filtering the solid. The growing peptide chain does not need to be purified at any intermediate stage. When the peptide is fully constructed, it is cleaved chemically from the solid support.

Typically the solid phase is a cross-linked polystyrene (Sec. 3.17) in which some fraction of the aromatic rings contain chloromethyl ($ClCH_2$—) groups:

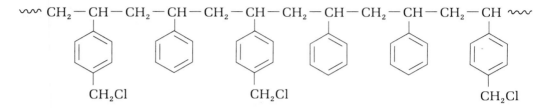

The polymer is treated with an N-protected amino acid. The first amino acid that is attached to the polymer will eventually become the C-terminal amino acid of the synthetic peptide.

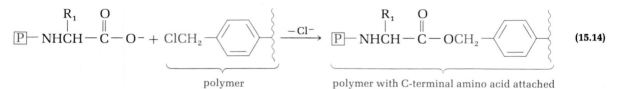

(15.14)

polymer   polymer with C-terminal amino acid attached

The protecting group is removed and the next N-protected amino acid is added to the chain:

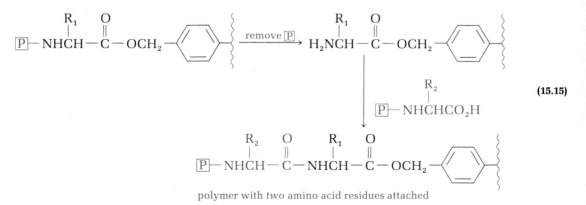

(15.15)

polymer with *two* amino acid residues attached

The process may be repeated to add a third amino acid, a fourth, and so on. Finally, when the correct number of amino acids have been connected in the desired sequence and the N-terminal amino group has been "de-protected," the polypeptide chain is detached from the polymer. Treatment with anhydrous hydrogen fluoride or with HBr in trifluoroacetic acid cleaves the benzyl ester without hydrolyzing the amide bonds in the polypeptide:

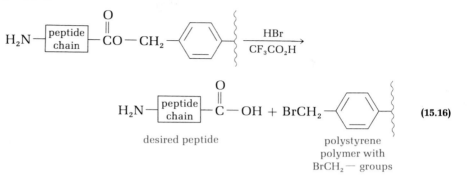

$$H_2N-\boxed{\begin{array}{c}\text{peptide}\\\text{chain}\end{array}}-\overset{\displaystyle O}{\overset{\|}{C}}O-CH_2-\text{(ring)} \xrightarrow[\text{CF}_3\text{CO}_2\text{H}]{\text{HBr}}$$

$$H_2N-\boxed{\begin{array}{c}\text{peptide}\\\text{chain}\end{array}}-\overset{\displaystyle O}{\overset{\|}{C}}-OH + BrCH_2-\text{(ring)} \qquad (15.16)$$

desired peptide

polystyrene polymer with $BrCH_2-$ groups

The most frequently used N-protecting group in solid-phase peptide synthesis is the **t-butoxycarbonyl (Boc)** group:

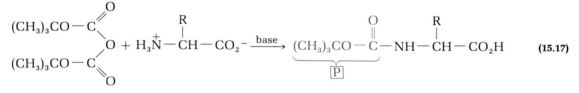

$$(CH_3)_3CO-\overset{\displaystyle O}{\overset{\|}{C}}\Big\rangle O + H_3\overset{+}{N}-\underset{\displaystyle R}{\overset{|}{C}H}-CO_2^- \xrightarrow{\text{base}} \underbrace{(CH_3)_3CO-\overset{\displaystyle O}{\overset{\|}{C}}-NH-\underset{\displaystyle R}{\overset{|}{C}H}-CO_2H}_{\boxed{P}} \qquad (15.17)$$

di-t-butyl dicarbonate

It has the virtue that it can be removed by treatment with acid under very mild conditions that do not affect peptide bonds or the ester bond that links the peptide to the polymer:

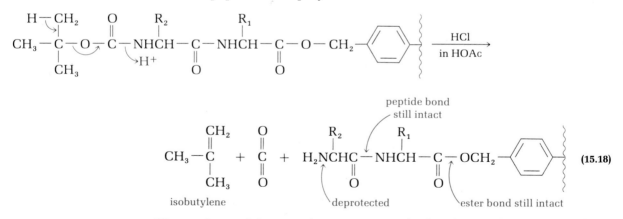

peptide bond
still intact

$$CH_3-\underset{\displaystyle CH_3}{\overset{\displaystyle CH_2}{\overset{\|}{C}}} + \overset{\displaystyle O}{\overset{\|}{C}} + H_2NCHC-NHCH-\overset{\displaystyle R_1}{\overset{|}{C}}-OCH_2-\text{(ring)} \qquad (15.18)$$

isobutylene   deprotected   ester bond still intact

The products of deprotection are gaseous (isobutylene and carbon dioxide) and are thus easily removed from the reaction mixture.

The attachment of each successive amino acid after the first one (that is, step 2 in eq. 15.15) is accomplished with the aid of **dicyclohexylcarbodiimide (DCC)**. It is able to link carboxyl and amino groups in a peptide bond; in the process, it is hydrolyzed to dicyclohexylurea.

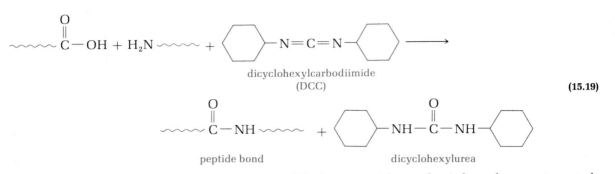

dicyclohexylcarbodiimide
(DCC)

(15.19)

peptide bond   dicyclohexylurea

The operations in solid-phase peptide synthesis have been automated. All the reactions occur in a single reaction vessel, with reagents and wash solvents added automatically from reservoirs by means of mechanical pumps. Working around the clock, the programmer can incorporate about eight amino acids into a polypeptide in a day. Merrifield synthesized the nonapeptide bradykinin (see A Word About Peptides, page 380) in just 27 hours using this technique. And in 1969 he used the automated synthesizer to prepare the enzyme ribonuclease (124 amino-acid residues), the first enzyme to be prepared synthetically from its amino-acid components. Automated peptide synthesis is now a fairly routine matter.

We have seen how the primary structure of peptides and proteins can be determined and how peptides can be synthesized in the laboratory. Now let us examine further details of protein structure.

**15.12
SECONDARY
STRUCTURE OF
PROTEINS**

Because proteins consist of long chains of amino acids strung together, one might think that their shapes would be rather amorphous or "floppy" and ill-defined. This is incorrect. Many proteins have been isolated in pure crystalline forms, where the polymer has a very well-defined shape. Indeed, even in solution the shapes seem to be quite regular. Let us examine some of the structural features of peptide chains that are responsible for their definite shapes.

15.12a Geometry of the Peptide Bond In Sec. 10.19 we pointed out that simple amides have a planar geometry, that the amide C—N bond is shorter than usual, and that rotation around that bond is restricted. Bond planarity and restricted rotation, which were rationalized by resonance, are also important in peptide bonds.

X-ray studies of crystalline peptides by Linus Pauling and his colleagues determined the precise geometry of peptide bonds. The characteristic dimensions, which are common to all peptides and proteins, are shown in Figure 15.7. Important features of this structure are the following:

1. The six atoms lie in a plane:

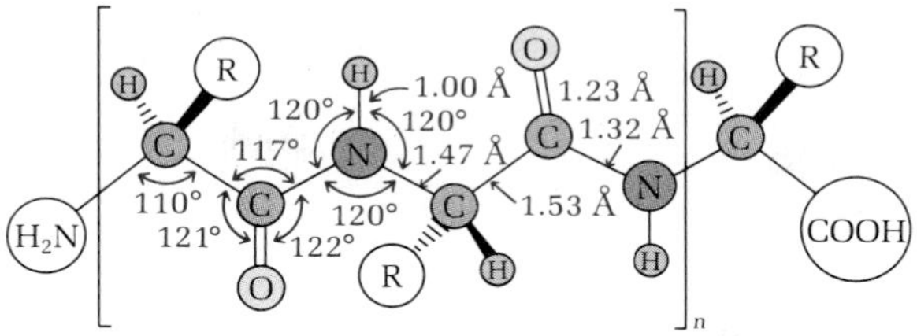

FIGURE 15.7

The characteristic bond angles and bond lengths in peptide bonds.

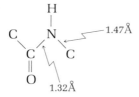

2. The angles around the nitrogen are 120°, showing it to be essentially sp^2-hybridized (as though the bond from the nitrogen to the carbonyl carbon were a double bond).

3. The bond from the nitrogen to the carbonyl carbon is much shorter than the other N—C bond, again as though the former were a double bond.

The rather rigid geometry and restricted rotation of the peptide bond impart a definite shape to proteins.

15.12b Hydrogen Bonding In Sec. 10.19 we pointed out that amides readily form intermolecular hydrogen bonds between the carbonyl group and the N—H group, bonds of the type C=O····H—N. Such bonds are also possible in peptide chains. The chain may coil in such a way that the N—H of one peptide bond may hydrogen-bond with a carbonyl group of another peptide bond further down the chain, thus rigidifying the coiled structure. Alternatively, carbonyl groups and N—H groups on different peptide chains may hydrogen-bond, thus binding the two chains. Although a single hydrogen bond is relatively weak (perhaps only 5 Kcal/mol in energy), the possibility of forming many intrachain or interchain hydrogen bonds makes this a very important factor in protein structure, as we will now see.

15.12c The α-Helix and the Pleated Sheet X-ray studies of α-keratin, a structural protein present in hair, wool, horns, and nails, showed that some feature of the structure repeats itself every 5.4 Å. From a study of molecular models using the geometry of the peptide bond, Linus Pauling suggested a structure that would explain this and other features of the x-ray studies. Pauling proposed that the polypeptide chain coils about itself in a spiral manner to form a helix, held rigid by intrachain hydrogen bonds. The **α-helix**, as it is called, is right-handed and has a pitch of 5.4 Å,

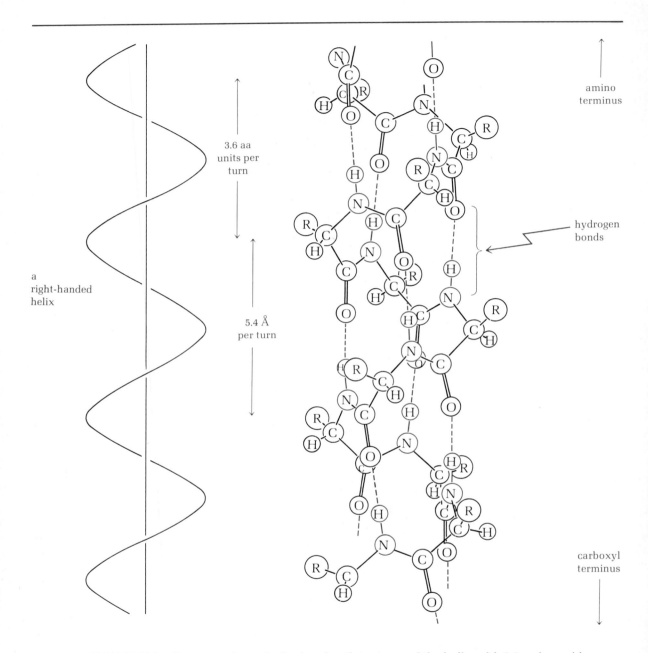

3.6 aa
units per
turn

a
right-handed
helix

5.4 Å
per turn

amino
terminus

hydrogen
bonds

carboxyl
terminus

FIGURE 15.8 Segment of an α-helix showing three turns of the helix, with 3.6 amino acid units per turn. H-bonds are shown as dashed colored lines.

FIGURE 15.9   Linus Pauling (California Institute of Technology and Stanford University),
who has made many contributions to our knowledge of organic structures.
He did fundamental work on the theory of resonance, on the measurement
of bond lengths and energies, and on the structure of proteins and the
mechanism of antibody action. He received the Nobel Prize in chemistry in
1954 and the Nobel Peace Prize in 1962.

or 3.6 amino acid units (Figure 15.8). Note several features of the α-helix.
Proceeding from the N-terminus (at the top of the structure as drawn in
Figure 15.8), each carbonyl group points ahead or down toward the C-
terminus and is hydrogen-bonded to an N—H bond farther down the
chain. The N—H bonds all point back toward the N-terminus. All the
hydrogen bonds are roughly aligned with the long axis of the helix. The
very large number of hydrogen bonds (one for each amino acid unit)
strengthen the helical structure. Note also that the R groups of the individual amino acid units are all directed outward and do not disrupt the
central core of the helix. It turns out that the α-helix is a natural pattern
into which many proteins fold. Figure 15.9 shows Professor Pauling seated
by a scale model of the α-helix.

The structural protein β-keratin, from silk fibroin, shows a different
repeat pattern (7 Å) in its x-ray structure. To explain the data, Pauling suggested a **pleated sheet** arrangement of the peptide chain (Figure 15.10).
In the pleated sheet, peptide chains lie side-by-side and are held together
by interchain hydrogen bonds. Adjacent chains run in opposite directions.
The repeat unit within each chain, which is stretched out compared with

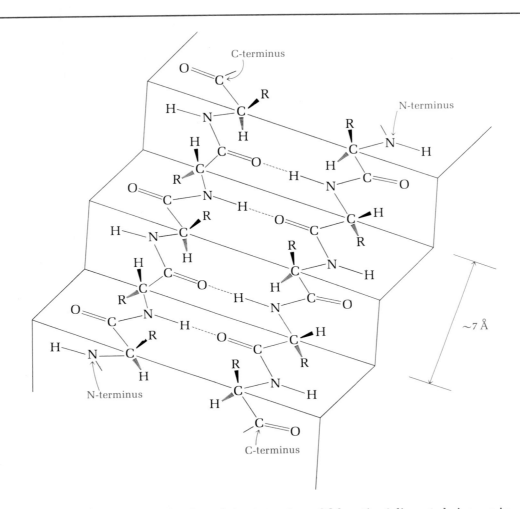

FIGURE 15.10   Segment of a pleated sheet structure of β-keratin. Adjacent chains run in
opposite directions and are held together by hydrogen bonds (shown in
color). R groups project above or below the mean plane of the sheet.

the α-helix, is about 7 Å. Note that, in the pleated sheet structure, the R
groups of each amino acid unit in any one chain alternate above and below
the mean plane of the sheet. If they are large, there will be appreciable
steric repulsion between R groups on adjacent chains. For this reason the
pleated sheet structure is important only in proteins that have a high percentage of amino acid units with small R groups. In β-keratin of silk
fibroin, for example, 36% of the amino acid units are glycine (R = H) and
another 22% are alanine (R = CH_3). Because this type of repulsion between

R groups is not encountered in the α-helix, the α-helix is by far the more common structure of the two.

15.13 FIBROUS AND GLOBULAR PROTEINS; TERTIARY STRUCTURE

Proteins generally fall into one of two main classes, the fibrous proteins and the globular proteins. These classes differ in shape and function, although they both contain the same primary and secondary structural features, polypeptide chains shaped by hydrogen bonds.

15.13a Fibrous Proteins Fibrous proteins are animal structural materials and hence are water-insoluble. We have already described some proteins of this type. They fall into three general categories: the **keratins,** which make up protective tissue such as skin, hair, feathers, claws, and nails; the **collagens,** which form connective tissue such as cartilage, tendons, and blood vessels; and the **silks,** such as the fibroin of spider webs and cocoons.

Fibrous proteins have different shapes that depend on further details of their structure. For example, the basic structural unit of the α-keratin of hair is the α-helix, a unit that is somewhat flexible and stretchable because of its intrachain hydrogen-bonded structure. In hair, three α-helices are braided to form a rope, the helices being held together by disulfide cross links (see A Word About Hair, page 189). The ropes are further packed side-by-side in bundles that ultimately form the hair fiber. The α-keratin of more rigid structures, such as nails and claws, is similar to that of hair except that there is a higher percentage of cysteine amino acid units in the polypeptide chain. This results in more disulfide cross links and a firmer, less flexible overall structure.

Collagens also consist of somewhat different helixes from the α-helix. Their particular amino acid content gives them a high tensile strength and allows a special shape. Roughly every third amino acid in the polypeptide chain is glycine, and collagen also has an unusually high proline content. (Recall that proline is the only common amino acid with a secondary amino group.) Collagen forms a triple-stranded helical rod, called *tropocollagen,* that is about 3000 Å long but only 15 Å in diameter. The high and regular glycine content is important, because the R group is small (H) and allows the three strands to hydrogen-bond tightly to one another. The prolines, with their five-membered rings, prohibit the close-packed α-helical structure but allow a less tightly coiled helix to be formed.

In silks, the β pleated sheet structures are stacked one on top of another. Thus the material is flexible but not easily stretched, because within each sheet the polypeptide chains are already in an extended form.

15.13b Globular Proteins Globular proteins are very different from the fibrous proteins. They tend to be water-soluble, and they have roughly spherical shapes, as their name suggests. Globular proteins in general fulfill one of four types of biological function.

1. Enzymes (such as trypsin, chymotrypsin, and hundreds of others) are biological catalysts. They catalyze the many organic reactions in the cell with great speed and remarkable specificity under mild reaction conditions.

2. Hormones are chemical messengers that regulate biological processes. Not all hormones are proteins, but several proteins and peptides, such as insulin and oxytocin, are hormones.

3. Transport proteins are carriers of small molecules from one part of the body to another. Examples include hemoglobin and myoglobin, which transport oxygen in the blood stream and muscles, respectively.

4. Storage proteins, such as casein of milk and ovalbumin of egg white, act as food stores.

We may well ask how materials as rigid as horses' hoofs, as springy as hair, as soft as silk, as slippery and shapeless as egg white, as inert as cartilage, and as reactive as enzymes can all be made of the same building blocks: amino acids and proteins. We have seen in fibrous proteins how hydrogen bonding and disulfide links can help form rather rigid structures. What structural features permit the more folded structures of the globular proteins?

So far we have concentrated on the protein backbone and its shape. But what about the marvelously unique and diverse R groups of the various amino acids? How do they affect protein structure?

Some amino acids have nonpolar R groups, simple alkyl or aromatic groups (for example, numbers 1–5, 10, 11, and to a lesser extent 8, 9, 12, and 13 in Table 15.1). Others have highly polar R groups, with carboxylate or ammonium ions and hydroxyl or other polar groups (for example, numbers 6, 7, and 14–20 in Table 15.1). Of course, these groups affect the gross properties of the protein. A water-soluble globular protein will, on the average, have more amino acids with polar or ionic side chains than will a water-insoluble fibrous protein. If an enzyme or other globular protein carries out its function mainly in the aqueous medium of the cell, it will adopt a structure in which the nonpolar, hydrophobic groups point in toward the center and the polar or ionic groups point out toward the water (much like micelles; see A Word About Soaps, page 277).

Many globular proteins are mainly helical but have folds that permit the overall shape to be globular. One of the 20 amino acids, proline, has a secondary amino group. Wherever a proline unit occurs in the primary peptide structure, there will be no N—H bond available for intrachain hydrogen bonding:

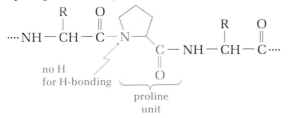

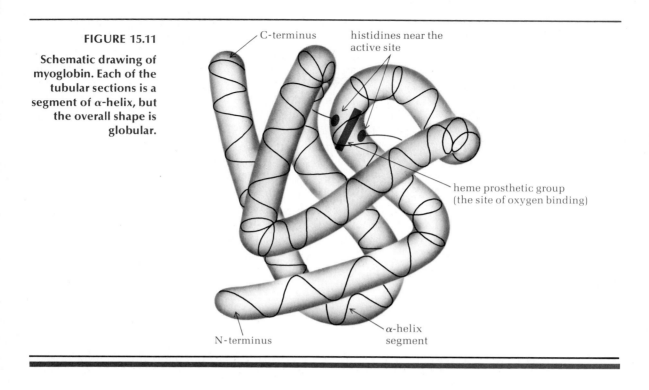

FIGURE 15.11

Schematic drawing of myoglobin. Each of the tubular sections is a segment of α-helix, but the overall shape is globular.

C-terminus

histidines near the active site

heme prosthetic group (the site of oxygen binding)

α-helix segment

N-terminus

Proline units tend to disrupt the α-helix, and we frequently find a proline unit at "turns" in a protein structure.

15.13c Myoglobin As a specific example of a globular protein, let us examine the structure of **myoglobin,** the oxygen-transport protein of muscle. Figure 15.11 shows a schematic drawing of this protein. This protein contains 153 amino acid units, yet it is extremely compact. Its approximate dimensions are $45 \times 35 \times 25$ Å, and there is very little empty space in its interior.

Approximately 75% of the amino acid units in myoglobin are part of eight major right-handed α-helical sections. There are four proline units and these all occur at or near "turns" in the structure. But there are also three other "turns" caused by other structural features of the amino acid R groups. The interior of myoglobin consists almost entirely of nonpolar R groups such as leucine, valine, phenylalanine, and methionine. There are *no* highly polar R groups "inside" the structure (no glutamic acid, aspartic acid, lysine, or the like). All R groups that are simultaneously polar and nonpolar have the polar parts pointing out and the nonpolar parts pointing in. For example, the hydroxyl groups of threonine and tyrosine point out though the rest of these groups point in. The only interior polar groups are two histidines. These perform a necessary function at the *active site* of the protein where the nonprotein portion, a molecule of the porphyrin *heme* (Sec 12.13), binds the oxygen. The outer surface of the protein includes many highly polar amino acid residues (lysine, arginine, glutamic acid, and so on).

To summarize this section, we see that the particular amino acid content of a peptide or protein influences its shape. These interactions are mainly a consequence of disulfide bonds and of the polarity or nonpolarity of the R groups, their shape, and their ability to form hydrogen bonds. When we refer to the **tertiary structure** of a protein, we refer to all the contributions of these factors to its three-dimensional structure.

The distinction between secondary and tertiary structure is not very sharp. They both refer to protein shape. In general, secondary structure results from interactions between groups close together in the primary structure (for example, *intra*chain hydrogen bonding which leads to the α-helix). Tertiary structure usually results from interactions of groups that may be far apart in the primary structure.

**15.14
QUATERNARY
STRUCTURE**
Many high-molecular-weight proteins exist as aggregates of several subunits. These aggregates are referred to as the **quaternary structure** of the protein. This aggregation helps to keep nonpolar portions of the protein surface from being exposed to the aqueous cellular environment.

Hemoglobin, the oxygen-transport protein of red blood cells, is an example of such aggregation. It consists of 4 almost spherical units, 2 α-units with 141 amino acids and 2 β-units with 146 amino acids. The 4 units come together in a tetrahedral array, shown in Figure 15.12.

Many other proteins form similar aggregates. Some are active only in their aggregate state, whereas others are active only when the aggregate dissociates into subunits. Aggregation in quaternary structures, then, provides an additional control mechanism over biological activity.

FIGURE 15.12

**Schematic drawing of
the four hemoglobin
subunits.**

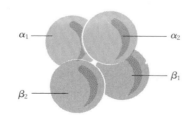

a word about

29. Proteins and Evolution

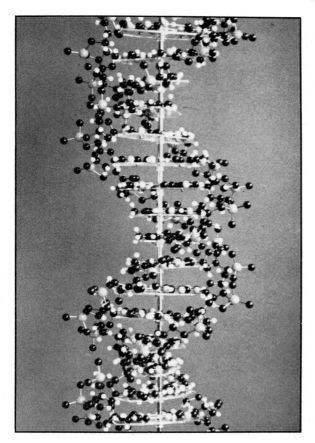

There are many reasons why it is important to determine the sequences of amino acids in proteins. First, we must know the detailed structures of proteins if we are to understand, at a molecular level, the way they function. The amino acid sequence is the link between the genetic message coded in DNA and the three-dimensional protein structure that forms the basis for its biological function.

There are medical reasons for knowing amino acid sequences. Certain genetic diseases, such as sickle-cell anemia, can result from the change of a single amino acid unit in a protein. (In this particular example, it is replacement of glutamic acid at position 6 in the β-chain of hemoglobin by a valine unit.) Sequence determination is an important part of medical pathology. One future possible application of genetic engineering is to devise ways of correcting these amino acid sequence errors.

Another important reason for determining amino acid sequences is that they provide a chemical tool for studying our evolutionary history. Proteins resemble one another in amino acid sequence if they have a common evolutionary ancestry. Let's look at a specific example.

Cytochrome c, an enzyme important in the respiration of most plants and animals, is an electron-transport globular protein with 104 amino acid residues. It is involved in oxidation–reduction processes. In these reactions, cytochrome c must react with and transfer an electron from one enzyme complex (cytochrome reductase) to another (cytochrome oxidase).

Cytochrome c probably evolved more than 1.5 billion years ago, before the evolutionary divergence of plants and animals. The function of this protein has been preserved all these years. We know this because the cytochrome c isolated from any eucaryotic microorganism (one that contains a cell nucleus) reacts *in vitro* with the cytochrome oxidase of any other species tested so far. For example, wheat germ cytochrome c reacts with human cytochrome c oxidase. Also, the three-dimensional structures of cytochrome c isolated from such diverse species as tuna heart and photosynthetic bacteria are very similar.

Although the shape and functions are similar for cytochrome c samples isolated from different sources, the amino acid sequence varies somewhat from species to species. The amino acid sequence of cytochrome c isolated from humans differs from that of monkeys in *just 1* (out of 104) amino acid residues. On the other hand, cytochrome c from dogs, a more distant relative on the evolutionary tree, differs from the human protein by 11 amino acid residues.

15.16. Give a definition and/or illustration of each of the following terms.
a. peptide bond   **b.** dipolar ion
c. dipeptide   **d.** L configuration of amino acids
e. essential amino acid   **f.** amino acid with a nonpolar R group
g. amino acid with a polar R group   **h.** amphoteric compound
i. isoelectric point   **j.** ninhydrin

15.17. Using Figure 15.1 as a guide, draw a three-dimensional structure for L-alanine. What is the priority order of the groups attached to the chiral center? What is the configuration of L-alanine: R or S?

15.18. Write Fischer projection formulas for:
a. L-phenylalanine   **b.** L-proline

15.19. Illustrate the amphoteric nature of amino acids by writing equations for the reaction of alanine in its dipolar ion form with 1 equivalent of:
a. hydrochloric acid   **b.** sodium hydroxide

15.20. Which is the most acidic proton in each of the following species?

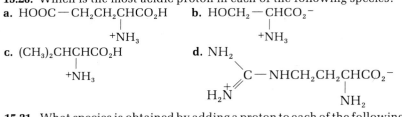

15.21. What species is obtained by adding a proton to each of the following species?
a. $CH_3CH-CHCO_2^-$ **b.** $^-O_2CCH_2CH-CO_2^-$
       | | |
      OH NH₃   +NH₃

15.22. Protonated alanine, $CH_3CH(\overset{+}{N}H_3)CO_2H$, has a pK_a of 2.34 and is appreciably more acidic than propanoic acid, $CH_3CH_2CO_2H$, whose pK_a is 4.85. Explain this increase in acidity on replacing an α-H by an $-\overset{+}{N}H_3$ substituent.

15.23. The pK_a's of glutamic acid are 2.19 (the α carboxyl group), 4.25 (the other carboxyl group), and 9.67 (the α ammonium ion). Write equations for the sequence of reactions that occurs when base is added to a strongly acidic (pH = 1) solution of glutamic acid.

15.24. The pK_a's of arginine are 1.82 for the carboxyl group, 8.99 for the ammonium ion, and 13.20 for the guanidinium ion. Write equations for the sequence of reactions that occurs when acid is gradually added to a strongly alkaline solution of arginine.

15.25. Write equations for the reaction of alanine with:
a. $CH_3CH_2OH + HCl$ **b.** $C_6H_5COCl + base$ **c.** acetic anhydride

15.26. Write equations for the following reactions.
a. serine + excess acetic anhydride →
b. tyrosine + bromine →
c. threonine + excess benzoyl chloride →
d. glutamic acid + excess methanol + HCl →
e. glutamine + aqueous NaOH + heat →

15.27. Write the equations that describe what occurs when phenylalanine is treated with ninhydrin.

15.28. Write structural formulas for the following peptides.

a. alanylalanine　　　　**b.** valyltryptophan
c. tryptophanylvaline　　**d.** glycylalanylglycine
e. serylleucylarginine .　**f.** histidylglycylglycylglutamic acid

15.29. Write formulas that show how the structure of alanylglycine changes as the pH of the solution changes from 1 to 10.

15.30. Use three-letter abbreviations to write out all possible tetrapeptides containing one unit each of Gly, Ala, Val, and Leu. How many structures are possible?

15.31. Write the structure of the product expected from the reaction of glycylcysteine with a mild oxidizing agent such as hydrogen peroxide.

15.32. Write an equation for the following reactions of Sanger's reagent.
a. 2,4-dinitrofluorobenzene + glycine →
b. excess 2,4-dinitrofluorobenzene + lysine →

15.33. A pentapeptide was converted to its DNP derivative (Sec. 15.10b), then completely hydrolyzed and analyzed quantitatively. It gave DNP-methionine, 2 mol of methionine, and 1 mol each of serine and glycine. The peptide was then partially hydrolyzed, the fragments were converted to their pure DNP derivatives, and each of them was hydrolyzed and analyzed quantitatively. Two tripeptides and two dipeptides isolated in this way gave the following products:

Tripeptide A: DNP-methionine and 1 mol each of methionine and glycine
Tripeptide B: DNP-methionine and 1 mol each of methionine and serine
Dipeptide C: DNP-methionine and 1 mol methionine
Dipeptide D: DNP-serine and 1 mol methionine

Deduce the structure of the original pentapeptide and explain your reasoning.

15.34. Write the equations for the removal of one amino acid from the peptide alanylglycylvaline by the Edman method. What is the name of the remaining dipeptide?

15.35. The following compounds are isolated as hydrolysis products of a peptide: Ala–Gly, Tyr–Cys–Phe, Phe–Leu–Try, Cys–Phe–Leu, Val–Tyr–Cys, Gly–Val, and Gly–Val–Tyr. Complete hydrolysis of the peptide shows that it contains one unit of each amino acid. What is the structure of the peptide, what are its N- and C-terminal amino acids, and what is its name?

15.36. Simple pentapeptides called *enkephalins* are abundant in certain nerve terminals. They have opiate-like activity and are probably involved in organizing sensory information pertaining to pain. An example is methionine enkephalin, Try–Gly–Gly–Phe–Met. Write out its complete structure, including all the side chains.

15.37. *Endorphins* were isolated in 1976 from the pituitary gland. They are potent pain-relievers. β-Endorphin is a polypeptide containing 32 amino acid residues. Digestion of β-endorphin with trypsin gave the following fragments:

Lys
Gly — Gln
Asn — Ala — His — Lys
Asn — Ala — Ile — Val — Lys
Tyr — Gly — Gly — Phe — Leu — Met — Thr — Ser — Glu — Lys
Ser — Gln — Thr — Pro — Leu — Val — Thr — Leu — Phe — Lys

From these data only, what is the C-terminal amino acid of β-endorphin? Treatment with cyanogen bromide gave the hexapeptide Tyr-Gly-Gly-Phe-Leu-Met and a 26-amino acid fragment. From these data only, what is the N-terminal amino acid of β-endorphin? Digestion of β-endorphin with chymotrypsin gave, among other fragments, a 15-unit fragment identified as Leu-Met-Thr-Ser-Glu-Lys-Ser-Gln-Thr-Pro-Leu-Val-Thr-Leu-Phe. You should now be able to locate 22 of the 32 amino acid units. Write out as much as you can of the sequence. What further information do you need to complete the sequence?

15.38. The attachment of the N-protected C-terminal amino acid to the polymer in solid phase peptide synthesis (eq. 15.14) is an S_N2 displacement reaction. What is the nucleophile? What is the leaving group? Rewrite eq. 15.14 in a way which clearly shows the reaction mechanism.

15.39. Eq. 15.16, the detachment of the peptide chain from the polymer in solid phase peptide synthesis, occurs by an acid-catalyzed S_N2 mechanism. Rewrite the equation to show this mechanism.

15.40. N-protection of an amino acid with a Boc group (eq. 15.17) involves nucleophilic substitution at a carbonyl group. Write out a mechanism for eq. 15.17).

15.41. Write the structure for glycylglycine, and show the resonance contributors to the peptide bond. At which bond is rotation restricted? What atoms will lie in a single plane?

15.42. *Glucagon* is a polypeptide hormone secreted by the pancreas when the blood sugar level is low. It increases the blood sugar level by stimulating the breakdown of glycogen in the liver. The primary structure of glucagon is

His—Ser—Glu—Gly—Thr—Phe—Thr—Ser—Asp—Tyr—Ser—Lys—Tyr—Leu—Asp—Ser—Arg—Arg—Ala—Gln—Asp—Phe—Val—Gln—Trp—Leu—Met—Asn—Thr

What fragments would you expect to obtain from digestion of glucagon with:
a. trypsin   **b.** chymotrypsin

15.43. In a globular protein, which of the following amino acid side chains are likely to point toward the center of the structure? Which will point toward the surface when the protein is dissolved in water?
a. arginine   **b.** phenylalanine   **c.** isoleucine
d. glutamic acid   **e.** asparagine   **f.** tyrosine

sixteen
Nucleotides and
Nucleic Acids

16.1
INTRODUCTION

DNA, the double helix, and the genetic code—through the popularization of science by the media, these have become household words, and they truly represent one of the greatest triumphs of all time for chemistry and biology. Yet the Watson–Crick hypothesis of the double helix and its role in the transfer of genetic information came on the scene only 30 years ago, and the genetic code was not "cracked" until a dozen years later.

In this chapter we will describe the structure of nucleic acids. We will first look at their building blocks, the nucleosides and nucleotides, and then describe how these building blocks are linked together to form giant nucleic acid molecules. Later we will consider the three-dimensional structures of these vital biopolymers and how the information they contain is transferred in the synthesis of proteins.

16.2
THE GENERAL
STRUCTURE OF
NUCLEIC ACIDS

The nucleic acids are linear, or chainlike, macromolecules that were first isolated from cell nuclei. Hydrolysis of nucleic acids gives **nucleotides**, which are the building blocks of nucleic acids, just as amino acids are the building blocks of proteins. A complete description of the primary structure of a nucleic acid would require knowledge of its nucleotide sequence, which would be comparable to knowing the amino acid sequence in a protein.

Further hydrolysis of a nucleotide gives one mole each of phosphoric acid and a **nucleoside**. The latter can be hydrolyzed further, to one equivalent each of a sugar and a heterocyclic base.

$$\boxed{\text{nucleic acid}} \xrightarrow[\text{enzyme}]{H_2O} \boxed{\text{nucleotides}} \xrightarrow[OH^-]{H_2O} \boxed{\text{nucleoside}} + H_3PO_4$$
$$\xrightarrow[H^+]{H_2O} \text{sugar} + \text{heterocyclic base}$$

(16.1)

The overall structure of a nucleic acid, then, is a macromolecule with a backbone of sugar molecules connected by phosphate links and with a base attached to each sugar unit.

FIGURE 16.1   The Pyrimidines   The Purines

The DNA bases.

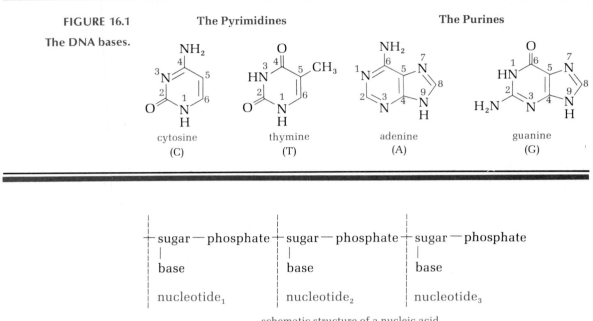

cytosine   thymine   adenine   guanine
  (C)   (T)   (A)   (G)

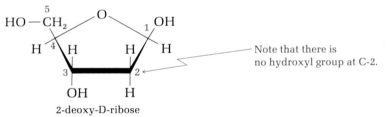

schematic structure of a nucleic acid

16.3
THE COMPONENTS
OF DEOXY-
RIBONUCLEIC
ACID (DNA)

Complete hydrolysis of DNA gives, in addition to phosphoric acid, a single sugar and four heterocyclic bases. The sugar is **2-deoxy-D-ribose**.

2-deoxy-D-ribose

The heterocyclic bases fall into two categories, the pyrimidines (**cytosine** and **thymine**) and the purines (**adenine** and **guanine**) as shown in Figure 16.1. When we refer to these bases later, especially in connection with the genetic code, we will use the capitalized first letters of their names as an abbreviation for their structures.

Now let us see how the sugar and bases are linked.

16.4
NUCLEOSIDES

A nucleoside is an *N-glycoside*. The pyrimidine or purine base is connected to the anomeric carbon (C-1) of the sugar. The pyrimidines are connected at N-1 and the purines at N-9. Here are two of the four nucleosides of DNA (the structures are numbered in the same manner as the component bases and the sugar, except that primes are used to designate the carbons of the sugar).

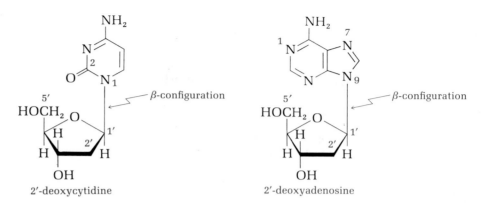

2'-deoxycytidine   2'-deoxyadenosine

N-glycosides, you will note, have similar structures to the O-glycosides discussed in Sec. 14.10. In O-glycosides, the -OH at the anomeric carbon was replaced by -OR, whereas in the N-glycosides, it is replaced by an -NR$_2$ group.

Example 16.1   **Draw the structures of**
a. the β-O-glycoside of 2-deoxy-D-ribose and methanol
b. the β-N-glycoside of 2-deoxy-D-ribose and dimethylamine

Solution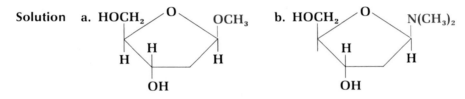

Note the similarity between N- and O-glycosides.

Problem 16.1   **Draw the structures for the remaining two nucleosides of DNA: 2'-deoxy-thymidine and 2'-deoxyguanosine.**

Because of their many polar groups, nucleosides are quite water-soluble. Like other glycosides, they can be hydrolyzed readily by acid (or by enzymes) to the sugar and the base. For example,

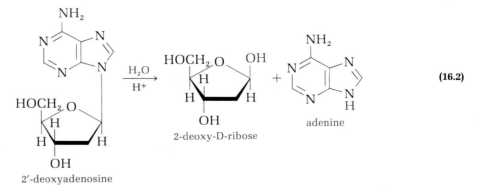

(16.2)

Equation 16.2 gives a specific example of the very last part of eq. 16.1, the hydrolysis of a nucleoside to a sugar and a heterocyclic base.

Problem 16.2   **Write equations that explain the mechanism of hydrolysis in eq. 16.2. Begin by adding a proton (the acid catalyst) to N-9 of the adenine part of the nucleoside.**

16.5   Nucleotides are phosphate esters in which a hydroxyl group in the sugar
NUCLEOTIDES   part of a nucleoside is esterified with phosphoric acid. In DNA nucleotides, either the 5′ or the 3′ hydroxyl group of the 2-deoxy-D-ribose can
be esterified.

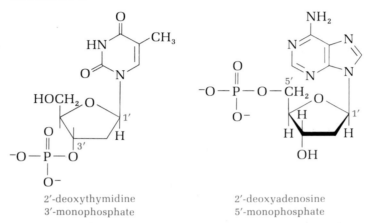

2′-deoxythymidine   2′-deoxyadenosine
3′-monophosphate   5′-monophosphate

Nucleotides may be named in several different ways. They may be named
as 3′- or 5′-monophosphate esters of a nucleoside, as shown. Since they
are long, these names are frequently abbreviated as shown in Table 16.1.
In these abbreviations, the small d stands for 2-deoxy-D-ribose, the next
letter refers to the base, and MP stands for monophosphate. (Later, we will
see that some nucleotides are diphosphates, abbreviated DP, or triphosphates, TP.) Unless otherwise stated, the abbreviations usually refer to the
5′-phosphates. Nucleotides may also be named as acids (for example,
dAMP is also known as deoxyadenylic acid). Table 16.1 lists the names of
the DNA nucleotides.

Example 16.2   **Write the structure for dTMP.**

Solution   **It is the same as the structure shown above for 2′-deoxythymidine 3′-
monophosphate, except that the phosphate group is at the 5′ position.**

Problem 16.3   **Write out the structures for:**
a. dCMP   b. dGMP

The phosphoric acid groups of nucleotides are quite acidic, and at pH 7
these groups exist mainly as the dianion, as shown in the structures.
Nucleotides can be hydrolyzed by aqueous base (or enzymatically) to

TABLE 16.1   The common 2-deoxyribonucleotides:

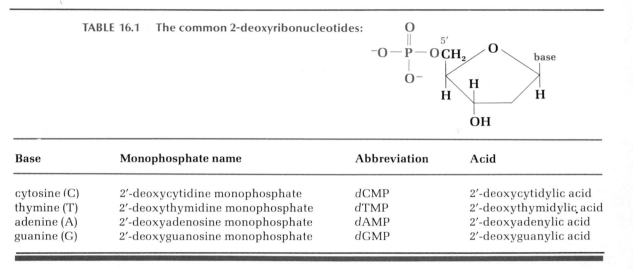

Base	Monophosphate name	Abbreviation	Acid
cytosine (C)	2′-deoxycytidine monophosphate	dCMP	2′-deoxycytidylic acid
thymine (T)	2′-deoxythymidine monophosphate	dTMP	2′-deoxythymidylic acid
adenine (A)	2′-deoxyadenosine monophosphate	dAMP	2′-deoxyadenylic acid
guanine (G)	2′-deoxyguanosine monophosphate	dGMP	2′-deoxyguanylic acid

nucleosides and phosphoric acid. Phosphoric acid is sometimes abbreviated P_i, meaning inorganic phosphate. This abbreviation is noncommital about the ionization state of the phosphate.

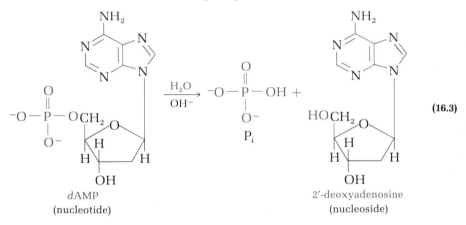

(16.3)

dAMP
(nucleotide)

2′-deoxyadenosine
(nucleoside)

This equation gives a specific example of the middle portion of eq. 16.1. Now let us go back another step in that equation and see how the nucleotides are linked to one another in DNA.

**16.6
THE PRIMARY
STRUCTURE OF
DNA**

Now we can be more precise than we were in Sec. 16.2 about nucleic acid structures. In *deoxyribonucleic acid* (DNA), 2-deoxy-D-ribose and phosphate units alternate in the backbone, the 3′ hydroxyl of one ribose unit being linked to the 5′ hydroxyl of the next ribose unit by a phosphodiester bond. Of course, a heterocyclic base is connected to the anomeric carbon of each deoxyribose unit by a β-N-glycosidic bond. Figure 16.2 shows a schematic drawing of a DNA segment. Note that in DNA there are

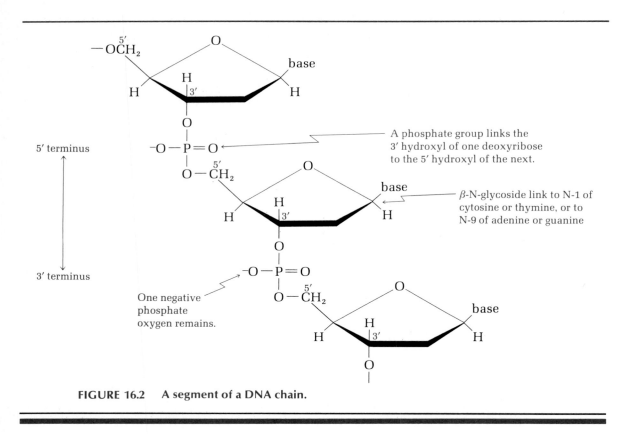

5′ terminus

A phosphate group links the
3′ hydroxyl of one deoxyribose
to the 5′ hydroxyl of the next.

β-N-glycoside link to N-1 of
cytosine or thymine, or to
N-9 of adenine or guanine

3′ terminus

One negative
phosphate
oxygen remains.

FIGURE 16.2   A segment of a DNA chain.

no remaining hydroxyl groups on each deoxyribose unit. Each phosphate,
however, still has one acidic proton that is usually ionized at pH 7, as
shown in Figure 16.2. If this proton were present, the substance would be
an acid; hence the name nucleic acid.

A complete description of any particular DNA molecule, which might
contain thousands of nucleotide units joined as shown in Figure 16.2,
would have to include the exact sequence of heterocyclic bases (A, C, G,
and T) along the chain. Suppose, for the moment, that we knew the
sequence in a segment of a DNA molecule. To write out the structure
completely, as in Figure 16.2, would be most cumbersome, so a shorthand
method is used.

**16.7
A SHORTHAND
CONVENTION FOR
NUCLEIC ACIDS**

By convention, we write the chain in the 5′→3′ direction, using capital
letters for the four bases. For example, the abbreviation dpApCpGpTp
stands for a tetranucleotide containing, in sequence, the four bases A, C,
G, and T. The small *d* tells us that the sugar is 2-deoxy-D-ribose. It can be
omitted if it is understood that we are discussing DNA and not some other
kind of nucleic acid. The small p's stand for phosphates. Because we write

from the 5′ end (from the top down, in Figure 16.2), we can say that the 3′ hydroxyl of the 2-deoxyadenosine is connected to the 5′ hydroxyl of the 2-deoxycytosine, and so on down the chain. Usually we abbreviate even further, to dpACGTp (the linking phosphates being understood). The p in front tells us that the 5′ hydroxyl of the first unit is present as a phosphate, and the final p tells us that the 3′ hydroxyl of the last unit is also present as a phosphate. If these hydroxyls were not esterified with phosphoric acid, the p's would be omitted.

Example 16.3   **Draw the complete structure of the dinucleotide dApCp.**

Solution

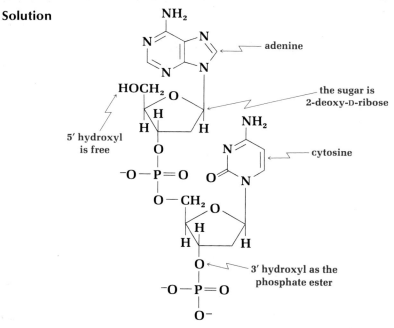

Problem 16.4   **Draw the complete structural formulas for:**
a. dCCA   b. dpACGTp

16.8 SEQUENCING NUCLEIC ACIDS

The problem of sequencing nucleic acids is in principle similar to that of sequencing proteins. At first, the job might appear to be easier because there are only 4 bases whereas there are 20 common amino acids. In fact, it is much more difficult. Even the smallest DNA molecule contains at least 5000 nucleotide units and some DNA molecules may contain as many as a million nucleotide units. To determine the exact base sequence in such a molecule is a task of considerable magnitude indeed.

Without trying to discuss nucleic acid sequencing in detail, we can describe the strategy. It is possible by various methods, to cleave nucleic acids selectively at a particular base—for example, on the 5′ side of an A or G nucleotide. The resulting fragments can then be separated, sequenced, and examined for overlapping sequences. Often nucleic acids labeled at the 5' terminus with a radioactive phosphate (^{32}P) can be helpful (just as labeling the N-terminal amino acid with a 2,4-dinitrophenyl group was helpful). A recently devised method involves the controlled interruption of nucleic acid replication by including nucleotides of the sugar 2,3-dideoxyribose in the growth medium. Having no 3' hydroxyl, these nucleotides block chain growth at whatever particular base they contain.

Progress in nucleic acid sequencing has been very rapid lately. When the previous edition of this text was published five years ago, the longest known nucleic acid sequences had about 200 nucleotide units. (These were in RNAs, which have shorter chains than DNAs.) But recently F. Sanger invented **controlled interruption of replication** and used this method to completely determine the base sequence in the DNA of a virus chromosome, OX 174, with 5375 nucleotide units. This achievement earned him his second Nobel Prize in chemistry.

This is the present state of determining primary DNA structure. But much can also be learned from the secondary structure of DNA, so let us go on to the double helix and the genetic code.

**16.9
SECONDARY DNA
STRUCTURE; THE
DOUBLE HELIX**

It has been known since 1938 that DNA molecules must have a discrete shape, because x-ray studies on DNA threads showed a regular stacking pattern with some periodicity. A key observation by E. Chargaff (Columbia University) in 1950 provided an important clue to the structure. Chargaff analyzed the base content of DNA from many different organisms and found that the mole percents of A and T are equal, and the mole percents of G and C are equal. For example, human DNA contains about 30% each of A and T and 20% each of G and C.

The meaning of these equivalences was not evident until 1953, when James D. Watson (United States) and Francis H. C. Crick (Great Britain), working together in Cambridge, England, proposed the double helix model for DNA. They received simultaneous supporting x-ray data for their proposal from Rosalind Franklin and Maurice Wilkins in London. The important features of their model are as follows:

1. DNA consists of two helical polynucleotide chains coiled around a common axis.

2. The helixes are right-handed and run in opposite directions with regard to their 3' and 5' termini.

3. The purine and pyrimidine bases lie *inside* the helix in planes that are perpendicular to the helical axis, whereas the deoxyribose and phosphate groups form the outside of the helix.

4. The two chains are held together by purine–pyrimidine base pairs connected by hydrogen bonds. Adenine is always paired with thymine, and guanine is always paired with cytosine.

5. The diameter of the helix is 20 Å. Adjacent base pairs are separated by 3.4 Å and related by a helical rotation of 36°. Hence there are 10 base pairs for every turn of the helix (360°), and the structure repeats every 34 Å.

6. There is no restriction on the sequence of bases along a polynucleotide chain. The exact sequence carries the genetic information.

FIGURE 16.3

Models and representation of the DNA double helix. The space-filling model at the left clearly shows the base pairs in the helix interior, in planes perpendicular to the main helix axis. The center drawing shows the structure more schematically, including the dimensions of the double helix. At the right is a schematic method for showing base pairing in the two strands.

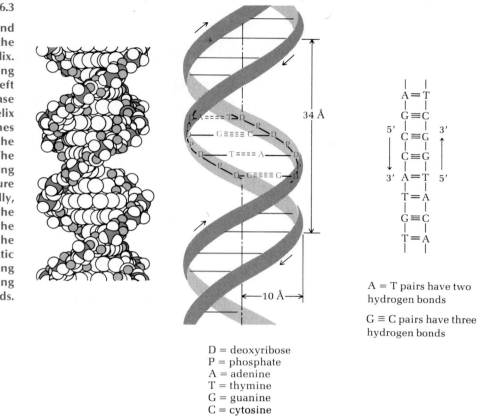

A = T pairs have two hydrogen bonds

G ≡ C pairs have three hydrogen bonds

D = deoxyribose
P = phosphate
A = adenine
T = thymine
G = guanine
C = cytosine

Figure 16.3 shows schematic models of the double helix. The key feature of the structure is the complementarity of the base pairing: **A—T** and **G—C**. Only purine–pyrimidine base pairs fit into the helical structure. There is insufficient room for the two purines and too much room for two pyrimidines, which would be too far apart to form hydrogen bonds. Of the purine–pyrimidine pairs, the hydrogen bonding possibilities are best for A—T and G—C pairing.

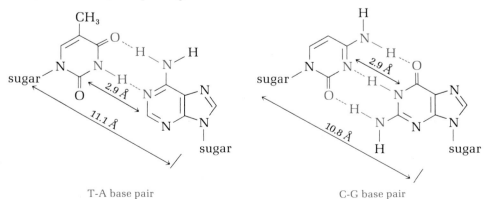

T-A base pair   C-G base pair

The A—T pair is joined by two hydrogen bonds and the G—C pair by three. The geometries of the two pairs are nearly identical.

Example 16.4   **Draw out the structure for the A—C base pair, and explain why it is less favorable than the A—T and G—C pairs.**

Solution   **Two possibilities for an A—C pair are shown:**

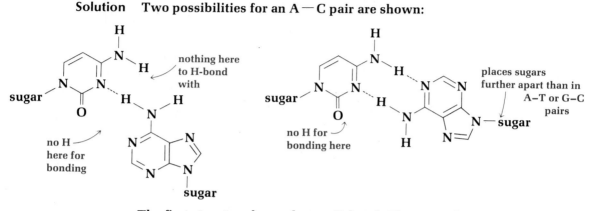

The first structure has only one H-bond. The second would substantially increase the sugar-sugar distance and distort the helix.

Problem 16.5   **Repeat Example 16.4 for the G—T pair.**

Problem 16.6   **Consider the following sequence of bases on one strand of DNA: —AGCCATGT— (written from 5′ → 3′). What will the sequence of bases be on the other strand?**

16.10
DNA REPLICATION

The beauty of the DNA double helix is that it suggests a molecular basis for transmitting information from one generation to the next: **DNA replication**. In 1954 Watson and Crick proposed that, as the two strands of a double helix separate, a new complementary strand is synthesized from nucleotides in the cell, using one strand as a template for the other. Their proposal is illustrated in Figure 16.4.

An ingenious experiment by Meselson and Stahl (United States, 1958) confirmed the Watson–Crick proposal. In the **Meselson–Stahl experiment**, bacteria were grown in an ^{15}N-rich medium (^{15}N is a heavy isotope of ordinary ^{14}N) to give a DNA in which the bases on both strands were rich in the heavy isotope. This "heavy–heavy" DNA was then allowed to replicate in a medium containing only ^{14}N (or "light") nutrients. First-generation DNA contained one "heavy" and one "light" strand, but second-generation DNA contained equal amounts of "heavy–light" and "light–light" molecules. This experiment is illustrated schematically in Figure 16.5.

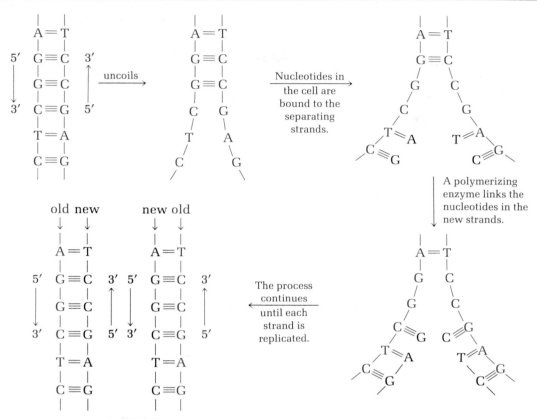

FIGURE 16.4   Schematic representation of DNA replication. As the double helix uncoils, nucleotides in the cell bond to the separate strands, following the base-pairing rules. A polymerizing enzyme links the nucleotides in the new strands to one another.

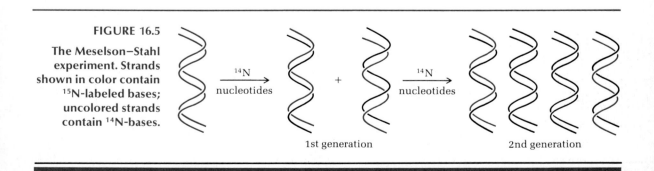

FIGURE 16.5

The Meselson–Stahl experiment. Strands shown in color contain ^{15}N-labeled bases; uncolored strands contain ^{14}N-bases.

Though simple in principle, replication is quite a complex process in practice. The nucleotides must be present as triphosphates (not mono-phosphates), an enzyme (DNA-polymerase) adds the nucleotides to a primer chain, other enzymes link DNA chains (DNA-ligase), there are specific places at which replication starts and stops, and so on. Our knowl-edge of the details of this process has increased considerably since the DNA double helix was proposed a few decades ago and, as in all science, it seems that the more we know, the more we need to learn.

Before we turn to the other important role of DNA, protein synthesis, we must first consider another type of nucleic acid that plays a key role in this process.

**16.11
RIBONUCLEIC
ACIDS, RNA**

Ribonucleic acids differ from DNA in three important ways: (1) the sugar is **D-ribose**, (2) one of the four heterocyclic bases is **uracil** (in place of thymine), and (3) most RNA molecules are single-stranded, although helical regions may be present by looping of the chain back on itself.

The RNA sugar D-ribose differs from the DNA sugar 2-deoxy-D-ribose in that it has a hydroxyl group at C-2. Otherwise, the nucleosides and nucleotides of RNA have structures similar to those of DNA (Sec. 16.4 and Sec. 16.5).

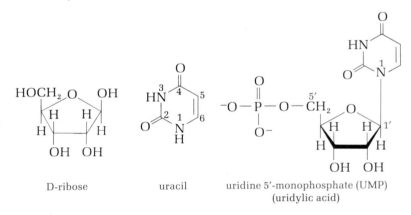

D-ribose   uracil   uridine 5'-monophosphate (UMP)
                                           (uridylic acid)

Note that uracil differs from thymine only in lacking the C-5 methyl group. Like thymine, it forms nucleotides at N-1, and their names are similar to those in Table 16.1.

Problem 16.7 **Write out the full structures for:
a. AMP   b. the RNA trinucleotide UCG**

Cells contain three major types of RNA. **Messenger RNA** (*mRNA*) is involved in **transcription** of the genetic code and is the template for protein synthesis. There is a specific mRNA for every protein synthesized by the cell. The base sequence of mRNA is complementary to the base sequence in a single strand of DNA, with U being the complement of A:

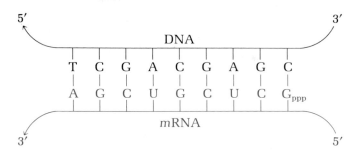

Transcription proceeds in the $3' \rightarrow 5'$ direction along the DNA template. That is, the mRNA chain grows from its own $5'$ end. The $5'$ nucleotide in RNA is usually present as a triphosphate, not monophosphate, and is commonly pppG or pppA. An enzyme called RNA polymerase is essential for transcription. Usually only one strand of DNA is transcribed. It contains base sequences, called promoter sites, which initiate transcription. It also contains certain termination sequences which signal the completion of transcription.

At the $3'$ end of mRNA, there is usually a special sequence of about 200 successive nucleotide units of the same base, adenine. This sequence plays a role in transporting the mRNA from the cell nucleus to the ribosomes, which are cellular structures where proteins are synthesized.

Transfer RNA (tRNA) carries amino acids in an activated form to the ribosome for peptide bond formation, in a sequence determined by the mRNA template. There is at least 1 tRNA for each of the 20 amino acids. Transfer RNAs are relatively small as nucleic acids go, with about 75 nucleotide units. Besides the 4 common heterocyclic bases, tRNAs contain several "rare" or uncommon pyrimidine and purine bases. Figure 16.6 shows the base sequence and general shape of a tRNA. Alanine tRNA, the species that delivers the amino acid alanine to its protein synthesis site, was the first RNA to have its base sequence worked out (R. W. Holley, Cornell University, 1965; Nobel Prize in 1968). Since then, the sequences of a large number of RNAs have been determined. Each tRNA has a three-base sequence CCA at the $3'$-hydroxyl end, where the amino acid is attached as an ester. Each tRNA also has an anticodon loop quite remote from the amino acid attachment site. This loop contains seven nucleotides, the middle three of which are complementary to the the three-base code word on the mRNA for that particular amino acid.

The third type of RNA is **ribosomal RNA** (rRNA). It comprises about 80% of the total cellular RNA (tRNA = 15%, mRNA = 5%) and is the main component of the ribosomes. Its molecular weight is large, and each

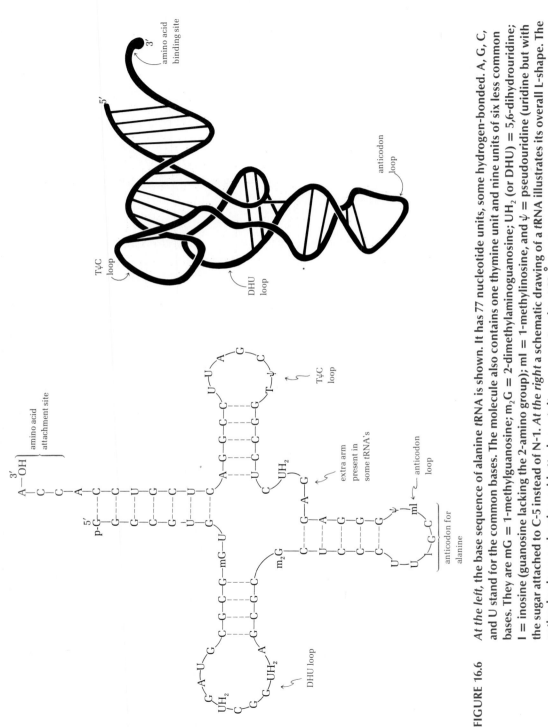

FIGURE 16.6 *At the left,* the base sequence of alanine tRNA is shown. It has 77 nucleotide units, some hydrogen-bonded. A, G, C, and U stand for the common bases. The molecule also contains one thymine unit and nine units of six less common bases. They are mG = 1-methylguanosine; m₂G = 2-dimethylaminoguanosine; I = inosine (guanosine lacking the 2-amino group); ml = 1-methylinosine, and ψ = pseudouridine (uridine but with the sugar attached to C-5 instead of N-1. *At the right* a schematic drawing of a tRNA illustrates its overall L-shape. The anticodon loop and amino acid attachment sites are remote, about 80 Å apart.

molecule may contain several thousand nucleotide units. Its precise role in protein synthesis is not yet known.

It is beyond the scope of this book to give a detailed account of the genetic code, how it was unraveled, and how over a hundred or more types of macromolecules must interact to translate that code into the synthesis of a protein. But we can present a few of the main concepts.

16.12a. The Genetic Code The **genetic code** is the relationship between the base sequence in DNA, or its RNA transcript, and the amino acid sequence in a protein. A 3-base sequence called a **codon** corresponds to 1 amino acid. Because there are 4 bases in RNA (A, G, C, and U), there are $4 \times 4 \times 4 = 64$ possible codons. However, there are only 20 common amino acids in proteins. Each codon corresponds to only 1 amino acid, but the code is *degenerate*. That is, several different codons may correspond to the same amino acid. Where this is the case, the codons usually differ only in the last letter. For example, GCU, GCC, GCA, and GCG are all codons for the amino acid alanine. Of the 64 codons, 3 are codes for "stop" (UAA, UAG, and UGA). They each signal that the particular protein synthesis is complete.

How was the genetic code "cracked"? The first successful experiment was done by Marshall Nirenberg (United States, 1961; Nobel Prize 1968). Nirenberg added a synthetic RNA, polyuridine (an RNA in which all the bases were uracil, U) to a cell-free protein-synthesizing system containing all the amino acids. He found a tremendous increase in the incorporation of phenylalanine in the resulting polypeptides. Since UUU is the only codon present in polyuridine, it must be the codon for phenylalanine. Similarly, polyadenosine led to the synthesis of polylysine, and polycytidine led to polyproline. Thus AAA = Lys and CCC = Pro. Later, other synthetic polyribonucleotides with other known repeating sequences were found to give polypeptides with repeating amino acid sequences.

Example 16.5 **A polyribonucleotide was prepared from the tetranucleotide UAUC. When it was subjected to peptide-synthesizing conditions, the polypeptide (Tyr–Leu-Ser-Ile)$_n$ was obtained. What are the codons for those four amino acids?**

Solution **The polyribonucleotide must have the sequence**

$$U \ A \ U \ C \ U \ A \ U \ C \ U \ A \ U \ C \ \cdots$$

If we divide the chain into codons, we get

$$\left.\begin{array}{c} UAU - CUA - UCU - AUC \cdots \\ Tyr \ - \ Leu \ - \ Ser \ - \ Ile \ \cdots \end{array}\right\} \text{ the sequence then repeats}$$

In this way, the meaning of four codons is cracked.

Problem 16.8   **A polynucleotide from the dinucleotide UA gave a polypeptide (Tyr–Ile)$_n$. How does this outcome confirm the results in Example 16.5? What is another codon for the amino acid isoleucine?**

Via experiments of this and other types, the entire genetic code has been worked out. *The code is universal for all organisms on earth and has remained invariant through all the years of evolution.* Consider what would happen if the "meaning" of a codon were changed. The result would be a change in the amino acid sequence of most proteins synthesized by that organism. Many of these changes would undoubtedly be disadvantageous. Hence there is a strong selection *against* changing the code.

16.12b. Protein Biosynthesis Coordinated reactions by many types of molecules are required for protein synthesis. The molecules include mRNA, tRNA, ribosomes, scores of enzymes, the amino acids, phosphate, and many others.

The amino acids must be attached, through enzyme catalysis, to the 3′ end of a tRNA (Figure 16.6) in preparation for delivery to the synthesis site. There is a unique tRNA for each amino acid and a unique enzyme for attaching the amino acid to it. The mRNA acts as the template for the synthesis, its code being "read" by the anticodons of the tRNAs.

Protein **biosynthesis** occurs in three phases: **initiation**, **elongation**, and **termination**. Synthesis is initiated when the mRNA and the first tRNA become aligned at a specific site on the ribosome. In bacteria, the "start" signal on the mRNA is usually AUG (the codon for methionine). The tRNA with the matching anticodon UAC becomes attached to the mRNA at this site through hydrogen-bonded base-pairing.

Elongation occurs in the following way. The tRNA corresponding to the next codon binds to the mRNA. The amino acid on the first tRNA is linked by an enzyme called peptidyl transferase to the amino acid on the adjacent tRNA with a peptide bond. The first tRNA, now lacking its amino acid, detaches from the mRNA. The same process is then repeated. Figure 16.7 illustrates the elongation process schematically. In this figure, the code on the mRNA is being read from bottom to top (in the 5′ → 3′ direction). Only two tRNAs are hydrogen-bonded to the mRNA at any one time. One of these tRNAs carries the growing peptide chain, and the other carries the next amino acid to be added. In the figure, it is valine. Below, a tyrosine tRNA is departing after having delivered its amino acid. Above, alanine tRNA and threonine tRNA are approaching the mRNA in preparation for delivery of their amino acids. The protein chain grows from the N-terminal to the C-terminal amino acid. For example, in Figure 16.7 the peptide bond is being made between the amino group of valine and the carboxyl group of serine.

Termination is signaled by one of three codons on the mRNA: UAA, UGA, or UAG. These codons are recognized by certain release factors

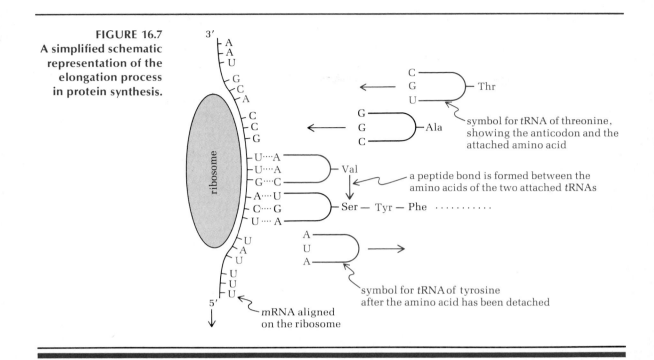

FIGURE 16.7
A simplified schematic representation of the elongation process in protein synthesis.

that bind to the "stop" codons and catalyze removal of the completed polypeptide chain. In Figure 16.7, threonine will become the C-terminal amino acid in the polypeptide, because the next codon on the mRNA is a "stop" codon (UAA).

Almost all polypeptides and proteins are synthesized by the scheme just described, including the enzymes that actually participate in the scheme. Some polypeptides may be modified after they are synthesized, for example, by forming disulfide bonds or by specific cleavage into shorter fragments.

To summarize, protein synthesis from DNA involves the following important steps:

1. Transcription of the message in the DNA codons to mRNA.
2. Alignment of the mRNA on the ribosome.
3. Initiation by binding of the first tRNA to the "start" codon on the mRNA.
4. Elongation, which consists of three steps that are repeated over and over again. They are (a) binding of a tRNA, carrying its particular amino acid; (b) peptide bond formation between the carboxyl group on the amino acid carried by the already bound tRNA and the amino group of the amino acid carried by the newly bound tRNA; and (c) release of the tRNA that has thus delivered its amino acid.

5. Termination, signalled by a "stop" codon on the mRNA. A termination factor binds to the "stop" codon and catalyzes the release of the completed peptide from the last tRNA molecule. In this step, the ribosome is then released and free to bind another mRNA molecule to repeat the entire process.

**16.13
OTHER
BIOLOGICALLY
IMPORTANT
NUCLEOTIDES**

The nucleotide structure finds utility not only in nucleic acids, but also in several other biologically active substances. We will describe the more important of these applications here.

Adenosine exists in several different phosphate forms. The 5'-monophosphate, diphosphate and triphosphates, as well as the 3',5'-cyclic phosphates, are key intermediates in numerous biological processes.

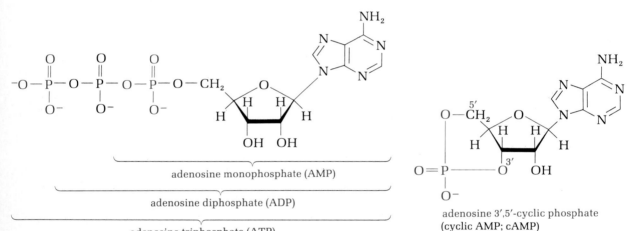

adenosine monophosphate (AMP)

adenosine diphosphate (ADP)

adenosine triphosphate (ATP)

adenosine 3',5'-cyclic phosphate
(cyclic AMP; cAMP)

ATP contains two phosphoric anhydride bonds, and considerable energy is released when ATP is hydrolyzed to ADP and further to AMP. These reactions often provide the energy for other biological reactions. Cyclic AMP is a mediator of certain hormonal activity. When a hormone outside a cell interacts with a receptor site on the cell membrane, it may stimulate cAMP synthesis within the cell. The cAMP in turn acts *within* the cell to regulate some biochemical process. In this way, a hormone need not penetrate the cell to exert its effect.

Four important coenzymes contain nucleotides as part of their structures. We have already mentioned **coenzyme A** (page 254), which contains ADP as part of its structure. It is a biological acyl-transfer agent and plays a key role in fat metabolism. **Nicotinamide adenine dinucleotide (NAD)** and its phosphate **NADP** are coenzymes that dehydrogenate alcohols to aldehydes or ketones, or the reverse (page 179). They consist of two nucleotides linked by the 5' hydroxyls of ribose:

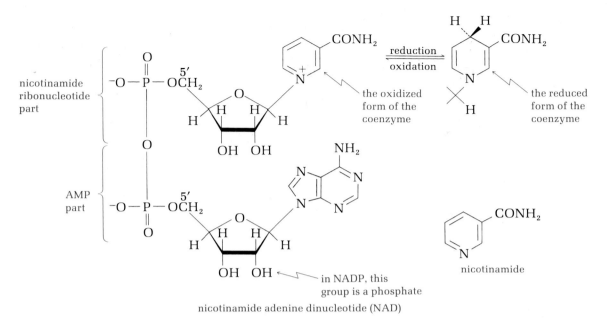

nicotinamide adenine dinucleotide (NAD)

Flavin adenine dinucleotide (FAD) is a yellow coenzyme involved in many biological oxidation–reduction reactions. It consists of a riboflavin (vitamin B_2) part connected to ADP. The reduced form has two hydrogens attached to the riboflavin part.

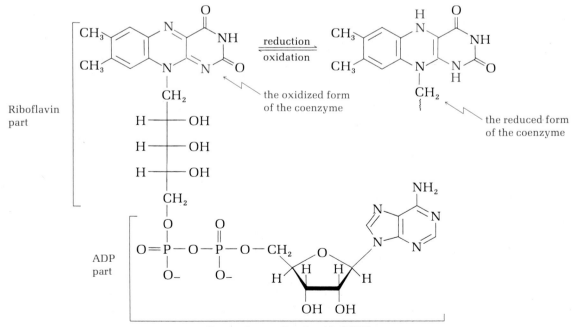

flavin adenine dinucleotide (FAD)

Vitamin B$_{12}$ (cobalamine), which is essential for the maturation and development of red blood cells, has an incredibly complex structure that includes a nucleotide part. **Coenzyme B$_{12}$** has two nucleotide parts in its structure. These molecules have a central cobalt atom surrounded by a macrocyclic molecule containing four nitrogens, similar to a porphyrin (Sec 12.13). But the cobalt has two additional ligands attached to it, above and below the mean plane of the nitrogen-containing rings. One of these is a ribonucleotide of the base **5,6-dimethylbenzimidazole.** The other ligand is a cyanide group in the vitamin and 5-deoxyadenosyl group in the coenzyme. In each case there is a direct carbon—cobalt bond. The reactions catalyzed by coenzyme B$_{12}$ usually involve replacement of the Co—R group by a Co—H group.

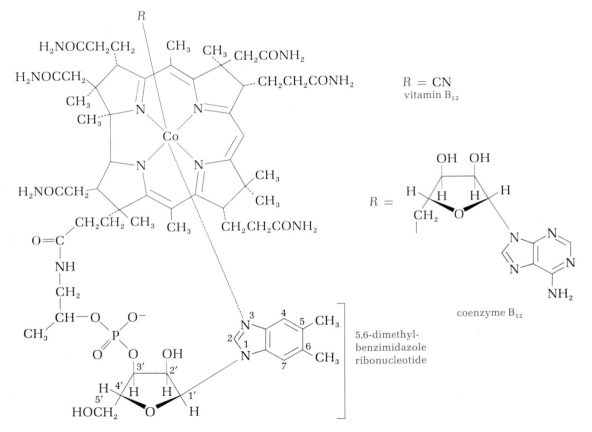

Vitamin B$_{12}$, which is produced by certain microorganisms but cannot be synthesized by humans, must be ingested. Only minute amounts are required, but pernicious anemia can result if the protein essential for absorption of the vitamin is lacking from gastric juices.

We see, then, that nucleotides play many roles in biological processes in addition to their role in DNA–RNA chemistry.

a word about ─────────────────────────────────────

30. The Future

How far have we come, and how much further do we have to go in our understanding of protein synthesis and the transfer of genetic information?

We know some of the basic ideas involved in these processes. We know the gross structures of DNA and RNA and how they replicate. We know the genetic code. We know something about mutations, which may occur when one base in a sequence is added or deleted or replaced by another, thus changing the "reading" of the base triplets. We know the base sequences of most tRNAs, which have about 75–80 nucleotides each. In 1981 the nucleotide sequence of the gene that codes for alanyl-tRNA synthetase, along with the complete amino acid sequence of the enzyme, was worked out. This was the first determination of the primary structure of a large synthetase, the enzyme that catalyzes the attachment of alanine to its tRNA. It contains 875 amino acid units.

We also know the entire base sequences of a handful of scientifically important small viruses about 5000 or so nucleotides in length. How much further do we have to go? The bacterium Escherichia coli is being worked on. Its genome has about 5 million bases! And the 46 human chromosomes, which may be contemplated next, contain about 500 million bases each. The magnitude of the task is huge, yet accomplishing it seems more conceivable now than it did a decade ago.

What good will this information do us? No one can tell for certain, but one obvious goal would be to correct genetic defects and alleviate the suffering they cause. Already, automatic DNA synthesizers are being developed in the laboratory. At present these machines can link nucleotides in a known and predetermined manner at the rate of about one every half hour, and the synthesis of oligonucleotides with a dozen or so units is commonplace. Recombinant-DNA techniques, whereby a new DNA molecule is formed by breakage and reunion of DNA strands, offer many possibilities for synthesizing large amounts of scarce genes or proteins. And no doubt there is more to be done than has yet been imagined.

ADDITIONAL
PROBLEMS

16.9. Write the structural formula for an example of each of the following.
a. a pyrimidine base   **b.** a purine base
c. a nucleoside   **d.** a nucleotide

16.10. The DNA bases in Figure 16.1 can exist in other tautomeric forms. Draw all possible tautomers (Sec. 9.14) of cytosine.

16.11. Examine the structures of adenine and guanine (Figure 16.1). Do you expect their structures to be planar or puckered? Explain.

16.12. Answer Problem 16.11 with respect to the pyrimidines cytosine and thymine.

16.13. Draw the structures of the following nucleosides.
a. cytidine (from β-D-ribose and cytosine)
b. deoxyadenosine (from β-2-deoxy-D-ribose and adenine)
c. uridine (from β-D-ribose and uracil)
d. deoxyguanosine (from β-2-deoxy-D-ribose and guanine)

16.14. *5-Fluorouracil 2-deoxyriboside* (FUdR) is used in medicine as an antiviral and antitumor agent. From its name, draw its structure.

16.15. *Psicofuranine* is a nucleoside used in medicine as an antibiotic and anti-tumor agent. Its structure differs from that of adenosine only in having a -CH_2OH attached with α geometry at carbon-l'. Draw its structure.

16.16. Write an equation for the complete hydrolysis of adenosine 5'-monophosphate (AMP) to its component parts.

16.17. Using Table 16.1 as a guide, write the structure of the following nucleotides.
a. guanylic acid (GMP)   **b.** 2'-deoxythymidylic acid (dTMP)

16.18. Write the steps in the mechanism for the base-catalyzed hydrolysis of AMP (eq. 16.3). What type of reaction mechanism is involved?

16.19. Draw the complete formula for the following RNA components.
a. pA   **b.** Ap   **c.** pApA

16.20. Draw the structure of the following DNA components.
a. dTpGp   **b.** dpppApC   **c.** dTAG

16.21. Draw the structure of the following RNA components.
a. UUU   **b.** UAA   **c.** GCA

16.22. Draw a structure showing the hydrogen bonding between uracil and adenine, and compare it with that for thymine and adenine (Sec. 16.9).

16.23. A segment of DNA contains the following base sequence:

5'—A—A—G—C—T—G—T—A—C—3'

Draw the sequence in its DNA complement and label its 3' and 5' ends.

16.24. For the DNA segment in Problem 16.23, write the mRNA complement and label its 3' and 5' ends.

16.25. Consider the following mRNA sequence:

5'—A—G—C—U—G—C—U—C—A—3'

Draw the DNA double helix from which this sequence was derived, using the schematic method at the right of Figure 16.3. Be sure to show the 5' and 3' ends of each strand.

16.26. Explain how the double-helical structure of DNA is consistent with Chargaff's analyses for the purine and pyrimidine content of DNA samples from various sources.

16.27. Draw the expected composition of the *third*-generation DNA expected in the Meselson–Stahl experiment (Figure 16.5).

16.28. The codon CAU corresponds to the amino acid histidine (His). How will this codon appear on the DNA strand from which it was transcribed? and on the complement of that strand? Be sure to label the 5' and 3' directions.

16.29. The anticodon in alanine-tRNA (Figure 16.6) contains the nucleoside unit inosine (I), which differs from guanosine only in lacking the 2-amino group. Draw the structure of inosine.

16.30. The alanine-tRNA anticodon CGI (Figure 16.6) can combine with the mRNA codons GCU, GCC, and GCA, all of which code for alanine. The reason is that the third base, inosine, can hydrogen-bond almost equally well with the third base in each codon (U, C, or A). Draw the hydrogen-bonded base pairs I—U, I—C, and I—A, and show that all three are about equally possible.

16.31. What products would you expect to obtain from the complete hydrolysis of nicotine adenine dinucleotide (NAD)? See Sec. 16.13 for its structure.

16.32. UDP–glucose is an activated form of glucose involved in the synthesis of glycogen. It is a nucleotide in which α-D-glucose is esterified at C-1 by the terminal phosphate of uridine diphosphate (UDP). From this description, draw the structure of UDP–glucose.

Index